PROBABILISTIC RELIABILITY ENGINEERING

PROBABILISTIC RELIABILITY ENGINEERING

BORIS GNEDENKO
Moscow State University and SOTAS, Inc.

IGOR USHAKOV
SOTAS, Inc. and George Washington University

Edited by **JAMES FALK**
George Washington University

A Wiley-Interscience Publication

JOHN WILEY & SONS, INC.

New York / Chichester / Brisbane / Toronto / Singapore

ISBN–0-471-30502-2

CONTENTS

9 Two-Pole Networks   **340**

The following Abbreviations are used frequently throughout this book:

Abbreviation	Meaning
BDP	birth and death process
d.f.	distribution function
g.f.	generating function
i.i.d.	independent and identically distributed
LST	Laplace-Stieltjes transform
m.g.f.	moment generating function
MRT	mean repair time
MTBT	mean time between failures
MTTF	mean time to failure
PFFO	probability of failure-free operation
p.r.v.	pseudo-random variable
r.v.	random variable
TTF	time to failure

PREFACE

This book was initially undertaken in 1987 in Moscow. We have found that the majority of books on mathematical models of reliability are very specialized: essentially none of them contains a spectrum of reliability problems. At the same time, many of them are overloaded with mathematics which may be beautiful but not always understandable by engineers. We felt that there should be a book covering as much as possible a spectrum of reliability problems which are understandable to engineers. We understood that this task was not a simple one. Of course, we now see that this book has not completely satisfied our initial plan, and we have decided to make it open for additions and a widening by everybody who is interested in it.

The reader must not be surprised that we have not touched on statistical topics. We did this intentionally because we are now preparing a book on statistical reliability engineering.

The publishing of this book became possible, in particular, because of the opportunities given by B. Gnedenko to visit the United States twice: in 1991 by George Washington University (Washington, DC) and in 1993 by SOTAS, Inc. (Rockville, Maryland). We both express our gratitude to Professor James E. Falk (GWU), Dr. Peter L. Willson (SOTAS), and Dr. William C. Hardy (MCI) for sponsoring these two visits of B. Gnedenko which permitted us to discuss the manuscript and to make the final decisions.

We would also like to thank Tatyana Ushakov who took care of all of the main problems in the final preparation of the manuscript, especially in dealing with the large number of figures.

We are waiting for the readers' comments and corrections. We also repeat our invitation to join us in improving the book for the future editions.

Professor of the Moscow State University   BORIS GNEDENKO
and Consultant to SOTAS, Inc.

Chief Scientist, SOTAS, Inc.   IGOR USHAKOV
and Visiting Researcher at the George Washington University

Moscow, Russia
Rockville, Maryland
December 1993

INTRODUCTION

The term *reliability*, in the modern understanding by specialists in engineering, system design, and applied mathematics, is an acquisition of the 20th century. It appeared because various technical equipment and systems began to perform not only important industrial functions but also served for the security of people and their wealth.

Initially, reliability theory was developed to meet the needs of the electronics industry. This was a consequence of the fact that the first complex systems appeared in this field of engineering. Such systems have a huge number of components which made their reliability very low in spite of their relatively highly reliable components. This led to the development of a specialized applied mathematical discipline which allowed one to make an a priori evaluation of various reliability indexes at the design stage, to choose an optimal system structure, to improve methods of maintenance, and to estimate the reliability on the basis of special testing or exploitation.

Reliability is a rich field of research for technologists, engineers, systems analysts, and applied mathematicians. Each of them plays a key role in ensuring reliability. The creation of reliable components is a very complex chemical–physical problem of technology. The construction of reliable equipment is also a very complex engineering problem. System design is yet another very complex problem of system engineering and systems analysis. We could compare this process to the design of a city: someone produces reliable constructions, another design and builds buildings, and a third plans the location of houses, enterprises, services, and so on. We consider mainly reliability theory for solving problems of system design. We understand all of the limitations of such a viewpoint.

To compensate for the deficiency in this book, we could recommend some books which are dedicated to reliability in terms of equipment and components. References can be found in the list of general publications at the end of this book. We understand that the problem of engineering support of reliability is very serious and extremely difficult. Most of this requires a concrete physical analysis and sometimes relates very closely to each specific type of equipment and component.

We are strongly convinced that the main problem in applied reliability analysis is to invent and construct an adequate mathematical model. Modeling is always an art and an invention. The mathematical technique is not the main issue. Mathematics is a tool for solution of the task.

Most modern mathematical models in reliability require a computer. Usually, reports prepared with the help of a computer hypnotize: accurate format, accurate calculations.... But the quality of the solution depends only on the quality of the model and input data. The computer is only a tool, not a panacea. A computer can never replace an analyst. The term "GIGO," which reminds one of FIFO and LIFO in queuing theory, was not conceived in vain. It means: garbage in, garbage out.

A mathematical model, first of all, must reflect the main features of a real object. But, at the same time, a model must be clear and understandable. It must be solvable with the help of available mathematical tools (including computer programs). It must be easily modified if a researcher can find some new features of the real object or would like to change the form of representation of the input data.

Sometimes mathematical models serve a simple purpose: to make a designed system more understandable for a designer. This use of modeling is very important (even if there are no practical recommendations and no numerical results) because this is the first stage of a system's testing, namely, a "mental testing." According to legend Napoleon, upon being asked why he could make fast and accurate decisions, answered that it is very simple: spend the night before the battle analyzing all conceivable turns of the battle—and you will gain a victory. The design of a mathematical model requires the same type of analysis: you rethink the possible uses of a system, its operational modes, its structure, and the specific role of different system's parts.

The reader will not find many references to American authors in this book. We agree that this is not good. To compensate for this deficiency, we list the main English language publications on the subject at the end of this book. We also supply a restricted list of publications in Russian which are close to the subject of this book.

As a matter of fact, we based our book on Russian publications. We also used our own practical experience in design and consulting. The authors represent a team of an engineer and a professional mathematician who have worked together for over 30 years, one as a systems analyst at industrial research and development institutes and the other as a consultant to the same institutes. We were both consultants to the State Committee of Standards of the former Soviet Union. For over 25 years we have been running the Moscow Consulting Center of Reliability and Quality Control which serves industrial engineers all over the country.

We had a chance to obtain knowledge of new ideas and new methods from a tide of contemporary papers. We have been in charge of the journal *Reliability and Quality Control* for over 25 years, and for more than 20 years we have been responsible for this section on reliability and queuing theory in the journal *Tehnicheskaya Kibernetika* (published in the United States as *Engineering Cybernetics* and later as the *Soviet Journal of Computer and Systems Sciences*).

This activity in industry and publishing was fruitful for us. Together we wrote several papers including review on the state of reliability theory in Russia.

We hope that the interested reader meets with terra incognita—Russian publications in the field, Russian names, and, possibly, new viewpoints, ideas, problems, and solutions. For those who are interested in a more in-depth penetration into the state of Russian results in reliability theory, we can suggest several comprehensive reviews of Russian works in the field: Levin and Ushakov (1965), Gnedenko, Kozlov, and Ushakov (1969), Belyaev, Gnedenko, and Ushakov (1983), and Rukhin and Hsieh (1987).

We tried to cover almost the entire area of applied mathematical models in the theory of reliability. Of course, we might only hope that the task is fulfilled more or less completely. There are many special aspects of the mathematical theory of reliability which appear outside the scope of this book. We suggest that our readers and colleagues join us in the future: the book is open to contributions from possible authors. We hope that the next edition of the book will contain new contributors. Please send us your suggestions and/or manuscripts of proposed new sections and chapters to the address of John Wiley & Sons.

BORIS GNEDENKO
IGOR USHAKOV

REFERENCES

Belyaev, Yu. K., B. V. Gnedenko, and I. A. Ushakov (1983). Mathematical problems in queuing and reliability theory. *Engrg. Cybernet.* (USA), vol. 22, no. 6.

Gnedenko, B. V., B. A. Kozlov, and I. A. Ushakov (1969). The role of reliability theory in the construction of complex systems (in Russian). In *Reliability Theory and Queuing Theory*, B. Gnedenko, ed. Moscow: Sovietskoe Radio.

Levin, B. R., and I. A. Ushakov (1965). Some aspects of the present state of reliability (in Russian). *Radiotechnika*, No. 4.

Rukhin, A. L., and H. K. Hsieh (1987). Survey of Soviet work in Reliability. *Statist. Sci.*, vol. 2, no. 4.

CHAPTER 1

FUNDAMENTALS

We decided to begin with a brief discussion of the more or less standard subject of probability theory and the theory of stochastic processes. Of course, we are trying to review all this from a reliability standpoint. We not only give a formal description of the main discrete and continuous distribution functions usually used in reliability analysis, but explain as well the nature of their appearance and their mutual interrelationships.

A presentation of stochastic processes does not pretend to cover this branch of probability theory. It is rather a recollection of some necessary background for the reader.

With the same purpose we decided to include an appendix to the chapter with a very short overview of the area of generating functions and Laplace–Stieltjes transforms.

1.1 DISCRETE DISTRIBUTIONS RELATED TO RELIABILITY

1.1.1 Bernoulli Distribution

In applications, one often deals with a very simple case where only two outcomes are possible—success or failure. For example, in analyzing the production quality of some production line, one may choose a criterion (an acceptable level or tolerance limit) to divide the entire sample into two parts: "good" and "bad."

Consider another example: during equipment testing one may predetermine some specified time and check if the random time-to-failure of the chosen item exceeds it or not. Thus, each event might be related to success or failure by this criterion.

1

We will denote a successful outcome as 1, and a failure as 0. This leads us to consider a random variable (r.v.) X for which $\Pr\{X = 1\} = p$ and $\Pr\{X = 0\} = 1 - p = q$. The value of p is called the parameter of the Bernoulli distribution. The distribution function (d.f.) of the r.v. X can be written in the form

$$f_B(x|p) = p^x q^{1-x} \qquad x = 0, 1 \tag{1.1}$$

where the subscript B signals the Bernoulli distribution. Clearly, $f_B(1|p) = p$ and $f_B(0|p) = 1 - p = q$. For the Bernoulli r.v. we know

$$E\{X\} = 1 \cdot p + 0 \cdot q = p \tag{1.2}$$

and

$$E\{X^2\} = 1^2 p + 0^2 q = p$$

The variance is expressed through the first and second moments:

$$\text{Var}\{X\} = E\{X^2\} - [E\{X\}]^2 = p - p^2 = p(1 - p) = pq \tag{1.3}$$

The moment generating function (m.g.f.) of the r.v. X can be written as

$$\varphi(s) = E\{e^{sX}\} = pe^s + q \qquad \text{for} \quad -\infty < s < \infty \tag{1.4}$$

The m.g.f. can also be used to obtain the moments of the distribution:

$$M^{(1)} = E\{X\} = \frac{d}{ds}(pe^s + q)\Big|_{s=0} = p$$

$$M^{(2)} = E\{X^2\} = \frac{d^2}{ds^2}(pe^s + q)\Big|_{s=0} = p$$

which coincide with (1.2) and (1.3).

A sequence of independent identically distributed (i.i.d.) Bernoulli r.v.'s is called a *sequence of Bernoulli trials with the parameter p*. For example, one may sequentially test n statistically identical items by setting $X_i = 1$ if the ith item operates successfully during the time period t, and $X_i = 0$ otherwise ($i = 1, \dots, n$). Thus, one has a random sequence of 1's and 0's which reflects the Bernoulli trial outcomes.

1.1.2 Geometric Distribution

Consider a unit installed in a socket. The unit is periodically replaced by a new one after time t. Thus, the socket's operation is represented by a sequence of cycles, each of which consists of the use of a new unit. Let X denote the trial's outcome: $X = 1$ if a unit has not failed during the time interval t, and $X = 0$ otherwise. The probability of a unit's successful operation during one cycle equals p. All units are identical and stochastically independent. The socket operates successfully for a random number of cycles X before a first failure. The distribution of the r.v. X is the subject of interest. This distribution of the length of a series of successes for the sequence of Bernoulli trials is called a geometrical distribution:

$$\Pr\{X = x\} = f_g(x|p) = p^x q \tag{1.5}$$

where the subscript g denotes the geometrical distribution. For (1.5) the d.f. is

$$\Pr\{X \le x\} = q \sum_{0 \le k \le x} p^k \tag{1.6}$$

Since (1.6) includes the geometric series, it explains the origin of the distribution's name.

Everybody knows how to calculate (1.6) in a standard way, but we would like to show an interesting way which can be useful in other situations. Let

$$z = 1 + p + p^2 + \cdots + p^{x+1} \tag{1.7}$$

and

$$y = 1 + p + p^2 + \cdots + p^x$$

Then (1.7) can be rewritten as

$$z = y + p^{x+1} = 1 + p(1 + p + p^2 + \cdots + p^x) = 1 + py$$

and, finally, if the sum converges

$$\sum_{0 \le k \le x} p^k = y = \frac{1 - p^{x+1}}{1 - p} = \frac{1}{q}[1 - p^{x+1}]$$

Now returning to (1.6), we obtain

$$\Pr\{X \le x\} = 1 - p^{x+1} \tag{1.8}$$

Thus, with the probability defined in (1.8), a failure has occurred before the xth cycle. The probability of a series of successes of length not less than x, that is, $\Pr\{X \ge x\}$, is, obviously,

$$\Pr\{X \ge x\} = 1 - \Pr\{X \le x - 1\} = p^x \tag{1.9}$$

Of course, the last result can be obtained directly. The set of all events, consisting of series of not less than x successes, is equivalent to x first successes and any other outcome afterwards.

For the geometric distribution, the m.g.f., $\overline{\varphi}$, can be written as

$$\overline{\varphi}_g(s) = \mathrm{E}\{e^{sX}\} = q \sum_{x \geq 0} p^x e^{sx} \tag{1.10}$$

This sum has a limit if $0 < pe^s < 1$. To compute (1.10), we can use the same procedure as above. With the same notation, we obtain

$$y = 1 + a + a^2 + a^3 + \cdots = 1 + a(1 + a + a^2 + \cdots) = 1 + ay$$

and then

$$y = (1 - a)^{-1}$$

Thus,

$$\overline{\varphi}_g(s) = \frac{q}{1 - pe^s} \tag{1.11}$$

The mean and variance of the geometric distribution can be found in a direct way with the use of bulky transformations. We will derive them using (1.11):

$$\mathrm{E}\{X\} = \frac{d}{ds}\overline{\varphi}_g(s)\Big|_{s=0} = \frac{d}{ds}\left(\frac{q}{1 - pe^s}\right)\Big|_{s=0} = \frac{p}{q} \tag{1.12}$$

and

$$\mathrm{E}\{X^2\} = \frac{d^2}{ds^2}\overline{\varphi}_g(s)\Big|_{s=0} = \frac{d^2}{ds^2}\left(\frac{q}{1 - pe^s}\right)\Big|_{s=0} = \frac{p(1 + p)}{q^2} \tag{1.13}$$

Thus, the variance by (1.11) is

$$\mathrm{Var}(X) = \frac{p(1 + p)}{q^2} - \left(\frac{p}{q}\right)^2 = \frac{p}{q^2} \tag{1.14}$$

Substituting e^s for z, we obtain the generating function (g.f.), $\hat{\varphi}$, of the distribution, that is, a sum of the form

$$\hat{\varphi}(z) = \sum_{k \geq 0} p_k z^k = \sum_{k \geq 0} p^k q z^k = \frac{q}{1 - pz} \tag{1.15}$$

In conclusion, we should emphasize that the geometric distribution possesses the memoryless, or Markovian, property: the behavior of a sequence of Bernoulli trials, taken after an arbitrary moment, does not depend on the evolution of the trials before this moment. This statement can be written as

$$\Pr\{X = k + t | X \geq k\} = \Pr\{X = t\}$$

Of course, this property of the geometric distribution follows immediately from the definition of a Bernoulli trial. At the same time, (1.14) follows from (1.7) and the definition of the conditional probability:

$$\Pr\{X = k + t | X \geq k\} = \frac{\Pr\{X = k + t \text{ and } X \geq k\}}{\Pr\{X \geq k\}} = \frac{qp^{k+t}}{p^k} = qp^t$$

For example, in the case with cycles of successful operations of a socket, the reliability index of the socket at an arbitrary moment of time does not depend on the observed number of successful cycles before this moment.

1.1.3 Binomial Distribution

In a sequence of Bernoulli trials, one may be interested in the total number of successes in n trials rather than in the series of successes (or failures). In this case the r.v. of interest is

$$X = X_1 + \cdots + X_n = \sum_{1 \leq i \leq n} X_i$$

For example, consider a redundant group of n independent units operating in parallel. The group operates successfully if the number of operating (or functioning) units is not less than m. Let X_i be 1 if the ith unit is functioning at some chosen time, and 0 otherwise. Then X is the number of successfully operating units in the group. Thus, the group is operating successfully as long as $X \geq m$.

When considering the distribution of the r.v. X, one speaks of the *binomial distribution* with parameters n and p.

By well-known theorems of probability theory, for any set of r.v.'s X_i,

$$\mathrm{E}\left\{ \sum_{1 \leq i \leq n} X_i \right\} = \sum_{1 \leq i \leq n} \mathrm{E}\{X_i\} \tag{1.16}$$

In this particular case

$$\mathrm{E}\{X\} = np \tag{1.17}$$

For independent r.v.'s the variance of X is expressed as

$$\text{Var}\left\{ \sum_{1 \le i \le n} X_i \right\} = \sum_{1 \le i \le n} \text{Var}\{X_i\} \tag{1.18}$$

For i.i.d. Bernoulli r.v.'s

$$\text{Var}\{X\} = npq \tag{1.19}$$

For this distribution the m.g.f. is

$$\overline{\varphi}(s) = (pe^s + q)^n \tag{1.20}$$

Both (1.17) and (1.19) can be easily obtained from (1.20).

Substituting $e^s = z$ transforms (1.20) into the g.f. of a binomial distribution

$$\hat{\varphi}(s) = (pz + q)^n \tag{1.21}$$

The reader can see that (1.21) is a Newton binomial so the origin of the distribution's name is clear.

If one writes (1.21) in expanded form, the coefficients at z^k is the probability of k successes in n trials

$$\hat{\varphi}(z) = p^n z^n + \binom{n}{1} p^{n-1} qz^{n-1} + \binom{n}{2} q^{n-2} q^2 z^{n-2} + \cdots \tag{1.22}$$

So the probability that there will be x successes in n trials equals the coefficient of z^x:

$$\text{Pr}\{X = x\} = \binom{n}{x} p^x q^{n-x} \tag{1.23}$$

Of course, (1.21) can be written in the form $\varphi(z) = (p + qz)^n$. In this case the coefficient of z^x will be the probability that exactly x failures have occurred.

1.1.4 Negative Binomial Distribution

The negative binomial distribution arises if one considers a series of Bernoulli trials before the appearance of the kth event of a chosen type. In other words, the r.v. is a sum of a fixed number, say k, of geometric r.v.'s. This distribution is sometimes called the *Pascal distribution*.

As an illustrative example consider a relay. With each switching the relay performs successfully with probability p. With probability $q = 1 - p$ the relay fails and then is replaced by another identical one. Let us assume that each switching is independent with a constant probability p, and the relay replaces the failed one is identical to the initial one. If there is one main and $x - 1$ spare relays, the time to failure of the socket has a negative binomial distribution.

Thus, a negative binomially distributed r.v. X can be expressed as

$$X = X_1 + \cdots + X_n = \sum_{1 \leq i \leq n} X_i$$

where each X_i, $i = 1, \ldots, n$, has a geometric distribution.

Of course, in a direct way one can easily find the mean and variance of the negative binomial distribution using the corresponding expressions (1.12) and (1.14) for the geometric distribution

$$E\{X\} = \sum_{1 \leq i \leq n} E\{X_i\} = \frac{np}{q} \tag{1.24}$$

and

$$\mathrm{Var}\{X\} = \sum_{1 \leq i \leq n} \mathrm{Var}\{X_i\} = \frac{np}{q^2} \tag{1.25}$$

The m.g.f. of the negative binomial distribution can be easily written with the help of the m.g.f. of the geometric distribution:

$$\bar{\varphi}_{nb}(s) = \left[\varphi_g(s)\right]^x = \left[\frac{q}{1 - qe^s}\right]^x \tag{1.26}$$

Obviously, the mean and variance can be obtained from (1.26) by a standard procedure, but less directly. The example above shows that the use of an m.g.f. can result in a more straightforward analysis.

Consider a geometric r.v. representing a series of successes terminating with a failure. Let us find the probability that n trials will terminate with the xth failure; that is, during n trials one observes exactly x geometric r.v.'s. This event can occur in the following way: the last event must be a failure by necessity (by assumption) and the remaining $n - 1$ trials contain $x - 1$ failures and $(n - 1) - (x - 1) = n - x$ successes, in some order. But the latter is exactly the case that we had when we were considering a binomial distribution: $x - 1$ failures (or, equivalently, $n - x$ successes) in $n - 1$ trials. The probability equals

$$\Pr\{X = n\} = \Pr\{x - 1 \text{ failures among } n - 1 \text{ trials}\}$$

$$\cdot \Pr\{\text{the } n\text{th trial is a failure}\} \tag{1.27}$$

The second term of the product in (1.27) equals q and the first term (considered relating to failures) is defined to be

$$f_b(x - 1|p, n - 1) = \binom{n-1}{x-1} q^{x-1} p^{n-x} \tag{1.28}$$

Now (1.28) can be rewritten as

$$\Pr\{X = n\} = \binom{n-1}{x-1} q^x p^{n-x} \tag{1.29}$$

The expression (1.29) can be written in the following form:

$$\Pr\{X = n\} = \binom{-n}{x}(-q)^x p^{n-x} \tag{1.30}$$

[We leave the proof of (1.30) for Exercise 1.1.]

Equation (1.26) explains the name of the distribution.

We mention that the negative binomial and the binomial distributions are connected in the following manner. The following two events are equivalent:

- In n Bernoulli trials, the kth success occurs at the n_1th trial where $n_1 \leq n$, and all remaining trials are unsuccessful.
- The negative binomially distributed r.v. is less than or equal to n.

The first and second events are described with the help of binomial and negative binomial distributions, respectively. In other words,

$$\binom{n}{k} p^k q^{n-k} = \sum_{0 \leq j \leq n-k} \binom{n-1-j}{k-1} p^k q^{n-k}$$

Thus, in some sense, a binomial d.f. plays the role of a cumulative d.f. for an r.v. with a negative binomial d.f.

1.1.5 Poisson Distribution

The Poisson distribution plays a special role in many practical reliability problems. The role of the Poisson distribution will be especially clear when we consider point stochastic processes, that is, processes which are represented by a sequence of point events on the time axis.

Before we begin to use this distribution in engineering problems, let us describe its genesis and its formal properties.

Again, let us consider a sequence of Bernoulli trials. One observes n_1 experiments each with a probability of success of p_1 and a probability of failure of q_1. The probability of no failures occurring during the experiment is

$$\Pr\{\text{no failure}|n_1, p_1\} = p_1^{n_1} \tag{1.31}$$

Let the probability (1.31), that is, the probability that there are no failures in n_1 trials, be equal to P. Now let us assume that each mentioned trial consists, in turn, of m identical and independent subtrials, or "trials of the second level." So now we consider $n_2 = n_1 m$ experiments at the second level. If at least one failure has occurred in this group of experiments at the second level, we will consider that a failure of the entire process has occurred. If the probability of success for this second level is p_2, then one has the obvious relationship $p_1 = p_2^m$ or, consequently,

$$\Pr\{\text{no failure}|n_2, p_2\} = p_2^{n_2} = P$$

We can continue this procedure of increasing the number of trials and correspondingly increasing the probability of success in such a manner that for any jth stage of the procedure

$$\Pr\{\text{no failure}|n_j, p_j\} = p_j^{n_j} = P$$

Now let us consider the probability of k failures for the same process at a stage with n trials and corresponding probabilities p and q. We can use the binomial distribution

$$
\begin{aligned}
\Pr\{k \text{ failures}|n, p\} \\
= \binom{n}{k} p^{n-k} q^k \\
= \frac{n \cdot (n-1) \cdot \cdots \cdot (n-k+1)}{1 \cdot 2 \cdot \cdots \cdot k} (1-q)^{n-k} q^k
\end{aligned}
$$

Now let us write the expression for the case when k is fixed but $n \to \infty$ and $p \to 1$ in correspondence with the above-described procedure:

$$
\begin{aligned}
\lim_{n \to \infty} \Pr\{k \text{ failures}|n, p\} &= \frac{q^k}{k!} \lim_{n \to \infty} [n \cdot (n-1) \cdot \cdots \cdot (n-k+1)](1-q)^{n-k} \\
&= \frac{(nq)^k}{k!} e^{-nq} \tag{1.32}
\end{aligned}
$$

Thus, the Poisson distribution can be considered as a limiting distribution for the binomial when the number of trials goes to ∞ (or, in practice, is very large) and the value nq is restricted and fixed.

For this case it is convenient to introduce a special parameter, say λ, which characterizes the intensity of a failure in a time unit for this limiting case.

For the limit (1.32) one can speak of the transformation of a discrete Bernoulli trials process into a continuous process. Then λt is the mean number of failures during a time interval t. (The memoryless property of Bernoulli trials is independent of when this interval begins.) So one can substitute nq in (1.32) for λt and obtain

$$\Pr\{k;\lambda t\} = \frac{(\lambda t)^k}{k!} e^{-\lambda t} \tag{1.33}$$

We will soon discuss the main applications of the Poisson distribution. Here we emphasize that this distribution is a very good approximation for the binomial distribution when the number of trials is very large and the probability of failure in a single trial is extremely small (but the mean number of events during a fixed time interval is finite).

Now let us consider different characteristics of this distribution. Based on the definition of the parameter λ, one can directly find the mean, that is, the average number of failures during a time interval t,

$$E\{X\} = \lambda t = \Lambda \tag{1.34}$$

The equation for the m.g.f. can be easily obtained with the use of (1.33)

$$\overline{\varphi}(z) = E\{e^{Xz}\} = \sum_{x>0} \frac{\Lambda^x}{x!} e^{-\Lambda} e^{xz} = e^{-\Lambda} \sum_{x>0} \frac{(\Lambda e^z)^x}{x!} = e^{-\Lambda(1-e^z)} \tag{1.35}$$

The expression can also be used to obtain the second moment

$$E\{X^2\} = \frac{d^2 e^{-\Lambda(1-e^z)}}{dz^2}\bigg|_{z=0} = \Lambda^2 + \Lambda \tag{1.36}$$

and hence from (1.34) and (1.36) we obtain

$$\mathrm{Var}\{X\} = \Lambda \tag{1.37}$$

1.2 CONTINUOUS DISTRIBUTIONS RELATED TO RELIABILITY

1.2.1 Exponential Distribution

The exponential distribution is the most popular and commonly used distribution in reliability theory and engineering. Its extreme popularity usually generates two powerful "lobbies" among the community of reliability specialists: "exponentialists" and "antiexponentialists." Both groups have many pro's and con's. Sometimes these groups remind one of the two political parties of egg eaters described by Jonathan Swift in his famous book *Gulliver's Travels*!

The "exponential addicts" in engineering will tell you that this distribution is very attractive because of its simplicity. This may or may not be a good reason! Many mathematical researchers love the exponential distribution because they can obtain a lot of elegant results with it. If, in fact, the investigated problem has at least some relation to an exponential model, this is an excellent reason!

Antagonists of the exponential distribution maintain that it is an unreasonable idealization of reality. There are no actual conditions that could generate an exponential distribution. This is not a bad reason for criticism. But on the other hand, it is principally impossible to find a natural process that is exactly described by a mathematical model.

The real question that must be addressed is: under which conditions it is appropriate to use an exponential distribution. It is necessary to understand the nature of this distribution and to decide if it can be applied in each individual case. Therefore, sometimes an exponential distribution can be used, and sometimes not. We should always solve practical problems with a complete understanding of what we really want to do.

Consider a geometric distribution and take the expression for the probability that there is no failure during n trials. If n is large and p is close to 1, one can use the approximation

$$\Pr\{X \geq n\} = p^n = (1 - q)^n \approx e^{-nq} \tag{1.38}$$

If we consider a small time interval dt, then the probability of failure for a continuous distribution must be small. In our case this probability is constant for equal intervals. Let

$$\Pr\{\text{failure during } \Delta\} = \lambda \Delta$$

Then, for the r.v. X, a continuous analogue of a geometric r.v., with $n \to \infty$ and $\Delta \to 0$, we obtain

$$\lim_{\Delta \to 0} (1 - \lambda t)^{t/\Delta} = e^{-\lambda t} \tag{1.39}$$

It is clear that the exponential distribution is a continuous analogue of the geometric distribution under the aforementioned conditions. Using the memoryless property, (1.39) can be obtained directly in another way. This property means that the probability of a successful operation during the time interval $t + x$ can be expressed as

$$P(t + x) = P(t) \cdot P(x|t) = P(t) \cdot P(x)$$

or

$$f(t + x) = f(t) + f(x) \tag{1.40}$$

where $f(y) = \ln P(y)$.

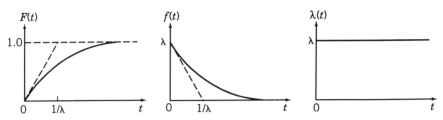

Figure 1.1. Exponential distribution $F(t)$, its density $f(t)$, and its hazard function $\lambda(t)$.

But the only function for which (1.40) holds is the linear function. Let $f(y) = ay$. Then $P(y) = \exp(ay)$. Now one uses the condition that $F(\infty) = 1 - P(\infty) = 1$ and finds that $a = -1$. Therefore, the probability of having no failure during the period t equals

$$P(t) = 1 - F(t) = \exp(-\lambda t) \tag{1.41}$$

The distribution function is

$$F(t) = 1 - \exp(-\lambda t)$$

and the density function is

$$f(t|\lambda) = \lambda \exp(-\lambda t) \tag{1.42}$$

The exponential distribution is very common in engineering practice. It is often used to describe the failure process of electronic equipment. Failures of such equipment occur mostly because of the appearance of extreme conditions during their operation. We will show below that such events can be successfully described by a Poisson process. In turn, the Poisson process very closely relates to the exponential distribution.

In addition, we should emphasize that the exponential distribution appears in several practical important cases when one considers highly reliable repairable (renewal) systems.

Both of these cases are related to the case where a continuous (or discrete) stochastic process crosses a high-level threshold. Indeed, intuitively we feel that a level might be considered as "high" because it is very seldom reached.

Now let us find the main characteristics of the exponential distribution. The easiest way to find the mean of the exponential r.v. is to integrate the function $P(t) = 1 - F(t)$:

$$E\{X\} = \int_0^\infty \lambda t e^{-\lambda t}\, dt = \frac{1}{\lambda} \tag{1.43}$$

The second initial moment of the distribution can also be found in a direct way

$$E\{X^2\} = \int_0^\infty \lambda t^2 e^{-\lambda t}\, dt = \frac{2}{\lambda^2} \tag{1.44}$$

and, consequently, from (1.43) and (1.44)

$$\mathrm{Var}\{X\} = \frac{2}{\lambda^2} - \frac{1}{\lambda^2} = \frac{1}{\lambda^2}$$

that is, the standard deviation of an exponential distribution equals the mean

$$\sigma = \sqrt{\mathrm{Var}\{X\}} = \frac{1}{\lambda}$$

The m.g.f. for the density can also be found in a direct way

$$\bar\varphi_e(s) = \int_0^\infty e^{st}\lambda e^{-\lambda t}\, dt = \frac{\lambda}{\lambda - s} \tag{1.45}$$

For future applications it is convenient to have the Laplace–Stieltjes transform (LST) of a density function. For the density of an exponential distribution, the LST equals

$$\varphi(s) = \int_0^\infty \lambda e^{-\lambda t} e^{-st}\, dt = \frac{\lambda}{\lambda + s} \tag{1.46}$$

As we considered above, the LST of the function $P(t) = 1 - F(t) = e^{-\lambda t}$, taken at $s = 0$, equals the mean. In this case

$$\varphi_p(s) = \int_0^\infty e^{-\lambda t} e^{-st}\, dt = \frac{1}{\lambda + s} \tag{1.47}$$

and, consequently,

$$\varphi_p(s = 0) = \frac{1}{\lambda}$$

One very important characteristic of continuous distributions is the *intensity function* which, in reliability theory, is called the *failure rate*. This function is determined as the conditional density at a moment t under the condition that the r.v. is not less than t. Thus, the intensity function is

$$\lambda(t) = \frac{f(t)}{1 - F(t)} = \frac{f(t)}{P(t)}$$

For the exponential distribution the intensity function can be written as

$$\lambda(t) = \frac{f(t)}{P(t)} = \lambda \qquad (1.48)$$

that is, the failure rate for an exponential distribution is constant. This follows as well from the memoryless property. In reliability terms it means, in particular, that current or future reliability properties of an operating piece of equipment do not change with time and, consequently, do not depend on the amount of operating time since the moment of switching the equipment on. Of course, this assumption seems a little restrictive, even for "exponential addicts." But this mathematical description is sometimes practically sufficient.

1.2.2 Erlang Distribution

The Erlang distribution is the continuous analogue of a negative binomial distribution. It represents the sum of a fixed number of independent and exponentially distributed r.v.'s. The principal mathematical model for the description of queuing processes in a telephone system is a Markov one. Consider a multiphase stage, or example, a waiting line of messages. An observed message can stand in line behind several previous ones, say N. Then for this message the waiting time can be represented as a sum of the N serving times of the previous messages. By assumption, for a Markov-type model, each of these serving times has an exponential distribution, and so the resulting waiting time of the message under consideration has an Erlang distribution.

The sum of N independent exponential r.v.'s forms an Erlang distribution of the Nth order. It is then clear that the mean of an r.v. with an Erlang distribution of the Nth order is a sum of N means of exponential r.v.'s, that is,

$$E\{X\} = \frac{N}{\lambda} \qquad (1.49)$$

and so the variance equals N times the variance of a corresponding exponential distribution

$$\text{Var}\{X\} = \frac{N}{\lambda^2} \tag{1.50}$$

Finally, the LST of the density of an Erlang distribution of the Nth order is

$$\varphi(s) = \left(\frac{\lambda}{\lambda + s}\right)^N \tag{1.51}$$

The last expression allows us to write an expression for the density function of this distribution

$$f_N(t) = \lambda \frac{\lambda^{N-1}}{(N-1)!} e^{-\lambda t} \tag{1.52}$$

(e.g., one can use a standard table of the Laplace–Stieltjes transforms). We will show the validity of (1.52) below when we consider a Poisson process.

Note that if the exponential r.v.'s which compose the Erlang r.v. are not identical, the resulting distribution is called a *generalized Erlang distribution*. Here we will not write the special expression for this case but one can find related results in Section 1.6.7 dedicated to the so-called death process.

1.2.3 Normal Distribution

This distribution occupies a special place among all continuous distributions because many complex practical cases can be modeled by it. This d.f. is often termed a *Gaussian distribution*.

The central limit theorem of probability theory states that the sum of independent r.v.'s under some relatively nonrestrictive conditions has an asymptotically normal distribution. This fundamental result has an intriguing history which has developed over more than two centuries.

A simple example of a practical application of the central limit theorem in engineering occurs in the study of the supply of spare parts. Assume that some unit has a random time to failure with an unknown distribution. We know only the mean and variance of the distribution. (These values can be estimated, even with very restricted statistical data.) If we are planning to supply spare parts over a long period of time, as compared to the *mean time to failure* (MTTS) of the unit, we can assume that the total time until exhaustion of n spare units has an approximately normal distribution. This approximation is practically irreproachable if the number of planned spare parts, n, is not less than 30.

In engineering practice the normal distribution is usually used for the description of the dispersion of different physical parameters. For example, the resistance or electrical capacity of a sample of units is often assumed to be normally distributed; the normal distribution characterizes the size of mechanical details; and so on. Incidentally, many mechanical structures exposed to wear are assumed to have a normal d.f. describing their time to failure.

The normal distribution of the random time to failure (TTF) also appears when the main parameter changes linearly in time and has a normal distribution of its starting value. (The latter phenomenon was mentioned above.) In this case the time to the excedance of a specified tolerance limit will have normal distribution. We will explain this fact in mathematical terms below.

The normal distribution has the density function

$$f_n(x|a, \sigma) = \frac{1}{\sigma\sqrt{2\pi}} e^{-(x-a)^2/2\sigma^2} \tag{1.53}$$

where a and σ^2 are the mean and the variance of the distribution, respectively. These two parameters completely characterize the normal distribution. The parameter σ is called the *standard deviation*. Notice that σ is always nonnegative. From (1.53) one sees that it is a symmetrical unimodal function; it has a bell-shaped graph (see Figure 1.2).

That a and σ^2 are, respectively, the mean and the variance of the normal distribution can be shown in a direct way with the use of (1.53). We leave this direct proof to Exercises 1.2 and 1.3. Here we will use the m.g.f.

$$\tilde{\varphi}_n(s) = \int_{-\infty}^{\infty} \frac{1}{\sigma\sqrt{2\pi}} e^{-(x-a)^2/2\sigma^2} e^{sx} \, dx = \exp\left(as + \tfrac{1}{2}\sigma^2 s^2\right) \tag{1.54}$$

(The proof of this is left to Exercise 1.4.)

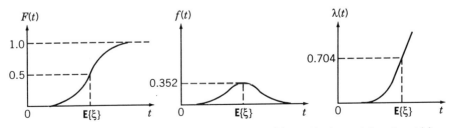

Figure 1.2. Normal distribution $F(t)$, its density $f(t)$, and its hazard function $\lambda(t)$.

From (1.54) one can easily find

$$E\{X\} = \frac{d\bar{\varphi}_n(z)}{dz}\bigg|_{z=0} = a \qquad (1.55)$$

$$E\{X^2\} = \frac{d^2\bar{\varphi}_n(z)}{dz^2}\bigg|_{z=0} = a^2 + \sigma^2 \qquad (1.56)$$

and

$$\text{Var}\{X\} = \sigma^2 \qquad (1.57)$$

[The proof of (1.56) is left to Exercise 1.5.]

In applications one often uses the so-called standard normal d.f. In this case $a = 0$ and $\sigma = 1$. It is clear that an arbitrary normal r.v. X can be reduced to a standard one. Consider the new r.v. $X' = X - a$ (obviously, the variances of X and X' are equal) and normalize this new r.v. by dividing by σ. In this way an arbitrary normal distribution can be reduced to the standard one (or vice versa) by means of a linear change of scale and changing the location of its mean to 0.

The density of a normal d.f. is (see Figure 1.2)

$$f_n(x|0, 1) = \frac{1}{\sigma\sqrt{2\pi}} e^{-x^2/2} \qquad (1.58)$$

The function (1.58) has been tabulated in different forms and over a very wide range (see Fig. 1.3). Using the symmetry of the density function, one can compile a table of the function

$$F_n^*(x) = \int_0^x f_n(x|0, 1)\, dx$$

The correspondence between the functions $F_n(x)$ and $F_n^*(x)$ is

$$F_n(-x) = \tfrac{1}{2} + F_n^*(x)$$

Often one can find a standard table of the so-called *Laplace function*: $1 - 2F(x)$. This kind of table is used, for instance, in artillery calculations to find the probability of hitting a target.

The distribution function of a normal distribution decreases very rapidly with increasing x. Most standard tables are, in fact, composed for $|x| < 3$ or 4, which is enough for most practical purposes. But sometimes one needs values for larger x. In this case we suggest the following iterative computational procedure.

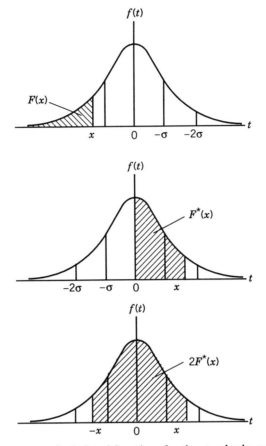

Figure 1.3. Three types of tabulated functions for the standard normal distribution.

Consider the integral

$$I = \int_t^\infty e^{-x^2/2} \, dx$$

It can be rewritten as

$$I = \int_t^\infty \frac{1}{x} \left[x e^{-x^2/2} \right] dx = -\int_t^\infty \frac{1}{x} d\left[e^{-x^2/2} \right]$$

Using integration by parts, one obtains

$$I = \frac{1}{t} e^{-t^2/2} - \int_t^\infty \frac{1}{x^2} \left[e^{-x^2/2} \right] dx = \frac{1}{t} e^{-t^2/2} - I_1 < \frac{1}{t} e^{-t^2/2}$$

Now we can evaluate I_1:

$$I_1 = \int_t^\infty \frac{1}{x^3}\left[\frac{1}{x}e^{-x^2/2}\right]dx = \int_t^\infty \frac{1}{x^3}\,d\left[e^{-x^2/2}\right]$$

and after integration by parts

$$I_1 = \frac{1}{t^3}e^{-t^2/2} - 3\int_t^\infty \frac{1}{x^4}e^{-x^2/2}\,dx = \frac{1}{t^3}e^{-t^2/2} - I_2 < \frac{1}{t^3}e^{-t^2/2}$$

Thus at this stage of the iteration

$$I > \left(\frac{1}{t} - \frac{1}{t^3}\right)e^{-t^2/2}$$

More accurate approximations can be obtained in an analogous manner.

1.2.4 Truncated Normal Distribution

A normal d.f. ranges from $-\infty$ to $+\infty$. But in reliability theory one usually focuses on the *lifetime* of some object, and so we need consider distributions defined over the domain $[0, +\infty)$. The new d.f. (see Fig. 1.4) is said to be "truncated (from the left)." The new density function, $\tilde{f}(x|a,\sigma)$, can be related to the initial one, $f(x|a,\lambda)$, as follows:

$$\tilde{f}(x|a,\sigma) = \frac{f(x|a,\sigma)}{1 - F(0)} = \frac{e^{-(x-a)^2/2\sigma^2}}{\displaystyle\int_0^\infty e^{-(x-a)^2/2\sigma^2}\,dx}$$

In practical problems this truncation often has a negligible influence if a/σ is greater than 4 or 5.

The mean of a truncated distribution is always larger than the mean of its related normal distribution. The variance, on the other hand, is always smaller. We will not write these two expressions because of their complex form.

1.2.5 Weibull – Gnedenko Distribution

One of then most widely used distributions is the Weibull–Gnedenko distribution. This two-parameter distribution is convenient for practical applications because an appropriate choice of its parameters allows one to use it to describe various physical phenomena. One of the parameters, λ, is called the scale parameter and another, β, is called the shape parameter of the

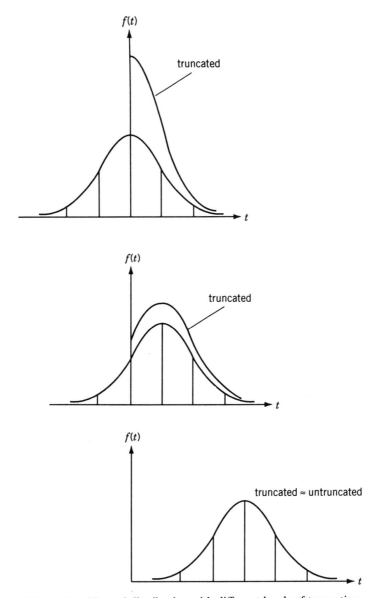

Figure 1.4. Normal distribution with different levels of truncation.

distribution. A Weibull–Gnedenko distribution has the form

$$F(t) = \begin{cases} 1 - e^{-(\lambda t)^{\beta}} & \text{for} \quad t \geq 0 \\ 0 & \text{for} \quad t < 0 \end{cases}$$

The density function is

$$f(t) = \begin{cases} \lambda^{\beta}\beta t^{\beta-1}e^{-(\lambda t)^{\beta}} & \text{for} \quad t \geq 0 \\ 0 & \text{for} \quad t < 0 \end{cases}$$

The density function for several different parameter values is presented in Figure 1.5.

The failure rate of the distribution is

$$\lambda(t) = \lambda^{\beta}\beta t^{\beta-1}$$

The behavior of the failure rate depending on the parameter values is depicted in Figure 1.6. For $\beta = 1$, the Weibull–Gnedenko d.f. transforms into a common exponential function (the failure rate is constant). For $\beta > 1$, one observes an increasing failure rate: for $1 < \beta \leq 2$, this is concave; for $\beta \geq 2$, this is convex. For $0 < \beta < 1$, the failure rate is decreasing.

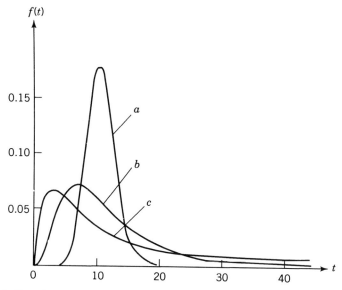

Figure 1.5. Density of the Weibull–Gnedenko distribution $f(t)$ for the following parameters: (a) $\beta = 1$, $\lambda = 1$; (b) $\beta = 2$, $\lambda = 1$; (c) $\beta = 4$, $\lambda = 1$; $\beta = 2$, and $\lambda = 0.7$.

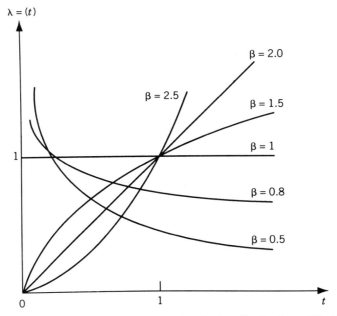

Figure 1.6. Hazard rate for the Weibull–Gnedenko distribution with $\lambda = 1$ and different parameter β.

The mean of this d.f. is

$$E\{\xi\} = \frac{1}{\lambda} \Gamma\left(1 + \frac{1}{\beta}\right)$$

and the variance is

$$\text{Var}\{\xi\} = \frac{1}{\lambda^2}\left[\Gamma\left(1 + \frac{2}{\beta}\right) - \left(\Gamma\left(1 + \frac{1}{\beta}\right)\right)^2\right]$$

where $\Gamma(\cdot)$ is the gamma function.

1.2.6 Lognormal Distribution

In mechanics one often sees that material fatigue follows a so-called lognormal distribution. This distribution appears if the logarithm of the time to failure has a normal distribution. For $t > 0$, one has

$$F(t) = \Phi\left(\frac{\log t - \mu}{\sigma}\right)$$

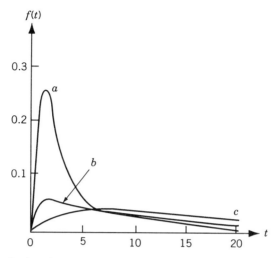

Figure 1.7. Density function for the lognormal distribution with different parameters: (a) $\mu = 1$, $\sigma = 1$; (b) $\mu = 3$, $\sigma = 1.7$; (c) $\mu = 3$, $\sigma = 1$.

and the density is

$$f(t) = \begin{cases} \dfrac{1}{\sqrt{2\pi}\,\sigma t}\,\exp\left[-\dfrac{(\log t - \mu)^2}{2\sigma^2}\right] & \text{for} \quad t > 0 \\ 0 & \text{for} \quad t > 0 \end{cases}$$

A sample of a lognormal distribution for several parameter values is depicted in Figure 1.7. The mean and the variance have the following forms, respectively:

$$E\{\xi\} = e^{\mu + \sigma^2/2}$$

and

$$\text{Var}\{\xi\} = e^{2\mu + \sigma^2}\left(e^{\sigma^2} - 1\right)$$

For a small coefficient of variation, one can use a normal approximation for a lognormal d.f.

1.2.7 Uniform Distribution

For this distribution the density function is constant over its domain $[a, b]$. The graphs of the density and distribution functions are presented in Figure 1.8. The density function is

$$f(x) = \begin{cases} \dfrac{1}{b-a} & \text{for} \quad a \leq x \leq b \\ 0 & \text{for} \quad x < a \text{ and } x > b \end{cases} \tag{1.59}$$

and the d.f. is

$$F(x) = \int_a^x \frac{1}{b-a} \, dx = \frac{x-a}{b-a} \tag{1.60}$$

Because of the symmetry of the density function, the mean is $(b-a)/2$. The variance can be calculated as

$$\text{Var}\{X\} = \int_a^b \frac{\left(x - \dfrac{b-a}{2}\right)^2}{b-a} \, dx = \frac{(b-a)^2}{12} \tag{1.61}$$

The uniform distribution on the interval $[0, 1]$ plays an important role in reliability and its related applications. It is determined by the fact that an r.v. $y = F^{-1}(x)$ [here F^{-1} is the inverse for $F(x)$] has a uniform distribution. This fact is often used for the generation of r.v.'s with a desired distribution on the basis of uniformly distributed r.v.'s. For example, to generate an r.v. ξ with a specified d.f. $F(x)$, we must take the generator of a uniformly distributed r.v. $y_1, y_2, \dots$ and arrange the inverse transforms: $\xi_1 = F^{-1}(y_1)$, $\xi_2 = F^{-1}(y_2), \dots$.

For computer simulations the so-called *pseudo-random variable* (p.r.v.) is usually generated. The first generator of uniformly distributed r.v.'s was

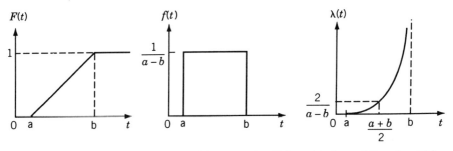

Figure 1.8. Uniform distribution $F(t)$, its density $f(t)$, and its hazard function $\lambda(t)$.

introduced by John von Neuman. The principle consists in the recurrent calculation of some function.

For example, one takes an exponential function with some two-digit power, chooses, say the 10th and 11th digits as the next power, and repeats the procedure from the beginning. Of course, such a procedure leads to the formation of a cycle: as soon as the same power appears, the continuation of the procedure will be a complete repetition of one of the previous links of p.r.v.'s. At any rate, it is clear that the cycle cannot be larger than 100 p.r.v.'s if the power of the exponent consists of two digits. Fortunately, modern p.r.v. generators have practically unrestricted cycle lengths.

At the same time, p.r.v.'s are very important for different numerical simulation experiments designed for comparison of different variants of a system design. Indeed, one can completely repeat a set of p.r.v.'s by starting the procedure from the same initial state. This allows one to put different system variants into an equivalent pseudo-random environment. This is important to avoid real random mistakes caused by putting one system in a more severe "statistical environment" than another.

1.3   SUMMATION OF RANDOM VARIABLES

The summation of random variables often comes up in engineering problems involving a probabilistic analysis. The observation of a series of time sequences or the analysis of the number of failed units arriving at a repair shop are examples. At the same time, the number of terms in the sum is not always given—sometimes it is random. Asymptotic results are also of practical interest.

1.3.1   Sum of a Fixed Number of Random Variables

General Case   Consider a repairable system which is described by cycles as "a period of operation" and "a period of repair." Each cycle consists of two r.v.'s ξ and η, a random time to failure (TTF) with distribution $F(t)$, and a random repair time with distribution $G(t)$, respectively. If the distribution of the complete cycle is of interest, we would analyze the sum $\theta = \xi + \eta$. The distribution of this new r.v., denoted as $D(t) = \Pr\{\theta \leq t\}$, is the convolution of the initial d.f.'s:

$$D(t) = \Pr\{\xi + \eta \leq t\} = F * G(t) = \int_0^t F(t - x)\, dG(x)$$

$$= G * F(t) = \int_0^t G(t - x)\, dF(x)$$

If the Laplace–Stieltjes transforms (LSTs) of these d.f.'s

$$\varphi_F(s) = \int_0^\infty F(t)e^{-st}\,dt$$

and

$$\varphi_G G(s) = \int_0^\infty G(t)e^{-st}\,dt$$

are known, the LST of the d.f. $D(t)$ is

$$\varphi_D(s) = \varphi_F(s)\varphi_G(s)$$

If one considers a sum of n i.i.d. r.v.'s, the convolution $F^{*n}(t)$ is

$$\Pr\left\{ \sum_{1\le k\le n} \xi_k \le t \right\} = \int_0^t \Pr\left\{ \sum_{1\le k\le n-1} \xi_i \le t - x \right\} dF(x)$$

$$= \int_0^t F^{*(n-1)}(t - x)\,dF(x) = F^{*(n)}$$

where all F^{*k}'s are determined recurrently. For a sum of i.i.d. r.v.'s each of which has LST equal to $\varphi(s)$,

$$\varphi_n(s) = [\varphi(s)]^n$$

For the sum of n r.v.'s with arbitrary distributions, one can write

$$E\{\xi_\Sigma\} = E\left\{ \sum_{1\le j\le n} \xi_j \right\} = \sum_{1\le j\le n} E\{\xi_i\} \tag{1.62}$$

and, for independent r.v.'s.

$$\sigma_\Sigma = \sqrt{ \sum_{1\le j\le n} \sigma_i^2 } \tag{1.63}$$

Now we begin with several important and frequently encountered special cases.

Sum of Binomial Random Variables   Consider two binomially distributed r.v.'s ν_1 and ν_2 obtained, respectively, by n_1 and n_2 Bernoulli trials with the same parameter q. From (1.21), the g.f. of the binomial distribution is

$$\hat{\varphi}(s) = (pz + q)^{n_i} \qquad i = 1, 2 \tag{1.64}$$

Thus,

$$\hat{\varphi}(z) = \hat{\varphi}_1(z)\hat{\varphi}_2(z) = (pz + q)^{n_1}(pz + q)^{n_2} = (pz + q)^{n_1 + n_2} \quad (1.65)$$

In other words, the sum of two bionomially distributed r.v.'s with the same parameter p will produce another binomial distribution. This can be easily explained: arranging a joint sample from two separate samples of sizes n_1 and n_2 from the same population is equivalent to taking one sample of size $n_1 + n_2$.

Obviously, an analogous result holds for an arbitrary finite number of binomially distributed r.v.'s. Thus, we see that the sum of binomially distributed r.v.'s with the same parameter p produces a new binomially distributed r.v. with the same p and corresponding parameter

$$n = \sum_{1 \le j \le N} n_j$$

For different binomial distributions, the result is slightly more complicated (see Exercise 1.8).

Sum of Poisson Random Variables   Consider the sum X_Σ of two independent Poisson r.v.'s X_1 and X_2 with corresponding parameters $\Lambda_i, i = 1, 2$. The m.g.f.'s for the two Poisson distributions are written as

$$\overline{\varphi}_i(z) = e^{\Lambda_i(1 - e^z)} \qquad i = 1, 2 \tag{1.66}$$

The m.g.f. for the distribution of the sum X_Σ can be written as

$$\varphi(z) = e^{\Lambda_1(e^z - 1)}e^{\Lambda_2(e^z - 1)} = e^{(\Lambda_1 + \Lambda_2)(e^z - 1)} \tag{1.67}$$

that is, the resulting m.g.f. is the m.g.f. of a new Poisson d.f. with parameter $\Lambda_\Sigma = \Lambda_1 + \Lambda_2$.

An analogous result can be obtained for an arbitrary finite number of Poisson r.v.'s. In other words, the sum of N Poisson r.v.'s is again a Poisson r.v. with parameter equal to the sum of the parameters:

$$\Lambda = \sum_{1 \le j \le n} \Lambda_i \tag{1.68}$$

Sum of Normal Random Variables   The sum of independent normally distributed r.v.'s has a normal distribution. Again consider a sum of two r.v.'s. Let X_i be a normal r.v. with parameters a_i and σ_i, $i = 1, 2$, and let

$X_\Sigma = X_1 + X_2$. Then the m.g.f. for X can be expressed as

$$\bar{\varphi}_\Sigma(z) = \bar{\varphi}_1(z)\bar{\varphi}_2(z) = \exp\left(a_1 z + \tfrac{1}{2}\sigma_1^2 z^2\right)\exp\left(a_1 z + \tfrac{1}{2}\sigma_2^2 z^2\right)$$
$$= \exp\left[z(a_1 + a_2) + \tfrac{1}{2}z^2(\sigma_1^2 + \sigma_2^2)\right] \tag{1.69}$$

Therefore, the sum of two normal r.v.'s produces an r.v. with a normal distribution. For n terms, the parameters of the resulting normal distribution are

$$a_\Sigma = \sum_{1 \le i \le N} a_i \tag{1.70}$$

and

$$\sigma_N = \sqrt{\sum_{1 \le i \le N} \sigma_i^2} \tag{1.71}$$

1.3.2 Central Limit Theorem

Many statisticians have worked on the problem of determining the limit distribution of a sum of r.v.'s. This problem has practical significance because, when a sum includes a large number of r.v.'s, the direct calculation of some characteristic of the sum becomes very complicated. The problem itself has aroused theoretical interest even outside of applications.

Above we showed that a sum of different normally distributed independent r.v.'s has a normal distribution, independent of the number of terms in the sum. The new resulting normal distribution has a mean equal to the sum of the means of the initial distributions and a variance equal to the sum of the variances. It is obvious that this property is preserved with the growth of n.

But what will be the limiting distribution of a sum of r.v.'s whose distributions are not normal? It turns out that, with increasing n, such a sum has a tendency to converge to a normally distributed r.v.

In simple engineering terms it appears that if we consider a sum of a large number n of independent r.v.'s ξ, then this sum has approximately a normal distribution. If we consider the sum of independent arbitrary distributed r.v.'s ξ with mean $a = E\{\xi\}$ and variance $v = \text{Var}\{\xi\}$, then the normal distribution of the sum will have mean $A = an$ and variance $V = vn$. (Of course, some special restrictions on the independence and properties of distributions must be fulfilled.)

Historically, limit theorems developed over several centuries. Different versions of them pertain to different cases. One of the first attempts in this direction is contained in the following theorem.

DeMoivre Local Theorem   Consider a sequence n of Bernoulli trials with a probability of success p. The probability of m successes $P_n(m)$ satisfies the relationship

$$\frac{\sqrt{npq}\, P_n(m)}{\frac{1}{\sqrt{2\pi}}e^{-x^2/2}} \xrightarrow[n\to\infty]{} 1$$

uniformly for all m such that

$$x = \frac{m - np}{\sqrt{npq}}$$

belongs to some finite interval.

This theorem, in turn, is the basis of the following theorem.

Integral DeMoivre – Laplace Theorem   If ν is the random number of successes among n Bernoulli trials, then for finite a and b the following relationship holds:

$$\Pr\left\{a \le \frac{\nu - np}{\sqrt{npq}} < b\right\} \xrightarrow[n\to\infty]{} \frac{1}{\sqrt{2\pi}}\int_a^b e^{-z^2/2}\, dz$$

The next step in generalizing the conditions under which the sum of a sequence of arbitrary r.v.'s converges to a normal distribution is formulated in the following theorem.

Liapounov Central Limit Theorem   Suppose that the r.v.'s X_i are independent with known means a_i and variances σ_i^2, and for all of them, $E\{|X_i - a_i|^3\} < \infty$. Also, suppose that

$$\lim_{n\to\infty} \frac{\sum\limits_{1\le i\le n} E\{|X_i - a_i|^3\}}{\sqrt[3]{\left(\sum\limits_{1\le i\le n}\sigma_i^2\right)^2}} = 0$$

Then, for the normalized and centered (with zero mean) r.v.,

$$Y_n = \frac{\sum\limits_{1\le i\le n} X_i - \sum\limits_{1\le i\le n} a_i}{\sqrt{\sum\limits_{1\le i\le n}\sigma_i^2}}$$

for any fixed number x,

$$\lim_{n \to \infty} \Pr\{Y_n \leq x\} = \frac{1}{\sqrt{2\pi}} \int_{-\infty}^{x} e^{-z^2/2} \, dz$$

Thus, this theorem allows for different r.v.'s in the sequence and the only restrictions are in the existence of moments of an order higher than 2. As a matter of fact, this statement is true even under weaker conditions (the restriction of a variance is enough) but all r.v.'s in the sum must be i.i.d.

For the sample mean, the related result is formulated in the following theorem.

Lindeberg – Levy Central Limit Theorem If the r.v.'s X_i are chosen at random from a population which has a given distribution with mean a and finite variance σ^2, then for any fixed number y,

$$\lim_{n \to \infty} \Pr\left\{ \frac{\sqrt{n}\left(\overline{X}_n - a\right)}{\sigma} \leq y \right\} = \frac{1}{\sqrt{2\pi}} \int_{-\infty}^{y} e^{-z^2/2} \, dz$$

where $\overline{X}_n$ is the sample mean.

Because

$$\overline{X}_n = \frac{1}{n} \sum_{1 \leq i \leq n} X_i$$

this theorem may be interpreted in the following way: the sum of i.i.d. r.v.'s approximately has a normal distribution with mean equal to na and variance equal to $n\sigma^2$.

A detailed historical review on the development of probability theory and statistics can be found in Gnedenko (1988).

1.3.3 Poisson Theorem

Considering the local DeMoivre theorem, we notice that this result works well for binomial distributions with p close to $1/2$. But the normal approximation does not work well for small probabilities or on the "tails" of a binomial distribution. An asymptotic result for small p (for the "tails" of the binomial distribution) is formulated in the following theorem.

Poisson Theorem If $p_n \to 0$ with $n \to \infty$, then

$$\binom{n}{m} p_n^m (1 - p_n)^{n-m} - \frac{a_n^m}{m!} e^{-a_n} \to 0$$

where $a_n = np_n$.

This means that for small p, instead of calculating the products of astronomically large binomial coefficients with extremely small p^n, we can use a simple approximation. A standard table of the Poisson distribution can be used.

1.3.4   Random Number of Terms in the Sum

Only a very general result can be given for the d.f. of the sum of r.v.'s, or for its LST when a random number of terms is distributed arbitrarily. Further, let us assume that ν is geometrically distributed. Then the distribution of the sum of arbitrarily distributed r.v.'s is

$$\Pr\{\xi_\Sigma \leq t\} = \sum_{1 \leq j < \infty} p^j q \Pr\left\{\sum_{1 \leq k \leq j} \xi_k \leq t\right\}$$

Consider a continuous d.f. The LST can be written as

$$\varphi_\Sigma(s) = \sum_{1 \leq k \leq N} p^k q [\varphi(s)]^k$$

In general, both of the latter expressions are practically useful only for numerical calculation.

To find the mean of ξ_Σ, we may use the Wald equivalence:

$$E\left\{\sum_{1 \leq k \leq \nu} \xi_k\right\} = E\{\nu\}\, E\{\xi\} \tag{1.72}$$

Below we consider two cases where the sum of finite r.v.'s will lead to simple results.

Geometrically Distributed Random Variables   We can investigate this case without using a mathematical technique. Consider an initial sequence of Bernoulli trials. The probability of success equals p and the probability of failure equals $q = 1 - p$. Now construct the new process consisting of only failures of the initial process and corresponding spaces between them. Consider a new procedure: each failure in the initial Bernoulli process creates a possibility for the appearance of a failure in the final process. (Failure cannot appear in the space between failures of the initial process.) A special moment concerning the "possibility" of a new process failure is considered. Let a failure of the initial process develop into a failure of the new (final) process with probability Q. Thus, if we consider the initial process, failure of the final process occurs there with probability $Q^* = qQ$.

We have obtained this result using only verbal arguments. Of course, it can be derived in strict mathematical terms.

Exponentially Distributed Random Variables Consider the sum of a random number of exponentially distributed identical and independent r.v.'s, with parameter λ. Assume that the number of terms in the sum has a geometric d.f. with parameter p. We will express the LST of the resulting density function through the LST of the density function of the initial d.f. From the formula for the complete mathematical expectation, we have

$$\varphi(s) = q\frac{\lambda}{\lambda + s} + pq\frac{\lambda^2}{(\lambda + s)^2} + p^2q\frac{\lambda^3}{(\lambda + s)^3} + \cdots$$

$$= \frac{q\lambda}{\lambda + s}\sum_{0 \le k < \infty}\frac{(p\lambda)^k}{(\lambda + s)^k} = \frac{q\lambda}{\lambda + s}\frac{1}{1 - \dfrac{p\lambda}{\lambda + s}} = \frac{q\lambda}{s + q\lambda} \quad (1.73)$$

Thus, we have an expression which represents the LST for an exponential distribution with parameter $\Lambda = \lambda q$.

We illustrate the usefulness of this result by means of a simple example. Imagine a socket with unit installed. Such a unit works for a random time, distributed exponentially, until a failure occurs. After a failure, the unit is replaced by a new one. The installation of each new unit may lead to a socket failure with probability q.

This process continues until the first failure of the socket. This process can be described as the sum of a random number of exponentially distributed random variables where the random number has a geometrical distribution.

Of course, in general, the final distribution of the sum strongly depends on the distribution of the number of terms in the sum. The distribution of the number of terms in the sum is the definitive factor for the final distribution.

1.3.5 Asymptotic Distribution of the Sum of a Random Number of Random Variables

In practice, we often encounter situations where, on the average, the random number of terms in the sum is very large. Usually, the number of terms is assumed to be geometric. If so, the following limit theorem is true.

Theorem 1.1 Let $\{\xi_i\}$ be a sequence of i.i.d. r.v.'s whose d.f. is $F(t)$ with mean $a > 0$. Let ν be the number of discrete r.v.'s of a sequence with a geometric distribution with parameter p: $\Pr\{\nu = k\} = qp^{k-1}$ where $q = 1 - p$. Then, if $p \to 1$, the d.f. of the normalized sum

$$\xi_\Sigma = q\sum_{1 \le k \le \nu}\xi_k$$

converges to the exponential d.f. $1 - e^{-qt}$.

Proof. Consider the normalized r.v.

$$\xi_\Sigma = \frac{\sum\limits_{1 \le k \le \nu} \xi_k}{\sum\limits_{1 \le k \le \nu} E\{\xi_k\}}$$

By the Wald equivalency,

$$E\left\{\sum_{1 \le k \le \nu}\right\} = E\{\nu\}\, E\{\xi\}$$

Without loss of generality, we can take $E\{\xi\} = 1$. Because ν has a geometric distribution, $E\{\nu\} = 1/q$. Hence,

$$\xi_\Sigma = q \sum_{1 \le k \le \nu} \xi_k$$

The LST of ξ_Σ is

$$\varphi_\Sigma(s) = E\{e^{-s\xi_\Sigma}\} = E\left\{\exp\left(-qs \sum_{1 \le i \le \nu} \xi_i\right)\right\}$$

$$= \sum_{1 \le k < \infty} p^{k-1} q \exp\left(-qs \sum_{1 \le j \le k} \xi_i\right)$$

Note that

$$\exp\left(-qs \sum_{1 \le i \le k} \xi_i\right) = [\varphi(sq)]^k$$

Then

$$\varphi_\Sigma(s) = \sum_{1 \le k < \infty} p^{k-1} q [\varphi(sq)]^k$$

$$= q\varphi(sq) \sum_{0 \le k < \infty} [p\varphi(sq)]^k = \frac{q\varphi(sq)}{1 - p\varphi(sq)}$$

Now with some simple transformations

$$\varphi_\Sigma(s) = \frac{q\varphi(sq)}{1 - p\varphi(sq)} = \frac{q\varphi(sq)}{1 - \varphi(sq) + q\varphi(sq)}$$

$$= \frac{\varphi(sq)}{s\dfrac{1 - \varphi(sq)}{sq} + \varphi(sq)}$$

Notice that $\varphi(s)|_{s=0} = 1$. Hence,

$$\lim_{q \to 0} \varphi(sq) = 1$$

and, consequently,

$$\lim_{q \to 0} \frac{1 - \varphi(sq)}{sq} = \lim_{q \to 0} \frac{\varphi(0) - \varphi(sq)}{sq} = -\varphi'(0) = -E\{\xi\}$$

Taking into account that $E\{\xi\} = 1$, we have finally

$$\varphi_\Sigma(s) = \frac{1}{s + 1}$$

that is, ξ_Σ has an exponential distribution with parameter $\lambda = 1$.

1.4 RELATIONSHIPS AMONG DISTRIBUTIONS

Various distributions have common roots, or are closely related. As we discussed previously, the normal and exponential distributions serve as asymptotic distributions in many practical situations. Below we establish some connections among different distributions that are useful in reliability analysis.

1.4.1 Some Relationships Between Binomial and Normal Distributions

The De Moivre–Laplace theorem shows that, for large n when $\min(nq, np) \gg 1$, the binomial distribution can be approximated by the normal distribution.

Example 1.1   A sample consists of $n = 1000$ items. The probability that the item satisfies some specified requirement equals $Pr\{success\} = p = 0.9$. Find $Pr\{880 \le \text{number of successes}\}$.

Solution.   For the normal d.f. which approximates this binomial distribution, we determine that $a = np = 900$ and $\sigma^2 = npq = 90$, that is, $\sigma = 9.49$. Thus,

$$Pr\{880 \le X\} = 1 - \Phi\left(\frac{880 - 900}{9.49}\right)$$

$$= 1 - \Phi(-2.11) = 1 - 0.0175 = 0.9825$$

Example 1.2 Under the conditions of the previous example, find the number of good items which the producer can guarantee with probability 0.99 among a sample of size $n = 1000$.

Solution. Using a standard table of the normal distribution, from the equation

$$\Pr\{m > x\} = \Phi\left(\frac{x - 900 + 0.5}{\sqrt{900}}\right) = \Phi\left(\frac{x - 900.5}{9.49}\right) = 0.01$$

we find

$$\frac{x - 900.5}{9.49} \approx -2.33$$

or $x = 978.6$. Thus, the producer can guarantee not less than 978 satisfactory items with the specified level of 99%.

We must remember that such an approximation is accurate for the area which is more or less close to the mean of the binomial distribution. This becomes clear if one notices that the domain of a normal distribution is $(-\infty, \infty)$, while the domain of a binomial distribution is restricted to $[0, n]$.

In addition, there is an essential difference between discrete and continuous distributions. Thus, we must use the so-called "correction of continuity":

$$\Pr\{\alpha \le X \le \beta\} \approx \Phi\left(\frac{\beta + \frac{1}{2} - a}{\sqrt{npq}}\right) - \Phi\left(\frac{\alpha - \frac{1}{2} - a}{\sqrt{npq}}\right)$$

1.4.2 Some Relationships Between Poisson and Binomial Distributions

By the Poisson theorem, a Poisson distribution is a good approximation for a bionomial distribution when p (or q) is very small.

Example 1.3 A sample consists of $n = 100$ items. The probability that an item is defective is equal to $p = 0.005$. Find the probability that there is exactly one defective item in the sample.

Solution. Compute $a = 100 \cdot (0.005) = 0.5$. From a standard table of the Poisson distribution, we find $p(1; 0.5) = 0.3033$. The computation with the use of a binomial distribution gives

$$p_b(1; 0.005, 100) = \binom{100}{1} 0.005 \cdot 0.995^{999} \approx 0.5 e^{-0.5}$$

$$\approx (0.5) \cdot (0.6065) \approx 0.3033$$

1.4.3   Some Relationships Between Erlang and Normal Distributions

The normal approximation can be used for the Erlang distribution when k is large, for instance, when k is more than 20. This statement follows from the "Lindeberg form" of the central limit theorem.

Let Y be an r.v. with an Erlang distribution of the kth order. In other words, $Y = X_1 + X_2 + \cdots + X_k$ where all X_i's are i.i.d. r.v.'s with an exponential distribution and parameter λ. Then, if $k \gg 1$, Y approximately has a normal distribution with mean $a = k/\lambda$ and standard deviation $\sigma = \sqrt{a}$.

Example 1.4   Consider a socket with 25 units which replace each other after a failure. Each unit's TTF has an exponential distribution with parameter $\lambda = 0.01$ [1/hour]. Find the probability of a failure-free operation of the socket during 2600 hours. (Replacements do not interrupt the system operation.)

Solution.   The random time to failure of the socket approximately has a normal d.f. with parameters $a = 25 \cdot 100 = 2500$ hours and $\sigma = \sqrt{(2500)} = 50$ hours. The probability of interest is

$$\Phi\left(\frac{2600 - 2500}{50}\right) = \Phi(2) = 0.9773$$

1.4.4   Some Relationships Between Erlang and Poisson Distributions

Consider the two following events:

(a) We observe a Poisson process with parameter λ. The probability that during the interval $[0, t]$ we observe k events of this process is

$$p_p(k; \lambda t) = \frac{(\lambda t)^k}{k!} e^{-\lambda t}$$

(b) We observe an r.v. ξ_k with an Erlang distribution of order k with parameter λ. Consider the event that ξ is smaller than t and, at the same time, ξ_{k+1} is larger than t. The probability of the latter event equals

$$\Pr\{\xi_k \le t | \xi_{k+1} > t\} = \int_0^t \lambda \frac{[\lambda(t - x)]^{k-1}}{(k - 1)!} e^{-\lambda(t-x)} e^{-\lambda x} \, dx$$

$$= \frac{(\lambda t)^k}{k!} e^{-\lambda t} = p_p(k; \lambda t)$$

Thus, events (a) and (b) are equivalent. It is important to remark that both the Erlang r.v. and the Poisson process are formed with i.i.d. r.v.'s with exponential d.f.'s.

Notice that the unconditional event $\xi_k > t$ is equivalent to the set of the following events in the Poisson process: {no events are observed} or {one event is observed} or {two events are observed} or . . . or {$k - 1$ events are observed}. This leads to the following condition:

$$\Pr\{\xi_k > t\} = \sum_0^{k-1} p_p(k; \lambda t) = \sum_0^{k-1} \frac{(\lambda t)^k}{k!} e^{-\lambda t} = P_p(k - 1; \lambda t)$$

or, for the probability of the event $\xi_k \le t$, that is, for the d.f. of the Erlang r.v. of the kth order, we have

$$\Pr\{\xi_k \le t\} = 1 - P_p(k - 1; \lambda t)$$

Therefore, in some sense, the Poisson d.f. is a cumulative function for an r.v. with an Erlang distribution.

1.4.5 Some Relationships Between Poisson and Normal Distributions

Note that a high-ordered Erlang r.v. can be approximated by a normal r.v. and, at the same time, it has a Poisson distribution as its cumulative distribution. This fact can be used as a heuristic justification for the possibility of approximating a Poisson distribution with the help of a normal distribution. The strict proof of this statement can be obtained with the help of a Gram–Charlie set (see below).

Here we take without proof that a Poisson distribution can be approximated by a normal distribution. For a Poisson d.f. with a large mean a, the approximation can be written as

$$\Pr\{m < x\} \approx \Phi\left(\frac{x - a - 0.5}{\sqrt{a}}\right)$$

Notice that this approximation is accurate in an area close to the mean and may be very bad for the "tails" of the Poisson distribution. This is explained by the fact that these two distributions have different domains: the normal distribution is defined on $(-\infty, \infty)$, while the Poisson d.f. has no meanings for $m < 0$.

Example 1.5 Assume that the number of failures of some particular unit of equipment has a Poisson distribution. The expected number of failures during a specified period of time equals 90. One has decided to supply the equipment with 100 spare units of this type. With what probability will there be no deficit of spare units?

Solution.

$$Pr = \{m \le 100\} \approx \Phi\left(\frac{100 - 90}{\sqrt{90}}\right) = \Phi(1.05)$$

From a standard table of the normal distribution, we find that this probability equals 0.853.

Example 1.6   Under the conditions of the previous example, find how many spare units should be supplied so that the probability exceeds 0.995.

Solution.

$$Pr\{m \ge x\} \approx \Phi\left(\frac{x - 90 + 0.5}{\sqrt{90}}\right) = 0.995$$

From a standard table of the normal distribution, we find that

$$\frac{x - 90.5}{\sqrt{90}} = 2.576$$

or $x = 114.9$. This means that one should have 115 spare units.

1.4.6   Some Relationships Between Geometric and Exponential Distributions

It is clear that an exponential distribution is an approximation to a geometric distribution with $q = \lambda \, \Delta t$:

$$\lim_{\Delta t \to 0} \frac{1}{\Delta t}\left[(1 - \lambda \, \Delta t)^{t/\Delta t} \lambda \, \Delta t\right] = \lambda \lim_{\Delta t \to 0} (1 - \lambda \, \Delta t)^{t/\Delta t} = \lambda e^{-\lambda t}$$

where $\Delta t \to 0$.

1.4.7   Some Relationships Between Negative Binomial and Binomial Distributions

The relationship between these distributions is similar to the relationship between the Erlang and Poisson distributions. Consider a sequence of Bernoulli trials that forms a negative binomially distributed r.v. ν_k consisting of the sum of k geometrically distributed r.v.'s. Let us pay attention to the first n trials where $n > k$. The event $\{v_k > n\}$ means that, in the first n trials, there are 0, or 1, or 2, ..., or $k - 1$ failures, that is,

$$\Pr\{v_k > n\} = \sum_{0 \le j \le k-1} \binom{n}{j} p^j q^{n-j}$$

1.4.8 Some Relationships Between Negative Binomial and Erlang Distributions

We noticed that the geometric distribution is related to the exponential distribution. In the same sense, the convolution of geometric distributions is related to the convolution of exponential distributions. No other comments are needed: the negative bionomial and Erlang distributions are these convolutions.

1.4.9 Approximation with the Gram – Charlie Distribution

Because of the wide applications of the normal distribution, many attempts were made to use various compositions of this distribution to express other distributions. Below is one of them.

Let $f(t)$ be the density function of a distribution other than the normal distribution. The mean a and the variance σ^2 of this distribution are known. Introduce a new variable

$$t = \frac{x - a}{\sigma} \tag{1.74}$$

The density function $f(t)$ can be represented with the help of the Gram–Charlie series

$$f(t) = A_0 \varphi(t) + A_1 \varphi'(t) + A_2 \varphi''(t) + \cdots$$

where $\varphi(t), \varphi'(t), \varphi''(t), \ldots$ are the density of the normal distribution and its subsequent derivatives. The standard normal density is expressed as

$$\varphi(t) = \frac{1}{\sqrt{2\pi}} e^{-t^2/2}$$

Introduce the Chebyshev–Hermit polynomials:

$$H_n(t) = (-1)^n \frac{\varphi^{(n)}(t)}{\varphi(t)} \tag{1.75}$$

where $\varphi^{(n)}(t)$ is the nth derivative of the normal density. By direct calculations we find

$$H_0(t) = 1$$
$$H_1(t) = t$$
$$H_2(t) = t^2 - 1 \tag{1.76}$$
$$H_3(t) = t^3 - 3t$$
$$H_4(t) = t^4 - 6t^2 + 3$$
$$\cdots$$

Usually, for practical problems, we do not need more than four terms of the Gram–Charlie set.

From (1.75) it follows that

$$\varphi^{(n)}(t) = (-1)^n \varphi(t) H_n(t) \tag{1.77}$$

These functions go to 0 for all n when $t \to \pm\infty$. The functions $\varphi(t)$, H_2, and H_4 are even, and the functions H_1 and H_3 are odd, so from (1.77) it follows that

$$\varphi'(-t) = \varphi(t)$$
$$\varphi''(-t) = \varphi(t)$$
$$\varphi^{(3)}(-t) = -\varphi^{(3)}(t)$$
$$\varphi^{(4)}(-t) = \varphi^{(4)}(t)$$

The Chebyshev–Hermit polynomials are orthogonal, that is,

$$\int_{-\infty}^{\infty} \varphi(t) H_n(t) H_m(t) \, dx = \begin{cases} 0 & \text{for} \quad m \neq n, \\ m! & \text{for} \quad m = n \end{cases}$$

This fact can be proven by direct calculation.

Now substitute the Chebyshev–Hermit polynomials into (1.75)

$$f(t) = A_0 H_0(t)\varphi(t) - A_1 H_1(t)\varphi(t) + A_2 H_2(t)\varphi(t) - \cdots \tag{1.78}$$

To find A_n, multiply both sides of (1.78) by $H_n(t)$ and integrate from $-\infty$ to ∞. Because of the above-mentioned orthogonality property of the Chebyshev–Hermit polynomials, we have

$$A_n = \frac{(-1)^n}{n!} \int_{-\infty}^{\infty} f(t) H_n(t) \, dt \tag{1.79}$$

After substituting (1.76) into (1.79), we obtain

$$A_0 = 1$$
$$A_1 = -m_1^0$$
$$A_2 = \tfrac{1}{2}m_2^0 - \tfrac{1}{2} \qquad\qquad (1.80)$$
$$A_3 = -\tfrac{1}{6}\left[m_3^0 - 3m_1^0\right]$$
$$A_4 = \tfrac{1}{24}\left[m_4^0 - 6m_2^0 + 3\right]$$

where m_n^0 is the central moment on the nth order of the r.v. t.

Thus, $m_1^0 = 0$, $m_2^0 = 1$, and, consequently, all initial moments are equal to centered moments. Then from (1.80)

$$A_1 = 0$$
$$A_2 = 0$$
$$A_3 = -\tfrac{1}{6}m_3(t) = -\tfrac{1}{6}S_k(t)$$
$$A_4 = \tfrac{1}{24}\left[m_4(t) - 3\right] = \tfrac{1}{24}k_4(t)$$

where k_3 and k_4 are known as the *coefficient of asymmetry* and the *coefficient of excess*, respectively,

$$k_3(x) = \frac{m_3(x)}{\sigma^3(x)}$$

$$k_4(x) = \frac{m_4(x)}{\sigma^4(x)} - 3$$

k_3 defines the deviation of the density function under consideration from a symmetrical function, and k_4 defines the sharpness of the mode of the density function. All symmetric densities have $k_3 = 0$, and a normal density has $k_4 = 0$.

Finally, we obtain

$$f(t) \approx \varphi(t) - \tfrac{1}{6}k_3\varphi^{(3)}(t) + \tfrac{1}{24}k_4\varphi^{(4)}(t)$$

or, after integration from $-\infty$ to t,

$$F(t) \approx \Phi(t) - \tfrac{1}{6}k_3\varphi^{(2)}(t) + \tfrac{1}{24}k_4\varphi^{(3)}(t)$$

Notice that t is the linear function of x. And so, $f(x)$ and $F(x)$ can be expressed as

$$f(x) \approx \frac{1}{\sigma} \left[\varphi(t) - \frac{1}{6} k_3 \varphi^{(3)}(t) + \frac{1}{24} k_4 \varphi^{(4)}(t) \right]$$

and

$$F(t) \approx \Phi(t) - \tfrac{1}{6} k_3 \varphi^{(2)}(t) + \tfrac{1}{24} k_4 \varphi^{(3)}(t)$$

Example 1.7   With the help of the Gram–Charlie series, the Poisson distribution can be approximately expressed as

$$\Pr\{x < c\} \approx \Phi(t) - \frac{1}{6\sqrt{a}} \varphi''(t) + \frac{1}{24a} \phi^{(3)}(t) \tag{1.81}$$

where

$$t = \frac{x - a - 0.5}{\sqrt{a}}$$

and a is the parameter of the Poisson distribution.

It is clear that for $a \gg 1$ one can disregard the last two terms of the right side of (1.81), and, consequently, the Poisson distribution can be approximated by the normal distribution for large a.

REMARK. The Gram–Charlie distribution can be successfully applied to the evaluation of d.f.'s. This takes place, for instance, in analyses of the distribution of a parameter of a piece of electronic equipment when the distributions of its components are known.

1.5 STOCHASTIC PROCESSES

Stochastic processes are used for the description of a system's operation over time. There are two main types of stochastic processes: discrete and continuous. Among discrete processes, point processes in reliability theory are widely used to describe the appearance of events in time (e.g., failures, terminations of repair, demand arrivals, etc.).

A well-known type of point process is the so-called *renewal process*. This process is described as a sequence of events, the intervals between which are i.i.d. r.v.'s. In reliability theory this kind of mathematical model is used to describe the flow of failures in time.

A generalization of this type of process is the so-called *alternating renewal process* which consists of two types of i.i.d. r.v.'s alternating with each other in turn. This type of process is convenient for the description of *renewal systems*. For such systems, periods of successful operation alternate with periods of idle time.

The more complex process is a process describing a system transition from state to state. The simplest kind of such a process is a *Markov process*. If the times that the process may change states are assumed to be discrete, the process is called the *Markov chain*.

We start with simplest cases and move in the direction of more complex mathematical models.

1.5.1 Poisson Process

In the theory of stochastic processes, the Poisson process plays a special role, comparable to the role of the normal distribution in probability theory. Many real physical situations can be successfully described with the help of a Poisson process. A classical example of an application of the Poisson process is the decay of uranium: radioactive particles from a nuclear material strike a certain target in accordance with a Poisson process of some fixed intensity.

In practice, the Poisson process is frequently used to describe the flow of failures of electronic equipment. In inventory control, the flow of random requests for replacement of failed units is also often assumed to be described by a Poisson process, especially if the system which generates these requests is large.

Sometimes the Poisson process is called "a process of rare events." Of course, the meaning of the word "rare" should be carefully defined in each particular case. Usually, we speak about rare events if they appear with a frequency which is lower than the frequencies of other accompanying processes. The Poisson process appears as the interaction of a large number of these processes and, consequently, has a frequency lower than the other processes.

In reliability, such "rare" events appear, for instance, when one considers a highly reliable *renewal redundant system* or a multicomponent *renewal series system*. This process also successfully describes the fluctuation over a high-level threshold.

This process is so named because the number of events in any fixed interval of length t has a Poisson distribution:

$$\Pr\{k \text{ events during } t\} = p_k(t) = \frac{(\lambda t)^k}{k!} e^{-\lambda t}$$

where λ is called the parameter of the Poisson process.

First of all, note that the Poisson process possesses the three following properties that are often referred to as *characterization properties*:

- Stationarity
- Memorylessness (Markov property)
- Ordinarity

The first property means that the d.f. of the number of observed events in a time interval depends only on the length of the interval and not on its position on the time axis.

The second property means that the d.f. of the number of observed events does not depend on the previous history of the process.

The third property means that the probability of an appearance of more than one event in an infinitesimally small interval h goes to 0:

$$\frac{1}{h} \lim_{h \to 0} \Pr\{k \text{ events appear during } h, k > 1\} \to 0$$

or, in another notation,

$$\Pr\{k \text{ events appear during } h, k > 1\} = o(h) \qquad (1.82)$$

In practical problems, these properties are often assumed. These properties, which seem to be—at a first glance—purely qualitative, allow us to obtain strict mathematical results.

First, for a better understanding, we present a semiintuitive proof of the fact that these properties generate a Poisson process. Consider a Bernoulli process with probability of success p and a sufficiently large number of trials n. The Bernoulli process satisfies the first two properties (and trivially satisfies the third one because of its discrete nature). As we considered in Section 1.1, the number of successes in a series of n Bernoulli trials has a binomial distribution. As we have shown in Section 1.1, for large n the binomial distribution can be successfully approximated by a Poisson distribution.

We now return to the exact mathematical terms. First, add one extra property to the above three properties, namely, assume that the probability that there is exactly one event in a time interval h:

$$P_1(h) = \lambda h + o(h) \qquad (1.83)$$

where λ is some constant and $o(h)$ was introduced in (1.82). As a matter of fact, (1.83) follows from the three properties characterizing a Poisson process.

Consider the probability of the appearance of k events in a time interval $t + h$. The formula for the probability can be easily written as

$$P_k(t + h) = \sum_{0 \le j \le k} P_j(t) P_{k-j}(h) \qquad (1.84)$$

Let

$$R_k = \sum_{0 \le j \le k-2} P_j(t) P_{k-j}(h) \qquad (1.85)$$

Obviously,

$$R_k \le \sum_{0 \le j \le k-2} P_{k-j}(h) \le \sum_{2 \le i \le k} P_i(h) \tag{1.86}$$

because all $P_j(t) < 1$. We only reinforce the inequality (1.86) by changing the limits of summation

$$R_k \le \sum_{2 \le i < \infty} P_i(h)$$
$$= \Pr\{\text{two or more events appear during interval } h\} \tag{1.87}$$

At the same time, by assumption, this probability equals $o(t)$.

As a result, we have the equality

$$P_k(t + h) = P_k(t)P_0(h) + P_{k-1}(t)P_1(h) + o(t) \tag{1.88}$$

In this equality, we can substitute $P_1(h) = \lambda h + o(h)$. Also, $P_0(h) + P_1(h) + o(h) = 1$, that is, $P_0(h) = 1 - \lambda h + o(h)$. Now (1.88) can be rewritten as

$$P_k(t + h) = P_k(t)(1 - \lambda h) + P_{k-1}(t)\lambda h + o(t) \tag{1.89}$$

and from (1.89) we obtain

$$\frac{1}{h}[P_k(t + h) - P_k(t)] = -\lambda P_k(t) + \lambda P_{k-1}(t) + o(1)$$

and, after $h \to 0$,

$$\frac{dP_k(t)}{dt} = -\lambda P_k(t) + \lambda P_{k-1}(t) \tag{1.90}$$

Thus, a system of equalities for $P_k(t)$, $k = 0, 1, \ldots$, has been obtained. We need to add one more equation to determine $P_0(t)$. Using the memoryless property, we can write

$$P_0(t + h) = P_0(t)P_0(h) = P_0(t)[1 - \lambda h + o(h)]$$

or, finally,

$$\frac{dP_0(t)}{dt} = -\lambda P_0(t) \tag{1.91}$$

To solve the system, we must determine the initial condition. Of course, at $t = 0$, the probability of no events equals 1; that is, the initial condition is $P_0(0) = 1$.

The system of differential equations (1.90) and (1.91) with the above initial condition can be solved by several different methods. We solve this system of equations with the use of the LST. Let $\varphi_0(s)$ be the LST of the function $P_0(t)$:

$$\varphi_0(s) = \int_0^\infty P_0(t) e^{-st} \, dt \tag{1.92}$$

Applying (1.92) to (1.91) and keeping in mind the properties of the LST, one obtains

$$-P_0(0) + s\varphi_0(s) = -\lambda\varphi_0(s) \tag{1.93}$$

which has the solution

$$\varphi_0(s) = \frac{1}{\lambda + s} \tag{1.94}$$

As it follows from a table of Laplace–Stieltjes transforms, the function $P_0(t)$ corresponding to (1.94) is exponential with parameter λ:

$$P_0(t) = e^{-\lambda t} \tag{1.95}$$

For arbitrary $k > 0$, from (1.90) the system of recurrent equations follows:

$$s\varphi_k(s) = -\lambda\varphi_k(s) + \lambda\varphi_{k-1}(s) \tag{1.96}$$

or

$$\varphi_k(s) = \frac{\lambda\varphi_{k-1}(s)}{\lambda + s} \tag{1.97}$$

Finally, using (1.94) systematically, we have

$$\varphi_k(s) = \frac{\lambda^k}{(\lambda + s)^{k+1}} \tag{1.98}$$

From a table of LSTs, the latter transformation corresponds to a Poisson distribution

$$P_k(t) = \frac{(\lambda t)^k}{k!} e^{-\lambda t} \tag{1.99}$$

For the Poisson distribution the mean number of events in a fixed interval of time is proportional to its length. The parameter λ is the mean number of events in a time unit, or, equivalently, it equals the inverse of the mean time between events. Also, as known (see the Appendix), a convolution of Poisson distributions produces a Poisson distribution. Thus, for several disjoined intervals of lengths $t_1, t_2, \ldots, t_m$, the distribution of the total number of events is Poisson with parameter $\lambda(t_1 + t_2 + \cdots + t_m)$. In other words, the Poisson process is a point stochastic process with exponentially distributed intervals between neighboring events.

1.5.2   Introduction to Recurrent Point Processes

We often encounter situations where some events occur sequentially in such a way that the times between occurrence (interarrival times) can be successfully described by a sequence of independent r.v.'s. For instance, consider a socket with an installed unit which is instantly replaced upon failure by a new unit; the times between replacement moments form such sequence. In general, the length of each interval might depend on the number of the event because of a changing environment, a wearing out of the socket, and so on. Here we ignore such phenomena. A process of this type is called a *point process with restricted memory*.

A point process with restricted memory is a sequence of r.v.'s. It is called a *renewal (recurrent) point process* if all interarrival intervals are i.i.d. r.v.'s with identical d.f.'s $F_k(t) = F(t)$, $k \geq 2$, with only the first interval having its own distribution $F_1(t)$.

The Poisson process represents a particular case of such a process in that the intervals between arrivals are independent and exponentially distributed.

We assume that a flow of failures is represented by a recurrent point process. This assumption is acceptable in many practical situations. At the same time, it allows us to obtain simple and understandable results.

For a renewal point process, there are two main characteristics: (1) the process intensity defined to be the mean number of process events arriving in a time unit and (2) the process parameter defined to be the limit probability of the arrival of at least one event.

Let $N(t)$ be the number of events arriving during an interval of length t. Then, for the stationary process,

$$\lambda^* = \lim_{t \to \infty} \frac{1}{\tau} E\{N(t, t + \tau)\} = \lim_{t \to \infty} \frac{1}{\tau} \sum_{0 \leq j < \infty} j p_j(t, t + \tau)$$

The parameter of the process is defined as

$$\tilde{\lambda} = \lim_{t \to \infty} \lim_{\tau \to 0} \frac{1}{\tau} \sum_{0 \leq j < \infty} p_j(t)$$

For an arbitrary stationary point process with a single arrival at a time and without so-called "points of condensation" (infinitesimally small intervals in which an infinite number of discrete events might appear), we have

$$\lambda^* \le \tilde{\lambda}$$

For a stationary and memoryless point process, the parameter coincides with the intensity. We can given an explanation of the parameter of a point process based on a more physical consideration:

$$\lambda^*(t)\Delta = \Pr\{\text{at least one failure occurs in } [t, t + \Delta]\}$$

Let $f^{*k}(t)$ stand for a convolution of the kth order of the function $f(t)$:

$$f^{*k}(t) = \int_0^t f^{*(k-1)}(x) f(t - x)\, dx$$

it is clear that at least one failure might occur if

- The first failure occurs with probability $f(t)\Delta$.
- The second failure occurs with probability $f^{*2}(t)\,\Delta, \ldots,$
- The kth failure occurs with probability $f^{*k}(t)\,\Delta$ and so on.

Thus, the probability that a failure will occur for any of these reasons is

$$\Pr\{\text{at least one failure occurs in the interval } [t, t + \Delta]\}$$

$$= \lambda^*(t)\,\Delta = \left[\sum_{k \ge 0} f^{*k}(t) \right] \Delta$$

where we use the conditional notation $f^{*0}(t) = f(t)$. Hence,

$$\lambda^*(t) = \left[\sum_{k \ge 0} f^{*k}(t) \right]$$

The function $\lambda^*(t)$ allows us to express the so-called *characterization point process function* which we denote by $\Lambda^*(t)$:

$$\Lambda^*(t, t + t_*) = \int_t^{t + t_*} \lambda^*(t)\, dt$$

Using this function, we can write

$$\Pr\{\text{no failures in } [t, t + t_*]\} = \exp\left[-\int_t^{t + t_*} \lambda^*(x)\, dx \right]$$

$$= e^{-\Lambda^*(t, t + t_*)} = e^{-[\Lambda^*(t + t_*) - \Lambda^*(t)]} = \frac{e^{-\Lambda^*(t + t_*)}}{e^{-\Lambda^*(t)}}$$

$$(1.100)$$

The function $\lambda(t)$ is defined to be the "instant" conditional density of the failure distribution $F(t)$. We emphasize that the functions $\lambda(t)$ and $\lambda^*(t)$ are quite different.

Now consider the main characteristics of a renewal process. One of the characteristics of a renewal process is the mean number of events occurring up to a moment t. Denote the random number of events by $N(t)$ and the mean number by $H(t) = E\{N(t)\}$. $H(t)$ is called the *renewal function*. The derivative $h'(t) = H(t)$ is called the *renewal density*. Consider a renewal process composed of i.i.d. r.v.'s with distribution $F(t)$. We can write

$$\Pr\{N(t) \geq k\} = \sum_{k \leq j < \infty} \Pr\{N(T) = j\} = \Pr\left\{ \sum_{1 \leq j \leq k} \xi_j < t \right\} = F^{*k}(t)$$

where $F^{*k}(t)$ is the kth-order convolution of $F(t)$:

$$F^{*k}(t) = \int_0^t F^{*(k-1)}(t-x) \, dF(x)$$

$$F^{*1} = F(t)$$

The expression can be easily written as

$$\Pr\{\text{any event occurs in interval } [t, t+dt]\}$$

$$= \sum_{1 \leq k < \infty} \Pr\{\text{the } k\text{th event occurs in interval } [t, t+dt]\}$$

$$= h(t) \, dt = f(t) \, dt + \sum_{2 \leq k < \infty} f^{*k}(t) \, dt \qquad (1.101)$$

Integrating (1.101) allows us to write an expression for $H(t)$:

$$H(t) = F(t) + \sum_{2 \leq k < \infty} F^{*k}(t) \qquad (1.102)$$

Of course, $H(t)$ can be found in a standard way as the mean number of events during time t. The probability that exactly k events happen up to moment t is expressed as

$$\Pr\{N(t) = k\} = \Pr\{N(t) \geq k\} - \Pr\{N(t) \geq k - 1\}$$

$$= F^{*k}(t) - F^{*(k+1)}(t)$$

Thus, the distribution of $N(t)$ is defined. $H(t)$ can be found by

$$H(t) = E\{N(t)\} = \sum_{1 \le k < \infty} k \Pr\{N(t) = k\}$$

$$= \sum_{1 \le k < \infty} k \left[F^{*k}(t) - F^{*(k+1)}(t) \right]$$

$$= F(t) + \sum_{2 \le k < \infty} k F^{*k}(t) - \sum_{2 \le k < \infty} (k-1) F^{*k}(t)$$

$$= \sum_{1 \le k < \infty} F^{*k}(t)$$

For $h(t)$ we can write

$\Pr\{$any event occurs in interval $[t, t + \Delta t]\}$

$= \Pr\{$the first event occurs in interval $[t, t + \Delta]\}$

$+ \Pr\{$the last event happens in interval $[t, t + \Delta t]$ (1.103)

and the following random time ξ is such that $x \le \xi \le x + \Delta x\}$

With $\Delta t \to 0$, (1.103) can be rewritten in differential form as

$$h(t) = f(t) + \int_0^t h(t - x) \, dF(x) \tag{1.104}$$

Naturally, the renewal density function at time t is the sum of the densities of the occurrence of all possible events of the renewal process: the first, or the second, or ..., or the k th ..., and so on. From (1.104), by integration, we obtain

$$H(t) = F(t) + \int_0^t H(t - x) \, dF(x) \tag{1.105}$$

It is important to note that $F^{*n} \ge [F(t)]^n$. Indeed,

$$F^{*n}(t) = \Pr\left\{ \sum_{1 \le k \le n} \xi_k < t \right\}$$

$$\le \Pr\left\{ \max_{1 \le k \le n} \xi_k < t \right\} = \prod_{1 \le k \le n} \Pr\{\xi_k < t\} = [F(t)]^n$$

This states the simple fact that the sum of n nonnegative values is not less than the maximal one. (Equality occurs only if at least $n - 1$ values are equal to 0.) From this fact it follows that

$$H(t) = \sum_{1 \le k < \infty} F^{*k} \le \sum_{1 \le k < \infty} [F(t)]^k = F(t) \sum_{0 \le k < \infty} [F(t)]^k = \frac{F(t)}{1 - F(t)}$$

Using (1.102) and observing that the integral on the right side of the equation is positive, we obtain two-sided bounds

$$F(t) \le H(t) \le \frac{F(t)}{1 - F(t)} \tag{1.106}$$

The next interesting bounds for a renewal process, built with "aging" r.v.'s ξ, can be obtained if we consider the following natural condition. Let $N(t)$ events be observed up to a moment t. Thus,

$$t \le \sum_{1 \le j \le N(t)} \xi_j$$

Using the Wald equivalency, we write

$$t \le E\{\xi\}[H(t) + 1]$$

which produces

$$H(t) \ge \frac{t}{E\{\xi\}} - 1$$

For an aging r.v. ξ, the residual time $\zeta(t)$ is decreasing, which allows us to write

$$H(t) \le \frac{t}{E\{\xi\}}$$

Thus, for a renewal process with aging r.v.'s ξ, we can write the two-sided bounds

$$\frac{t}{E\{\xi\}} - 1 \le H(t) \le \frac{t}{E\{\xi\}} \tag{1.107}$$

In practical reliability problems we are often interested in the behavior of a renewal process in a stationary regime, that is, when $t \to \infty$. This interest is understandable because repairable systems enter an "almost stationary" regime very quickly (see Section 6.1). Several important facts are established for this case.

Theorem 1.2   For any $F(t)$,

$$\lim_{t \to \infty} \frac{H(T)}{t} = \frac{1}{E\{\xi\}} \tag{1.108}$$

In a mathematical sense this theorem is close to the Wald theorem. In a physical sense it means that, for a large interval of size t, the mean number of events is inversely proportional to the mean interarrival time.

Theorem 1.3   If ξ is continuous, then

$$\lim_{t \to \infty} h(t) = \frac{1}{E\{\xi\}}$$

This theorem reflects the fact that with increasing t the renewal becomes stationary and its characteristics become independent of the current time.

Theorem 1.4 (Blackwell's Theorem)   For a continuous r.v. ξ and an arbitrary number τ,

$$\lim_{t \to \infty} [H(t + \tau) - H(t)] = \frac{\tau}{E\{\xi\}} \tag{1.109}$$

It is clear that this theorem is a simple generalization of the first one.

Theorem 1.5 (Smith's Theorem)   If ξ is a continuous r.v. and $V(t)$ is a monotone nonincreasing function, integrable on $(0, \infty)$, then

$$\lim_{t \to \infty} \int_0^t V(t - x) \, dH(t) = \frac{1}{E\{\xi\}} \int_0^\infty V(t) \, dt \tag{1.110}$$

The function $V(t)$ can be chosen arbitrarily between those which have a probabilistic nature. The choice of this function depends on the concrete applied problem. An interpretation of this theorem is provided in the following particular case.

Corollary 1.1   The stationary probability of a successful operation (the stationary interval availability coefficient) equals

$$R(t_0) = \frac{1}{E\{\xi\}} \int_{t_0}^\infty P(t) \, dt \tag{1.111}$$

Here t_0 is the time needed for a successful operation. The proof of this is left to Exercise 1.10.

1.5.3 Thinning of a Point Process

We often encounter where a unit failure leads to a system failure only if several additional random circumstances happen. For instance, in a system of a group of redundant units, a unit failure is the cause of a system failure if, at a particular moment, all of the remaining units have failed. Such a coincidence of random circumstances may be very rare. We may consider the flow of "possibilities" which generate a relatively rare flow of system failures. This procedure is called a *thinning procedure*.

Poisson Process The thinning of a Poisson process produces a Poisson process. To prove this fact, we consider the sum of a geometrically distributed random number of exponentially distributed r.v.'s. Indeed, thinning means that with some probability q an event remains in the final process and with probability $p = 1 - q$ it is removed from it. Thus, we have a sequence of Bernoulli trials.

Of course, in the particular case of the Poisson process, we can apply a simple deduction based on its three characteristic properties. Indeed, stationarity is not violated by the Bernoulli-like exclusion of events from the initial process: all p's are constant over the entire time axis. Ordinarity is also preserved because we only exclude events. The memorylessness property of the resulting process follows from the independent character of the event exclusion from the Bernoulli trial sequence.

General Case Consider a stationary recurrent point process for which the intervals between events have a distribution $F(t)$. Sometimes this distribution is called a "forming distribution." Apply the thinning procedure to this process. According to this procedure, each event remains in the process with probability q or is deleted with probability $p = 1 - q$. Thus, after such a procedure, the average number of points which remain in the newly formed process is $1/q$ times less than in the initial process. In other words, the time interval between points in the new process is $1/q$ times larger. The explanation of the procedure is depicted in Figure 1.9.

Each interval between events represents the sum of a random number of r.v.'s. Thus, the problem of renewal process thinning is equivalent to the

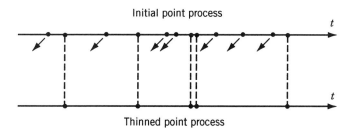

Figure 1.9. Example of the thinning procedure for a point process.

summation of a geometrically distributed random number of r.v.'s. (This was considered in Section 1.4.) In particular, in Section 1.4.5 we developed some asymptotic results. Here we use the standard terminology and methods of renewal theory, because this helps us to obtain some additional results.

Consider a special transformation of the renewal process, T_q: events are deleted from the process with probability p, and, simultaneously, the time scale is shrinking by a factor of $1/q$. This normalization of time keeps the length of the average interarrival interval the same as in the initial process. It is clear that the T_q transformation is equivalent to the summation of a geometrically distributed random number of r.v.'s with a d.f. $F(t)$ and the further normalization of the resulting r.v.

Sequential applications of the transformations T_{q_1} and T_{q_2} to the process are equivalent to the single transformation $T_{q_1 q_2}$. We ask the reader to prove this is Exercises 1.11 and 1.12.

Limit Rényi Theorem   The Rényi theorem is very important in many applications. These asymptotic results can be used if the thinning procedure is intensive enough. They are also very useful in developing heuristic approaches (see Chapter 13).

Theorem 1.6   If transformations $T_{q_1}, T_{q_2}, \ldots, T_{q_n}$ are such that, for $n \to \infty$,

$$Q_n = q_1 q_2 \cdots q_n \to 0 \qquad \text{as} \quad n \to 0$$

then their application to some point renewal process with an initial finite intensity $\tilde{\lambda}$ leads the resulting limit process to a Poisson process with the same intensity $\tilde{\lambda}$.

We omit the proof because, in general, it coincides with the corresponding proof of Section 1.3.5.

Later this result was generalized for the superposition of n different renewal processes with different thinning procedures.

We remark that a Poisson process is sometimes called "a process of rare events." From formulations of the above results, one can see that the T_q transformation generates the flow of "rare" events from a dense initial process.

1.5.4   Superposition of Point Processes

In reliability practice we frequently encounter a situation which might be described as the formation of a common point process from the superposition of several point processes (see Figure 1.10).

For example, consider the flow of failures of different units in a series system. Each unit generates its own renewal point process of failures: a failed

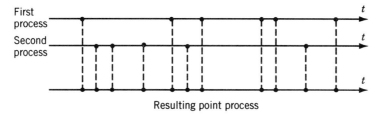

First
process

Second
process

Resulting point process

Figure 1.10. Example of the superposition of two point processes.

unit is replaced and the process continues. Unfortunately, even if we consider a small number of renewal processes, their superposition cannot be analyzed in terms of renewal processes! (The only exception is the superposition of Poisson processes.)

At the same time, fortunately, if the number of superimposed point processes is very large, the superposition of these processes produces a point process that is very close to being Poisson. In the theory of stochastic processes, the Poisson process plays a role which is analogous to that of the normal distribution in probability theory.

Poisson Process   For the superposition of n Poisson processes, the resulting process is Poissonian. If the initial processes have parameters $\lambda_1, \lambda_2, \ldots, \lambda_n$, the resulting process has the parameter $\lambda_\Sigma = \Sigma \lambda_i$.

To show this fact, consider an arbitrary moment of time t. Let ζ_k denote the residual time of the kth process, that is, the time from an arbitrary but fixed t until the appearance of the next event in the process. The memoryless property says that two r.v.'s ζ_k and ξ_k, which represent the time between events for the kth process, are statistically equivalent. Thus, for the kth initial process we can write the distribution of the residual time

$$\Pr\{\zeta_k \geq t\} = \Pr\{\xi_k \geq t\} = \exp(-\lambda t).$$

If we consider n processes, then, from a fixed (albeit arbitrary) moment t until the next arriving event, we observe an r.v.

$$\zeta_\Lambda = \min_{1 \leq k \leq n} \zeta_k$$

with d.f.

$$\Pr\{\zeta_\Lambda \geq t\} = \Pr\left\{ \min_{1 \leq k \leq n} \zeta_k \geq t \right\}$$

$$= \prod_{1 \leq k \leq n} \Pr\{\zeta_k \geq t\} = \prod_{1 \leq k \leq n} e^{-\lambda_k t} = \exp\left(-t \sum_{1 \leq k \leq n} \lambda_k\right)$$

Thus, the distribution of the time interval between neighboring events is exponential. As we know, only the Poisson process is characterized by such a property.

Of course, as above, we can prove this fact by checking that all three characteristic properties of the Poisson process are satisfied. Indeed, stationarity is kept because of the stationarity of all initial processes. Ordinarity is also preserved, because, for a continuous process, the probability of a coincidence of events equals 0. The memorylessness property of the resulting process follows from the independence of all the initial processes and their original memorylessness property.

General Case The proofs and even the formulation of the strict conditions of the related theorems are complex and lie outside the scope of this book. We only formulate the main results, sometimes even on a verbal level. The first strict result was formulated in the following theorem.

Khinchine – Ososkov Theorem If the limit

$$\lim_{n \to \infty} \sum_{1 \leq r \leq k_n} \lambda_{nr} = \Lambda$$

exists, a necessary and sufficient condition that the process $x_n(t)$ converges to a Poisson process with parameter Λ is that, for any fixed t and $n \to \infty$,

$$\sum_{1 \leq r \leq k_n} \lambda_{nr} \int_0^t \varphi_{nr}(0, x) \, dx \Rightarrow \Lambda t$$

Later general results, relating to the superposition of stochastic point processes, are contained in the Grigelionis–Pogozhev theorem. On a qualitative level, the theorem states that a limit point process which is formed by the superposition of independent "infinitesimally rare" point processes converges to a Poisson process. The parameter of this resulting process is expressed as a sum of the parameters of the initial processes.

1.6 BIRTH AND DEATH PROCESS

The *birth and death process* is an important branch of Markov processes. We will not give details for the general Markov processes, but we will consider some special models of renewal systems in a later chapter. This approach is also very useful for the analysis of renewal systems.

1.6.1 Model Description

The behavior of a number of practical systems can be portrayed with the help of the birth and death process (BDP). Birth and death processes are widely used for the construction of mathematical models in microbiology, zoology, and demography. They are also used in reliability and queuing theory.

Let us explain the nature of the BDP with three simple examples.

Consider a queueing system with one service unit and an unlimited number of input call sources. Suppose there are k calls on the line waiting for service at a moment t. We say that at this moment, the system is in state H_k. At the moment $t + \Delta t$, where Δt is infinitesimally small, the state changes to H_{k+1} if an additional call arrives during the interval Δt. If during Δt a call service in the system has been completed, at the moment $t + \Delta t$ the system changes its state to H_{k-1}. Recall that, for a Markov process, the probability of more than one state change is $o(\Delta t)$, which means that $o(\Delta t) \to 0$ as $\Delta t \to 0$. From our assumption of an unlimited number of input sources of calls, it follows the line length could be infinite. In other words, a BDP may have an infinite number of states. Suppose there is a specified criterion of system effectiveness; for example, a line of length more than m is considered to be inadmissible. Then the set of all system states may be divided into two subsets: the "up states" $H_0, \ldots, H_m$ and the "down states" $H_{m+1}, H_{m+2}, \ldots$.

As another example, consider a parallel system with one main unit and m identical active redundant units. There is one repair facility. This system may be thought of as a queuing system with a limited number (namely, $n + 1$) of input "call" sources. Indeed, if there are k units under repair, then only $m + 1 - k$ of the remaining units may fail. The state H_{m+1} corresponds to system failure, and only a transition from this state to state H_m is possible. State H_{m+1} is called *reflecting*.

As the final example, consider a parallel system with n main units and m identical active redundant units. Again, there is one repair facility. It is clear that the mathematical description of this system is very close to that of the above example. In this case there are, in total, $n + m$ sources of failure. The states $H_{m+1}, \ldots, H_{n+m}$ correspond to system failure states.

The last two examples are of BDPs with a finite number of states. In reliability theory it is sometimes reasonable to consider separately these two very similar cases.

The transition graphs for all three examples are shown in Figure 1.11.

If the system is in state H_j at a moment t, there are three possibilities during the next interval Δt:

- The process passes to state H_{j+1} with probability:

$$\Lambda_j \Delta t + o(\Delta t)$$

- The process passes to state H_{j-1} with probability:

$$M_j \Delta t + o(\Delta t)$$

- The process remains in state H_j with probability:

$$1 - (\Lambda_j + M_j) \Delta t + o(\Delta t)$$

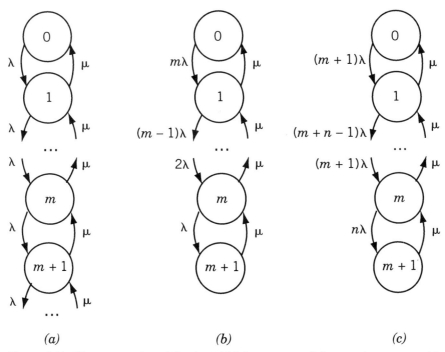

Figure 1.11. Three examples of death and birth processes: (*a*) a queuing system with an infinite source of demands; (*b*) a unit with m active redundant repairable units; (*c*) a series system of n units with m active redundant repairable units.

The failure state with the smallest index, say $m + 1$, may be considered absorbing. The other system failure states are of no interest because there are no transitions from them to the set of up states. In this case the process behavior in the subset of up states can be used to find the probability of a failure-free operation, the MTTF, and the mean time between failure (MTBF). Notice that if we are interested in finding repair (idle time) indexes, the state H_m must be chosen to be absorbing. In this case we consider the behavior of the process in the subset of system failure states.

If there are no absorbing states, the process is considered for finding the availability coefficients, both stationary and nonstationary. For a finite set of states, the state with the largest index is reflecting.

In reliability problems the state H_0 is always considered as reflecting (if the process is not of a special type with states $H_{-1}, H_{-2}, \ldots$).

For reliability problems it suffices to consider only BDPs with a finite number of states.

1.6.2 Stationary Probabilities

Consider a finite BDP with $N + 1$ states (see Figure 1.11a). For each state k and for two infinitesimally close time moments t and $t + \Delta t$, we can write the expression

$$p_k(t + \Delta t) = p_{k-1}(t)\big[\Lambda_{k-1}\,\Delta t + o(\Delta t)\big]$$
$$+ p_k(t)\big[1 - (\Lambda_k + M_k)\,\Delta t + o(\Delta t)\big]$$
$$+ p_{k+1}(t)\big[M_{k+1}\,\Delta t + o(\Delta t)\big] + o(t)$$

or

$$\frac{p_k(t + \Delta t) - p_k(t)}{\Delta t} = \Lambda_{k-1}p_{k-1}(t) - (\Lambda_k + M_k)p_k(t) + M_{k+1}p_{k+1}(t)$$

In the limit as $\Delta t \to 0$, we obtain

$$\lim_{\Delta t \to 0}\frac{p_k(t + \Delta t) - p_k(t)}{\Delta t} = p'(t)$$
$$= \Lambda_{k-1}p_{k-1}(t) - (\Lambda_k + M_k)p_k(t)$$
$$+ M_{k+1}p_{k+1}(t) \qquad (1.112)$$

Because we are considering a finite process, we must set $\Lambda_{-1} = M_0 = \Lambda_N = M_{N+1} = 0$. In other words,

$$p'_0(t) = -\Lambda_0 p_0(t) + M_1 p_1(t)$$

and

$$p'_N(t) = -\Lambda_{N-1}p_{N-1}(t) + M_N p_N(t)$$

We add to this system of equations the normalizing equation

$$\sum_{0 \le i \le N} p_k(t) = 1$$

and exclude any one of the above.

We note that (1.112) represents the equation of dynamic equilibrium. In other words, state k "loses each unit of its mass" $p_k(t)$ with intensity $\Lambda_k + M_k$ and "receives" a corresponding mass from states $k - 1$ and $k + 1$. If there is no absorbing state, the process has stationary states. Thus, there are limits

$$\lim_{t \to \infty} p_k(t) = p_k$$

Moreover, if for any k,

$$\Lambda_{k-1} + M_k \ne 0 \qquad (1.113)$$

the stationary probabilities of these states do not depend on the initial state at $t = 0$. The condition (1.113) means that there are no separated groups of states.

Consider this stationary case. If we take the limit as $t \to \infty$ in (1.112), we obtain the system of linear equations

$$
\begin{aligned}
0 &= -\Lambda_0 p_0 + M_1 p_1 \\
0 &= \Lambda_{k-1} p_{k-1} - (\Lambda_k + M_k) p_k + M_{k+1} p_{k+1} \\
0 &= -\Lambda_{N-1} p_{N+1} + M_N p_N.
\end{aligned}
\tag{1.114}
$$

We again must replace one of the above equations with the normalizing condition

$$
\sum_{0 \le i \le N} p_k = 1
\tag{1.115}
$$

Now let us recall that (1.114) represents an equilibrium. It means that if we consider any cut in the transition graph, for instance, a cut between states $k - 1$ and k, there is an equality of flows "up" and "down":

$$
M_k p_k = \Lambda_{k-1} p_{k-1}
\tag{1.116}
$$

From (1.116) we obtain the recurrent relationship

$$
p_k = \frac{\Lambda_{k-1}}{M_k} p_{k-1}
\tag{1.117}
$$

which allows us to obtain

$$
p_k = \frac{\Lambda_0 \Lambda_1 \cdot \cdots \cdot \Lambda_{k-1}}{M_1 M_2 \cdot \cdots \cdot M_k} p_0 = A_k p_0
\tag{1.118}
$$

From (1.115) it follows that

$$
p_k = \frac{A_k}{\displaystyle\sum_{0 \le j \le N} A_j} = \frac{\dfrac{\displaystyle\prod_{0 \le j \le k-1} \Lambda_j}{\displaystyle\prod_{1 \le j \le k} M_j}}{\displaystyle\sum_{0 \le j \le N} \dfrac{\displaystyle\prod_{0 \le i \le j-1} \Lambda_i}{\displaystyle\prod_{1 \le i \le j} M_i}}
\tag{1.119}
$$

where $A_0 = 1$.

1.6.3   Stationary Mean Time of Being in a Subset

Consider a BDP whose total set of n states is divided into two subsets: one subset of up states, $E_+ = \{H_0, \ldots, H_m\}$, and another of down states, $E_- = \{H_{m+1}, \ldots, H_N\}$.

To find the stationary mean time of the process present in a specified subset of states, we use a well-known result. Let us distinguish two subsets of so-called boundary states: e_+ which is a subset of E_+ and e_- which is a subset of E_-. The process may enter the subset E_-, leaving only a state belonging to the boundary subset e_+. The subset e_- plays an analogous role for the subset E_+. In considering case e_+ consists of only one state H_m. Briefly repeat the idea of the conclusion. The process may leave subset E_+ only from state H_m, and being in this state it leaves with intensity Λ_k. Hence, the intensity of leaving subset E_+ equals

$$\Lambda_+ = p_m^* \Lambda_m \tag{1.120}$$

where p_m^* is the conditional probability that the process is in state H_m under the condition that this is in subset E_+:

$$p_m^* = \frac{p_m}{\sum\limits_{0 \le j \le m} p_j}$$

where p_j can be found from (1.119). Finally, we find that the mean stationary time of the process being in subset E_+ is $T_+ = 1/\Lambda_+$:

$$T_+ = T_{M, M+1} = \frac{\sum\limits_{0 \le j \le m} p_j}{\Lambda_m p_m} \tag{1.121}$$

Obviously, the mean stationary time of the process being in subset E_- is similar to the above except for the following notation:

$$M_- = M_{m+1} p_{m-1}^{**}$$

where

$$p_{m+1}^{**} = \frac{p_{m+1}}{\sum\limits_{m+1 \le j \le N} p_j}$$

and, finally,

$$T_- = T_{M+1, M} = \frac{\sum\limits_{m+1 \le j \le N} p_j}{M_{m+1} p_{m+1}} \tag{1.122}$$

1.6.4 Probability of Being in a Given Subset

The BDP with an absorbing state H_{n+1} can be described with the help of the following system of differential equations:

$$\frac{d}{dt} p(t) = \Lambda_{j-1} p_{j-1}(t) - (\Lambda_j + M_j) p_j(t) + M_{j+1} p_{j+1}(t) \qquad 0 \le j \le n+1$$

$$\sum\limits_{0 \le j \le n+1} p_j(t) = 1 \tag{1.123}$$

$$\Lambda_{-1} = \Lambda_{n+1} = M_0 = M_{n+1} = M_{n+2} = 0$$

where $p_j(t)$ is the probability of state H_j at moment t. Let the probabilities $p_j(t)$ satisfy the initial conditions:

$$p_j(t = 0) = p_j(0) \qquad 0 \le j < n + 1 \qquad (1.124)$$

Let θ_{n+1} be the duration of time before the system has reached the absorbing state H_{n+1} for the first time. We need to find the distribution function $\Pr\{\theta_{n+1} \le t\} = p_{n+1}(t)$. This is the probability that the system has *not* reached the absorbing state H_{n+1} at time t. Let us apply the LST to (1.123). Then we find the system of linear algebraic equations:

$$\Lambda_{j-1}\varphi_{j-1}(s) - (\Lambda_j + M_j + s)\varphi_j(s) + \Lambda_{j+1}\varphi_{j+1}(s) = -p_j(0)$$
$$0 \le j \le n + 1$$
$$\Lambda_{j-1} = \Lambda_{n+1} = M_0 = M_{n+1} = M_{n+2} = 0 \qquad (1.125)$$

By using Cramér's rule,

$$\varphi_{n+1}(s) = \frac{D_{n+1}(s)}{s\,\Delta_{n+1}(s)} \qquad (1.126)$$

where

$\Delta_{n+1}(s)$

$$= \begin{vmatrix}
-(\Lambda_0 + s) & M_1 & 0 & \cdot & 0 & 0 \\
\Lambda_0 & -(\Lambda_1 + M_1 + s) & M_2 & \cdot & 0 & 0 \\
0 & \Lambda_1 & -(\Lambda_2 + M_2 + s) & \cdot & 0 & 0 \\
\cdot & \cdot & \cdot & \cdot & \cdot & \cdot \\
0 & 0 & 0 & \cdot & -(\Lambda_{n-1} + M_{n-1} + s) & M_n \\
0 & 0 & 0 & \cdot & \Lambda_{n-1} & -(\Lambda_n + M_n + s)
\end{vmatrix}$$

$$= (-1)^{n+1} \prod_{1 \le k \le n+1} \left(s + x_k^{n+1}\right) \qquad (1.127)$$

$D_{n+1}(s)$

$$\begin{vmatrix}
-(\Lambda_0 + s) & M_1 & 0 & \cdot & 0 & 0 & -p_0(0) \\
\Lambda_0 & -(\Lambda_1 + M_1 + s) & M_2 & \cdot & 0 & 0 & -p_1(0) \\
0 & \Lambda_1 & -(\Lambda_2 + M_2 + s) & \cdot & 0 & 0 & -p_2(0) \\
\cdot & \cdot & \cdot & \cdot & \cdot & \cdot & \cdot \\
0 & 0 & 0 & \cdot & -(\Lambda_{n-1} + M_{n-1} + s) & M_n & -p_{n-1}(0) \\
0 & 0 & 0 & \cdot & \Lambda_{n-1} & -(\Lambda_n + M_n + s) & -p_n(0) \\
0 & 0 & 0 & \cdot & 0 & \Lambda_n & -p_{n+1}(0)
\end{vmatrix}$$

$$(1.128)$$

Expanding the determinant $D_{n+1}(s)$ along the last row yields the recurrent equation:

$$D_{n+1}(s) = -p_{n+1}(0) \, \Delta_{n+1}(s) + \sum_{0 \leq i \leq n} (-1)^{n+1-i} p_i(0) \, \Delta_i(s) \prod_{i \leq =k \leq n} \Lambda_k$$

(1.129)

The probability $p_{n+1}(t)$ is found with the help of the inverse LST:

$$p_{n+1}(t) = \frac{1}{2\pi i} \int_L \varphi_{n+1}(s) e^{st} \, ds = \frac{1}{2\pi i} \int_L \frac{D_{n+1}(s) e^{st} \, ds}{-s \, \Delta_{n+1}(s)}$$

(1.130)

where $i = \sqrt{-1}$.

We are now faced with the problem of calculating the roots (eigenvalues) of (1.130). It can be shown that $\Delta_{n+1}(s)$ is a polynomial of power n, and all of its roots $s_k^{(n+1)}$, $1 \leq k \leq n+1$, are distinct and negative. Also, all roots of the polynomials of the neighboring orders n and $n+1$ are intermittent. This fact facilitates the computation of the recurrent equation (1.130).

We omit the cumbersome intermediate transformations and write the final result, taking into account that the probability of interest $P(t) = 1 - p_{n+1}(t)$:

$$P(t) = \sum_{0 \leq i \leq n} p_i(0) \prod_{i \leq j \leq n} \Lambda_j \sum_{i \leq k \leq n+1} \frac{e^{-s_k^{(n+1)}t}}{s_k^{(n+1)}}$$

$$\times \frac{\prod_{1 \leq m \leq i} \left(s_m^{(i)} - s_k^{(n+1)} \right)}{\prod_{\substack{1 \leq i \leq n \\ i \neq k}} \left(s_i^{(n+1)} - s_k^{(n+1)} \right)}$$

(1.131)

In (1.131) we need to insert the roots which are usually calculated by numerical methods.

Two cases are of special interest. They are given without special explanations:

1. When all units are in up states at moment $t = 0$.
2. When the system has just come out of a failure state at $t = 0$.

If the system begins its operation at the state with all units completely operational:

$$p_0(0) = 1 \qquad p_i(0) = 0 \qquad 1 \leq i \leq n+1$$

then

$$P^{(0)}(t) = \Lambda_0 \Lambda_1 \cdots \Lambda_n \sum_{1 \le i \le n+1} \frac{e^{-s_i^{(n+1)}t}}{s_i^{(n+1)} \prod_{\substack{1 \le k \le n+1 \\ k \ne i}} \left(s_k^{(n+1)} - s_i^{(n+1)} \right)} \quad (1.132)$$

If the system begins its operation just after coming out of a failure state.

$$p_n(0) = 1 \qquad p_i(0) = 0 \qquad i = 0, 1, \ldots, n-1, n+1$$

then

$$P^{(n)}(t) = (-1)^n \Lambda_n \sum_{1 \le i \le n+1} \Delta_n\left(-s_i^{(n+1)} \right) \frac{e^{-s_i^{(n+1)}t}}{s_i^{(n+1)} \prod_{\substack{1 \le i \le n+1 \\ k \ne i}} \left(s_k^{(n+1)} - s_i^{(n+1)} \right)}$$

$$(1.133)$$

1.6.5 Mean Time of Staying in a Given Subset

Now let us determine the mean time of the process staying in the subset $E_+ = \{H_0, \ldots, H_m\}$ starting from the state H_0. Of course, we may use a standard procedure for calculating this value: first, find the probability of staying in this subset with the initial condition $H_0(0) = 1$ and then integrate the obtained expression. But this approach is too difficult. (Also, we did not obtain the result in a form which is easily integrated!) Thus, we choose another method.

A transition of the process from the initial state H_0 to the absorbing state H_{n+1} can be considered as consisting of $n + 1$ steps:

- from H_0 to H_1, plus
- from H_1 to H_2 (with the probability of going back and forth to the state H_0), plus
- from H_2 to H_3 (with the possibility of going back and forth to the states H_0 and H_1), plus
  $\vdots$
- from H_n to H_{n+1}.

Let us find the mean time of passing from H_k to H_{k+1} where $k > 0$. Consider the auxiliary BDP with only $k + 1$ states where H_{k+1} is absorbing. We can use the stationary mean time of entrance in the absorbing state found in (1.131). Thus, the value of interest can be found to be

$$T_{0,n+1} = \sum_{0 \le k \le n} T_{k,k+1} = \sum_{0 \le k \le m} \frac{\displaystyle\sum_{0 \le i \le k} p_i}{\Lambda_k p_k}$$

1.6.6 Stationary Probability of Being in a Given Subset

Let us again consider two subsets, $E_+ = \{H_0, \ldots, H_m\}$ and $E_- = \{H_{m+1}, \ldots, H_N\}$. One can find the stationary probability of being in a given subset, say E_+, in two ways. Denote this probability K. The first way amounts to finding

$$K = \lim_{t \to \infty} \Pr\{H_j(t) \in E_+\} = \sum_{0 \le j \le m} p_j$$

The second way uses the means T_+ and T_- determined in (1.121) and (1.122), respectively. Let $K - 1 = k$. For the stationary process, the probability of being in a given state is proportional to the portion of time occupied by this state over the entire interval of observation on the time axis. This leads to the condition $(K/k) = (T_+/T_-)$. With the condition $K + k = 1$ we find

$$K = \frac{T_+}{T_+ + T_-}$$

1.6.7 Death Process

A particular class of birth and death processes is the so-called death process. From the name of the process, it is clear that this can be obtained from the BDP by putting all M_j, $1 \le j \le N$, equal to 0. We consider this mathematical model because it is very useful when dealing with certain redundant systems without repair. For example, a redundant system consisting of n identical dependent units might be analyzed with this technique. The units can be dependent in a special way when the failure rate of each of them depends on the number of failed (or, equivalently, the number of operating) units at the moment.

This process can be described by the linear transition graph (see Figure 1.12). Let the process have $N + 1$ states: $H_0, \ldots, H_N$. Let Λ_k denote the transition rate from state H_k to state H_{k+1}. Using the same technique as above, we can write the equation:

$$p_k'(t) = -\Lambda_k p_k(t) + \Lambda_{k-1} p_{k-1}(t) \tag{1.134}$$

for all $0 \le k \le N - 1$ and $\lambda_{-1} = \lambda_N = 0$.

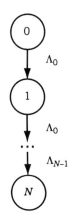

Figure 1.12. Example of a death process.

We add the initial condition to the system of linear differential equations. In reliability problems this is usually $P_0(0) = 1$; that is, the system is supposed to be in the state with all units up at the initial moment $t = 0$.

Using the Laplace–Stieltjes transform, we can represent (1.134 – in the form of the linear equations

$$-1 + s\varphi_0(s) = -\lambda_0\varphi_0(s)$$

$$s\varphi_k(s) = \lambda_{k-1}\varphi_{k-1}(s) - \lambda_k\varphi_k(s) \qquad (1.135)$$

$$s\varphi_N(s) = \lambda_{N-1}\varphi_{N-1}(s)$$

Solving (1.135) beginning with the first equation and sequentially substituting the obtained results in the next equation, we obtain

$$\varphi_0(s) = \frac{1}{s + \Lambda_0}$$

$$\varphi_k(s) = \frac{\Lambda_{n-1}}{s + \Lambda_k}\varphi_{k-1}(s) \qquad (1.136)$$

$$\varphi_N(s) = \frac{\Lambda_{N-1}}{s}\varphi_{N-1}(s)$$

The solution for $\varphi_n(s)$ is

$$\varphi_N(s) = \frac{\Lambda_0\Lambda_1 \cdot \,\cdots\, \cdot \Lambda_{N-1}}{s(s + \Lambda_0)(s + \Lambda_1) \cdot \,\cdots\, \cdot (s + \Lambda_{N-1})} \qquad (1.137)$$

For different Λ_i's the solution for $p_n(t)$, which is the probability that at moment t the process enters state H_N, can be found by using the inverse Laplace–Stieltjes transform

$$p_N(t) = 1 - \prod_{0 \le i \le N-1} \lambda_i \sum_{0 \le i \le N-1} \frac{1}{\lambda_k \omega'(-\lambda_k)} e^{-\lambda_k t} \qquad (1.138)$$

where $\omega(x)$ is a polynomial of the form

$$\omega(x) = (x + \lambda_0)(x + \lambda_1) \ldots (x + \lambda_{n-1}) \qquad (1.139)$$

and $\omega'(-\lambda_k)$ is the derivative with respect to x with the corresponding substitution.

If not all Λ_k are different, the expression for $p_N(t)$ becomes more complicated. But even (1.138) is not particularly convenient for practical use.

In a very important practical case, $\Lambda_k = \Lambda$ for all $0 \le k \le N - 1$. (Notice that this case corresponds to spare redundancy of identical units.) In this case (1.137) may be written in the form

$$\varphi_N(s) = \frac{\Lambda^N}{s(s + \Lambda)^N} \qquad (1.140)$$

In this case we find (with the use of a table of the LSTs) that

$$p_N(t) = 1 - \sum_{1 \le k \le N} \frac{(\Lambda t)^k}{k!} e^{-\Lambda t}$$

This fact becomes clear if we consider a sequence of N identical exponentially distributed r.v.'s which represents a sample of the process until the entrance into state H_N (see Figure 1.13). As we mentioned above, a sum of N such r.v.'s has an Erlang distribution. The Poisson distribution is the cumulative function for the Erlang density and the result follows immediately.

Figure 1.13. Time sequence of random values.

The mean time of the process entering into state H_N in the general case can easily be calculated as the sum of the time periods during which the process is remaining in each state

$$T_{0,N} = \sum_{1 \le k \le N} \frac{1}{\Lambda_k}$$

Some details concerning death processes will be discussed later.

CONCLUSION

Two distributions which are often used in engineering practice are the normal and the exponential. Each has its advantages and disadvantages. First of all, these distributions are very convenient for varied mathematical manipulations. But this argument is weak for practical applications. The question of their reasonable use, as with any modeling of real objects with the help of mathematical abstraction, always requires special "physical" verification based on experience and engineering intuition.

A Weibull–Gnedenko distribution is very convenient as a model for various physical phenomena because it is two parametrical. Besides, it has a clear physical sense as a distribution of extremal values. This distribution, as it relates to applied mechanical problems, was first mentioned in Weibull (1939). shortly after this, Gnedenko (1943) found classes of limit distributions of extreme values. A particular type of limit distribution has the form of the distribution discovered by Weibull

$$F(t) = 1 - - \exp\left(-\exp\left(\frac{t - a}{b}\right)\right)$$

where the new parameters are expressed as $b = 1/\beta$ and $a - \log \lambda$.

The reader interested in a deeper understanding of the probabilistic fundamentals of reliability theory should pay attention to special monographs. It is practically impossible to enumerate the books dedicated to this subject. An older, but nevertheless highly recommended book, is the book by Feller (1966). This book along with the book by Gnedenko (1967, 1988) were the main textbooks for several generations of statisticians and applied mathematicians.

For everyday use the books by DeGroot (1987) and Devore (1991) are recommended.

Concerning the limit theorems in the theory of stochastic processes, we must especially mention several works. Khinchine (1956a, 1956b, 1960) and Ososkov (1956) considered superposition of point processes, and later Grigelionis (1963) and Pogozhev (1964) generalized their result. Rényi (1962) formulated the theorem on "thinning" of point processes which later was generalized by Belyaev (1962). Summary of all of these results can be found in Gnedenko and Kovalenko (1987).

The reader can find details concerning generalized generating sequences in Ushakov (1986, 1987, 1988a, 1988b).

REFERENCES

Belyaev, Yu. K. (1962). Line Markov processes and their application to problems in reliability theory. *Trans. Sixth Conf. Probab. Statist.*, Vilnius.

DeGroot, M. H. (1987). *Probability and Statistics*, 2nd ed. Reading, MA: Addison-Wesley.

Devore, J. L. (1991). *Probability and Statistics for Engineering and the Sciences*, 3rd ed. Pacific Grove, CA: Brooks/Cole.

Feller, W. (1966). *An Introduction to Probability Theory and Its Applications*. New York: Wiley.

Gnedenko, B. V. (1943). Sur la distribution limit du terme maximum d'une serie aleatoir. *Ann. Math.*, no. 44.

Gnedenko, B. V. (1967). *The Theory of Probability*. New York. Chelsea.

Gnedenko, B. V., Kovalenko, I. N. (1987). Introduction in Queuing Theory, 2nd ed. (in Russian). Moscow: Nauka.

Gnedenko, B. V. (1988). Course of Probability Theory, 6th ed., revised (in Russian). Moscow: Nauka.

Grigelionis, B. (1963). On the convergence of sums of step stochastic processes to a Poisson process. *Theory Probab. Appl.*, vol. 8, no. 2.

Grigelionis, B. (1964). Limit theorems on sum of renewal processes. In *Cybernetics in the Service of Communism*, vol. 2, A. Berg, N. Bruevich, and B. Gnedenko, eds. Moscow: Energia.

Khinchine, A. Ya. (1956a). Streams of random events without aftereffect. *Theory Probab. Appl.*, No. 1.

Khinchine, A. Ya. (1956b). On Poisson streams of random events. *Theory Probab. Appl.*, no. 1.

Khinchine, A. Ya. (1960). *Mathematical Methods in the Theory of Queueing*. London: Charles Griffin.

Ososkov, G. A. (1956). A limit theorem for flows of similar events. *Theory Probab. Appl.*, vol. 1, no. 2.

Pogozhev, I. B. (1964). Evaluation of deviation of the equipment failure flow from a Poisson process. In *Cybernetics in the Service of Communism*, vol. 2, (in Russian) A. Berg, N. Bruevich, and B. Gnedenko, eds. Moscow: Energia.

Rènyi. A. (1956). Poisson-folyamat egy jemllemzese (in Hungarian). *Ann. Math. Statist.*, vol. 1, no. 4.

Rényi (1962).

Ushakov, I. A. (1986). A universal generating function. *Soviet J. Comput. Systems Sci.*, vol. 24, no. 5.

Ushakov, I. A. (1987). Optimal standby problem and a universal generating function. *Soviet J. Comput. Systems Sci.*, vol. 25, no. 4.

Ushakov, I. A. (1988a). Reliability analysis of multi-state systems by means of a modified generating function. *J. Inform. Process. Cybernet.*, vol. 24, no. 3.

Ushakov, I. A. (1988b). Solving of optimal redundancy problem by means of a generalized generating function. *J. Inform. Process. Cybernet.*, vol. 24, no. 4–5.

Weibull, W. (1939). A statistical theory of the strength of materials. *Ing. Vetenskaps Akad. Handl.*, no. 151.

Weibull, W. (1951). A statistical distribution of wide applicability. *J. Appl. Mech.*, no. 18.

APPENDIX: AUXILIARY TOOLS

1.A.1 Generating Functions

Let ν be a discrete random variable (r.v.) with distribution

$$\Pr\{\nu = k\} = p_k \qquad k = 0, 1, 2, \ldots$$

$$\sum_{\forall k} p_k = 1$$

The *generating function* (g.f.) of ν denoted by $\hat{\varphi}(z)$ is defined as

$$\hat{\varphi}(z) = \sum_{\forall k} p_k z^k$$

Thus, the coefficient of z^k equals the probability that ν equals k. The g.f. is very convenient when one deals with the summation of discrete r.v.'s. Generating functions are especially effective when algorithms for computer calculations are involved.

For example, suppose we have two discrete r.v.'s α and β, with distributions a_k and b_k, respectively. We are interested in the distribution g_k of the new r.v. $\gamma = \alpha + \beta$. Of course, we could find the desired distribution directly:

$$\Pr\{\gamma = k\} = \Pr\{\alpha = 0\} \Pr\{\beta = k\} + \Pr\{\alpha = 1\} \Pr\{\beta = k - 1\}$$

$$+ \cdots + \Pr\{\alpha = k\} \Pr\{\beta = 0\}$$

$$= \sum_{0 \le j \le k} a_j b_{k-j} = \sum_{0 \le j \le k} a_{k-j} b_j \qquad (1.141)$$

But such an approach is not always simple or convenient. For computational purposes it is often better to use the g.f.'s of α and β. Let $\hat{\varphi}_\alpha(z)$, $\hat{\varphi}_\beta(z)$, and $\hat{\varphi}_\gamma(z)$ be the g.f.'s of the respective distributions. Then we have

$$\hat{\varphi}_\gamma(z) = \hat{\varphi}_\alpha(z) \hat{\varphi}_\beta(z)$$

In the new polynomial the coefficient of z^k is automatically equal to expression (1.141). This example does not exhibit all of the advantages of generating functions, but below we will show other cases where the use of g.f.'s is very effective.

Suppose we wish to find $\Pr\{\nu \le k\}$. We note that

$$\Pr\{\nu \le k\} = \left[\frac{1}{z^k} \hat{\varphi}(z) \right]_{z=0}$$

where $[\cdot]_{z=0}$ is the operator that turns any negative power t of the term z^t into 1. Thus, after the substitution $z = 0$,

$$\Pr\{\nu \le k\} = \sum_{0 \le j \le k} p_k$$

Furthermore, it is clear that

$$\frac{d}{dz} \hat{\varphi}(z) \bigg|_{z=1} = \left[\sum_{\forall k \ge 1} kp_k z^{k-1} \right]_{z=1} = \sum_{\forall k \ge 1} kp_k = E\{\nu\}$$

To obtain higher moments, it is more convenient to use the so-called *moment generating function* (m.g.f.) $\tilde{\varphi}(s)$ of the r.v. ν. This function can be written formally by simply substituting $z = e^s$ into the generating function, that is, $\hat{\varphi}(e^s) = \hat{\varphi}(z)$. Then

$$\frac{d}{ds} \hat{\varphi}(e^s) \bigg|_{s=0} = \frac{d}{ds} \tilde{\varphi}(s) \bigg|_{s=0} = \left[\sum_{\forall k \ge 1} kp_k e^{sk} \right]_{s=0} = \sum_{\forall k \ge 1} kp_k = E\{\nu\}$$

$$\frac{d^2}{ds^2} \hat{\varphi}(e^s) \bigg|_{s=0} = \left[\sum_{\forall k \ge 1} k^2 p_k e^{sk} \right]_{s=0} = \sum_{\forall k \ge 1} k^2 p_k = E\{\nu^2\} = m^{(2)}$$

1.A.2 Laplace – Stieltjes Transformation

With continuous r.v.'s the Laplace–Stieltjes transformation (LST) is often used. This transformation allows one to solve integral–differential equations with the "reduced" mathematical technique. The essence of the LST is depicted in Figure 1.14.

In this book we usually consider distributions of nonnegative r.v.'s. The transforms of such r.v.'s are defined for the distribution function (d.f.) $F(t)$ as

$$\varphi_F(s) = \int_0^\infty F(t) e^{-st} dt$$

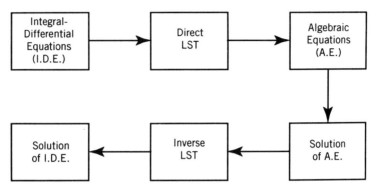

Figure 1.14. Scheme of the Laplace–Stieltjes transform usage.

and for the density function $f(t)$ as

$$\varphi_f(s) = \int_0^\infty f(t) e^{-st}\, dt = \int_0^\infty e^{-st}\, dF(t)$$

If we consider the LST corresponding to the density function, the LST can be rewritten in the form

$$\varphi_f(s) = \int_0^\infty e^{-st}\, dF(t) = \mathrm{E}\{e^{-\xi s}\}$$

The correspondence between the original function $f(t)$ and its LST $\varphi(s)$ is usually denoted as

$$f(t) \leftrightarrow \varphi_f(s)$$

We now consider some properties of LSTs.

Sum of Functions   the transformation of the sum of functions is the sum of the transforms:

$$f_1(t) + f_2(t) \leftrightarrow \varphi_1(s) + \varphi_2(s) \tag{1.142}$$

This follows directly from the property of the integration. Obviously, (1.142) is true for any number of functions in the sum.

Convolution of Functions   The convolution of two functions $f_1(t)$ and $f_2(t)$ is the function $f(t)$ defined by

$$f(t) = \int_0^t f_1(t-x)f_2(x)\,dx = \int_0^t f_2(t-x)f_1(x)\,dx$$

This operation over the functions $f_1(t)$ and $f_2(t)$ is also denoted by

$$f(t) = f_1 * f_2(t)$$

The transform of the convolution of a pair of functions is the product of the transforms:

$$f_1 * f_2(t) \leftrightarrow \varphi_1(s)\varphi_2(s) \tag{1.143}$$

The proof of this is left as Exercise 1.14. Obviously, the correspondence is true for any number of functions in the convolution:

$$f_1 * f_2 * \cdots * f_n(t) \leftrightarrow \varphi_1(s)\varphi_2(s)\ldots\varphi_n(s)$$

Derivative of a Function   The transform of the derivative of a function can be expressed in terms of the transform of the function as

$$f'(t) \leftrightarrow s\varphi(s) - f(0) \tag{1.144}$$

The proof of this is left as Exercise 1.15.

Integral of a Function   The transform of the integral of a function can be expressed by the transform of the function as

$$\int_0^t f(t)\,dt \leftrightarrow \frac{1}{s}\varphi(s) \tag{1.145}$$

The proof of this is left as Exercise 1.16.

Property of the LST of the Density Function   If the function $f(t)$ is the density of the distribution of the r.v. ξ, that is, $f(t) = [dF(t)]/dt$, then

$$\left[\int_0^\infty f(t)e^{-st}\,dt\right]_{s=0} = \int_0^\infty f(t)\,dt = 1 \tag{1.146}$$

and

$$\left[-\int_0^\infty tf(t)e^{-st}\,dt\right]_{s=0} = -\int_0^\infty tf(t)\,dt = -\mathrm{E}\{\xi\}$$

Property of the LST of the PFFO If $P(t)$ is the probability of a failure-free operation, that is, $P(t) = 1 - F(t)$, then the corresponding LST at 0 is

$$\left[\int_0^\infty P(t) e^{-st} \, dt \right]_{s=0} = \int_0^\infty P(t) \, dt = T \qquad (1.147)$$

where T is the mean of the distribution $F(t) = 1 - P(t)$. This value is called the *mean time to failure* (MTTF).

Initial Moments of a Distribution The Laplace–Stieltjes transformation of the density function allows us to obtain the moments as

$$m^{(k)} = (-1)^{k+1} \frac{d^k}{ds^k} \varphi_f(s) \bigg|_{s=0} \qquad (1.148)$$

These moments are obtained more conveniently with the help of the continuous m.g.f. which coincides with the LST except in the sign of the power in the exponential:

$$\overline{\varphi}^{(s)} = \int_0^\infty f(t) e^{st} \, dt$$

In this case there is no change in the sign:

$$m^{(k)} = \frac{d^k}{ds^k} \overline{\varphi}^{(M)}(s) \bigg|_{s=0}$$

The Laplace–Stieltjes transformation represents a very useful mapping from one functional space into a new one where the original functions are replaced with transformed ones. Operations over these new functions are often simpler in the transformed space. The general idea is reflected in Figure 1.14.

1.A.3 Generalized Generating Sequences

The method of *generalized generating sequences* (GGS) is based on a new approach which is genetically tied to generating functions. It is very convenient for a computerized realization of different enumeration problems which often arise in discrete optimization. We begin with a simple example to illustrate the main features of the GGS.

Consider a series connection of n resistors. Each unit in the series has a resistance which has a random value (for various reasons, e.g., manufacturing, storage, environmental influence, etc.). This random value of the unit's resistance is characterized by some distribution. We assume that this distribution is discrete and the resistance of the ith resistor equals the value r_{ij} with probability p_{ij}, so that

$$\sum_{1 \le j \le M_i} p_{ij} = 1$$

where M_i is the number of discrete values of the ith resistor. For each unit we can construct the generating function of the distribution of the resistance values:

$$G_i(z) = \sum_{1 \le j \le M_j} p_{ij} z^{r_{ij}}$$

To find the distribution of the resistance of the entire series connection, we can compute its g.f.

$$G(z) = \prod_{1 \le i \le f} G_i(z) = \prod_{1 \le i \le n} \sum_{1 \le j \le M_i} p_{ij} z^{r_{ij}} \qquad (1.149)$$

After simple algebraic transformations, we write the final expression in the form of a polynomial

$$G(z) = \sum_{1 \le s \le n} P_s z^{R_s} \qquad (1.150)$$

where the coefficient P_s of the term z^{R_s} equals the probability that the series system's resistance is R_s.

We remark that, in a computational sense, the introduction of the auxiliary variable z permits us to separate the variables of interest: p and r. (We omit other useful properties of the g.f. for this discussion because they are irrelevant here.) To compute P_s and R_s, one needs only to multiply the p's and to add the r's.

This example is very clear and contains no new information for those who know how to work with generating functions. Of course, if the problem is to calculate the resistance of a parallel connection of resistors, it is impossible to use (1.149) and (1.150) in any direct way. To use the g.f., one has to consider r.v.'s which measure conductivity (instead of resistance) and then find the desired result in terms of conductivity. Finally, the result can be transformed from units of conductivity to units of resistance.

Now suppose it is necessary to analyze the pipeline capacity of a set of pipes connected in series. In this example the collective capacity is the *minimum* of the capacities of the individual units. The usual generating function does not work here at all! We suggest a new approach which we call the *generalized generating sequence* (GGS).

To explain how the GGS works, we use the above example with resistors in series. First, we analyze the computations involved in moving $G(z)$ as expressed by (1.149) to $G(z)$ as expressed by (1.150). For the moment, consider a series system of two resistors labeled A and B. In terms of calculations, we perform the following operations.

1. The probability distributions of the resistances are stored as sequences of ordered pairs. We can associate these sequences with the symbols A and B and so write

$$A = (\{p_{11}, r_{11}\}, \{p_{12}, r_{12}\}, \ldots, \{p_{1v}, r_{1v}\})$$

and

$$B = (\{p_{21}, r_{21}\}, \{p_{22}, r_{22}\}, \ldots, \{p_{2w}, r_{2w}\})$$

where, for example, the pair $\{p_{1j}, r_{1j}\}$ exhibits the probability p_{1j} that the resistance of resistor A will have the value r_{1j}.

2. Now introduce a new operator Ω which operates on the pair of sequences A and B and produces a new sequence C of ordered pairs $\{p_{3k}, r_{3k}\}$. The sequence C represents the probability distribution of the resistance of the series connection of A and B. Thus,

$$\Omega(A, B) = C$$

or, since each term of the sequence C is a pair of numbers, it can also be rewritten as

$$\Omega(A, B) = \big(\Omega_p(A, B), \Omega_r(A, B)\big)$$

The sequence C is formed under Ω from the pair (A, B) as follows:
(a) For each pair (p_{1i}, r_{1i}) and (p_{2j}, r_{2j}) compute the pair $(p_{1i} p_{2j}, r_{1i} + r_{2j})$.
(b) Order the obtained pairs according to increasing values of their second components.
(c) When two or more pairs in the newly obtained sequence are tied in their second components, combine all such pairs into the single pair. The first component of the new pair is the *sum* of all first components of the tied pairs, and the second component of the new pair is the (common) *product* of the tied second components.

Note that the operators Ω_p and Ω_r have a very specific meaning in this example. But this meaning can be substituted by others in different situations. For example, for the pipeline consisting of a series connection of units with different capacities, one can write $\Omega_c(c_1, c_2) = \min(c_1, c_2)$ where c_i is the capacity of the ith pipe. All of the remaining formal operations and the order of their performance are similar. Therefore, the above-described computational algorithm, in general, can be used with no restrictions on the polynomial form. The new approach can be used for enumeration problems involving different physical parameters. We will show the effectiveness of this operator for computational problems of complex system reliability analysis and discrete optimization problems.

Now let us describe the procedure in more general terms. Keeping in mind the use of a computer, we introduce a more formal description.

For a more vivid presentation we will use a special terminology to distinguish the GGS from the g.f. This will relieve us of having to use traditional terms in a new sense, which often leads to confusion. Moreover, we hope that this new terminology can help us, in a mnemonical sense, to remember and even to explain the procedure.

In an ancient Roman army, a *cohort* was the main combat unit. Each cohort consisted of *maniples* which were independent and sometimes specialized simple combat units. Several cohorts composed a *legion*. The use of this essentially military terminology appears to be convenient in this essentially peaceful applied mathematical field. We set up a one-to-one correspondence between the above-mentioned military units and the GGS with its attributes.

Consider a system consisting of n units. Each unit j is characterized by its GGS. Let the GGS of a unit be called a legion. Each legion j includes v_j cohorts:

$$L_j = \left(C_{j1}, C_{j2}, \ldots, C_{jv_j} \right)$$

Each cohort C_{jk} is composed of some set of the unit's parameters, special characteristics, and auxiliary attributes. We call these components of the cohort maniples. Therefore,

$$C_{jk} = \left(M_{jk1}, M_{jk2}, \ldots, M_{jv_js} \right)$$

where M_{jv_jl} is the corresponding maniple and s is the number of different maniples (assumed to be the same for each cohort).

The operation of interaction between legions is denoted by Ω^L. This operator is used to obtain the resulting legion

$$L = \Omega^L_{1 \leq j \leq n} L_j$$

The operator Ω^L denotes a kind of "n-dimensional Cartesian product" and a special "reformatting" of the resulting cohorts. This reformatting depends on the specific nature of the problem [see, e.g., item (c) of the series resistors example].

As a result of this interaction of the legions, one obtains

$$N = \prod_{1 \le i \le n} v_i$$

new cohorts. For each cohort the following notation is used where Ω^C denotes the cohort's interaction, k is the subscript of this cohort in the set obtained as a result of the procedure

$$C_k = \Omega^C_{\substack{1 \le j \le n \\ i_j \leftrightarrow k}} C_{ji_j}$$

(before using the formatting procedure), and i_j are corresponding subscripts of the cohorts taking part in the interaction (this fact is conditionally reflected in the notation $i_j \leftrightarrow k$). The new cohort can be represented as

$$C_k = (M_{k1}, M_{k2}, \ldots, M_{ks})$$

Each new cohort is obtained as a result of a vector product–type interaction of maniples: n maniples of the first type interact between themselves, n maniples of the second type interact between themselves, and so on. The interaction between maniples of a specified type can be called a "natural" interaction because they involve a real physical sense of the corresponding parameters:

$$M_{kl} = \Omega^M_{\substack{1 \le j \le n \\ i_j \in k}} M_{ji_j l}$$

Here the subscript l defines the type of maniple interaction.

The resulting legion consists of a set of cohorts obtained by using the formatting procedure. It can be written as

$$L = (C_{(1)}, C_{(2)}, \ldots, C_{(N^*)})$$

where $N^* \le N$. This formatting procedure can consist of special operations over N cohorts. For example, several cohorts can be joined into an equivalent one in which some specified maniple equals the sum of others: we have the same solution with the g.f. when we add the probabilities of the terms with the same power of z. It may also be the selection of a "priority" (or "domination") of one cohort over another. Such a formatting procedure will be encountered in Chapter 10. The essential ideas of the proposed method of generating sequences can best be explained with the help of concrete examples. Such examples will be provided in Chapters 3, 8, and 10.

EXERCISES

1.1 Prove the equivalency of expressions (1.29) and (1.30), that is, prove that

$$\binom{n-1}{x-1}q^x p^{n-x} = \binom{-n}{x}(-q)^x p^{n-x}$$

1.2 Prove that a is the mean of a normal distribution with density function

$$f_N(x|a,\sigma) = \frac{1}{\sigma\sqrt{2\pi}}e^{-(x-a)^2/2\sigma^2}$$

1.3 Prove that σ^2 is the variance of a normal distribution with density function

$$f_N(x|a,\sigma) = \frac{1}{\sigma\sqrt{2\pi}}e^{-(x-a)^2/2\sigma^2}$$

1.4 Using the m.g.f. for the normal distribution, find the expression for the first moment (the mean).

1.5 Using the m.g.f. for the normal distribution, find the expression for the variance.

1.6 One observes two Bernoulli sequences with n_1 and n_2 trials, respectively. A successful trial appears at the first sequence with probability p_1 and at the second sequence with probability p_2.
(a) Find the probability, $R_k(n_1, n_2)$, that there will be k successes in the entire $n = n_1 + n_2$ trials.
(b) Show that for $p_1 = p_2 = p$ the probability of interest equals

$$R = \binom{n}{k}p^k(1-p)^{n-k}$$

1.7 Prove that

$$\sum_{1 \le k \le n} \binom{n}{k} = 2^n$$

1.8   Prove that

$$\sum_{1 \le k \le n} (-1)^k \binom{n}{k} = 0$$

1.9   Prove that

$$\binom{n_1 + n_2}{k} = \sum_{0 \le j \le n} \binom{n_1}{k} \binom{n_2}{k-j}$$

1.10   There are two variants of equipment: one performs its operation during time t and another performs the same operation during time $2t$. Both units have an exponentially distributed time to failure. The systems under consideration have different reliability: the first one has a failure rate equal to 2λ, and the second one has a failure rate equal to λ. What variant of equipment will perform its task with larger probability?

1.11   A production line manufactures good quality items with probability 0.9. Find the probability that in a sample of size $n = 500$ the number of failed items does not exceed 80.

1.12   The average portion of deficient items equals 0.01. Find the probability that in a sample of size $n = 100$ the number of failed items does not exceed 2.

1.13   A flow of the equipment failures is formed by superposition of the flows of different types of units. Each type of unit produces a failure flow which can be described as a Poisson process. During a given period of time, the average number of failures of some specified type of unit, equals 36. How many spare units should be supplied for this period of time to support failure-free operation of this type of unit with probability 0.95?

1.14   Prove that the LST of the convolution of a pair of functions is the product of the LSTs of the transforms of the initial functions in convolution.

1.15   Prove that the LST of the derivative of a function can be expressed as $f'(t) \leftrightarrow s\varphi(s) - f(0)$.

1.16   Prove that the LST of the integral of a function can be expressed as

$$\int_0^t f(t)\, dt \leftrightarrow \frac{1}{s}\varphi(s)$$

SOLUTIONS

1.1 For a binomial coefficient one can write the well-known expression

$$\binom{n}{x} = \frac{1 \cdot 2 \cdot \cdots \cdot n}{(1 \cdot 2 \cdot \cdots \cdot x)\left[1 \cdot 2 \cdot \cdots \cdot (n - x - 1)\right]}$$

$$= \frac{(n - m + 1) \cdot (n - m + 2) \cdot \cdots \cdot (n - 1) \cdot n}{1 \cdot 2 \cdot \cdots \cdot m}$$

As one knows from mathematical combinatorics, the latter expression is true for any n—even for a negative noninteger. Thus, setting n negative, one obtains

$$\binom{-n}{x} = \frac{(-n - m - 1) \cdot (-n - m + 2) \cdot \cdots \cdot (-n - 1) \cdot (-n)}{1 \cdot 2 \cdot \cdots \cdot m}$$

or, after trivial transformations,

$$\binom{-n}{x} = (-1)^n \frac{(n + m - 1) \cdot (n + m - 2) \cdot \cdots \cdot (n + 1) \cdot n}{1 \cdot 2 \cdot \cdots \cdot m}$$

$$= (-1)^n \binom{n + m - 1}{n - 1}$$

Because

$$\binom{n}{x} = \binom{n}{n - x}$$

one can finally write

$$\binom{-n}{m} = (-1)^n \binom{n + m - 1}{m}$$

1.2 Introduce a new variable

$$y = \frac{x - a}{\sigma\sqrt{2}}$$

Then the initial expression takes the form

$$\frac{1}{\sqrt{\pi}} \int_{-\infty}^{\infty} \left(\sigma\sqrt{2}y + a\right)e^{-y^2} \, dy$$

$$= \sigma \frac{\sqrt{2}}{\sqrt{\pi}} \int_{-\infty}^{\infty} ye^{-y^2} \, dy + \frac{a}{\sqrt{\pi}} \int_{-\infty}^{\infty} e^{-y^2} \, dy$$

The first term of the latter sum equals 0 because the function under integral is symmetrical in respect to $y = 0$. The second term is the well-known Euler–Poisson integral

$$\int_{-\infty}^{\infty} e^{-y^2} \, dy = 2 \int_{0}^{\infty} e^{-y^2} \, dy = \sqrt{\pi}$$

Thus, the final expression of the integral of interest equals a.

1.3   Introduce a new variable

$$y = \frac{x - a}{\sigma\sqrt{2}}$$

Then the initial expression takes the form

$$\frac{2\sigma^2}{\sqrt{\pi}} \int_{-\infty}^{\infty} y^2 e^{-y^2} \, dy$$

which can be represented as

$$\frac{\sigma^2}{\sqrt{\Pi}} \int_{-\infty}^{\infty} y \, d\left(e^{-y^2}\right)$$

Taking the latter integral by parts, one obtains

$$\frac{\sigma^2}{\sqrt{\pi}} \left(-ye^{-y^2}\Big|_{-\infty}^{\infty} + \int_{-\infty}^{\infty} e^{-y^2} \, dy \right)$$

The first term of the latter sum equals 0 because an exponential function grows faster than a linear one. The second term is the Euler–Poisson integral obtained above. Thus, the final expression of the integral of interest equals σ^2.

1.4   Consider (1.54). Assume that the first derivative with the substitution $s = 0$ derives the mean:

$$\frac{d}{ds}\exp\left(as + \frac{1}{2}\sigma^2 s^2\right)\bigg|_{s=0} = (a + \sigma^2 s)\exp\left(as + \frac{1}{2}\sigma^2 s^2\right)\bigg|_{s=0} = a$$

1.5   Using the intermediate result of Exercise 1.4, one obtains

$$\frac{d^2}{ds^2}\tilde{\varphi}(s) = \frac{d}{ds}\left[\frac{d}{ds}\tilde{\varphi}(s)\right] = \frac{d}{ds}(a + \sigma^2 s)\exp\left(as + \frac{1}{2}\sigma^2 s^2\right)$$

$$= \sigma^2 \exp\left(as + \frac{1}{2}\sigma^2 s^2\right) + (a + \sigma^2 s)^2 \exp\left(as + \frac{1}{2}\sigma^2 s^2\right)$$

$$= \left[\sigma^2 + (as + \sigma^2)^2\right]\exp\left(as + \frac{1}{2}\sigma^2 s^2\right)$$

This gives the second initial moment which is equal to the sum of the variance σ^2 and the mean squared a^2. Substituting $s = 0$ gives the desired result.

1.6   Denote by $b(k, n)$ the probability that there will be k successes in n trials. Then

(a) $$R_k(n_1, n_2) = \sum_{1 \le j \le k} b(j, n_1)b(k - j, n_2)$$

and, finally,

$$R_k(n_1, n_2) = \sum_{0 \le j \le k} \binom{n_1}{j} p_1^j q_1^{n_1 - j} \binom{n_2}{k - j} p_2^{k-j} q_2^{n_2 - k - j} \quad \text{(E1.1)}$$

(b)   If $p_1 = p_2 = p$ one can consider two experiments as one experiment with a total number of trials equal to $n = n_1 + n_2$. For this case one has

$$R = \binom{n}{k} p^k q^{n-k} \quad \text{(E1.2)}$$

1.7   The solution follows immediately if one considers a binomial of the form $(1 + 1)^n$:

$$(1 + 1)^n = \sum_{0 \le i \le n} \binom{n}{i} 1^i 1^{n-i} = \sum_{0 \le i \le n} \binom{n}{i}$$

1.8 The solution follows immediately if one considers a binomial of the form $(1 - 1)^n$.

1.9 Compare solutions (E1.1) and (E1.2) obtained in Exercise 1.6. Substitution of $p_1 = p_2 = p$ into (E1.1) gives

$$R = p^k q^{n-k} \sum_{0 \le j \le k} \binom{n_1}{j} \binom{n_2}{k - j}$$

Comparison of the latter expression with (E1.2) leads to the desired result.

1.10 Both systems are equivalent in terms of the chosen criteria.

1.11 Apply the normal approximation with mean = 450 and a standard deviation $\sigma = \sqrt{45} \approx 6.7$. Use a standard table of the normal d.f. for an argument $= (420 - 450)/6.7 \approx -4.48$.

1.12 Apply the Poisson approximation with parameter $(0.01)(100) = 1$. Use a standard table of the Poisson d.f.

1.13 Apply the normal approximation with mean = 36 and a standard deviation $\sigma = \sqrt{36} = 6$. Use a standard table of the normal d.f.

1.14 The convolution of two functions is defined as

$$f^*(t) = f_1 * f_2(t) = \int_0^t f_1(t - x) f_2 \, dx$$

By definition, the LST is

$$\varphi(s) = \int_0^\infty \left[\int_0^t f_1(t - x) f_2 \, dx \right] e^{-st} \, dt$$

Using the Dirichlet formula, we obtain

$$\varphi(s) = \int_0^\infty dx \int_x^\infty e^{-st} f_1(x) f_2(t - x) \, dt$$

$$= \int_0^\infty dx \int_x^\infty e^{-sx} f_1(x) e^{-s(t-x)} f_2(t - x) \, dt$$

Substituting $y = t - x$, we obtain

$$\varphi(s) = \left[\int_0^\infty f_2(x) e^{-sx} \, dx \right] \left[\int_0^\infty f_1(y) e^{-sy} \, dy \right] = \varphi_1(s) \varphi_2(s)$$

Thus, $f_1 * f_2(t) \leftrightarrow \varphi_1(s) \varphi_2(s)$ which corresponds to (1.3).

1.15 By definition, the LST $\varphi^*(s)$ of the derivative $f'(t) = df/dt$ is

$$\varphi^*(s) = \int_0^\infty f'(t)e^{-st}\, dt$$

The following simple transformations need no explanation:

$$\int_0^\infty f'(t)e^{-st}\, dt = \int_0^\infty e^{-st}\, df(t) = f(t)e^{-st}\big|_0^\infty - \int_0^\infty f(t)\, d\left(e^{-st}\right)$$

$$= -f(0) + s\int_0^\infty f(t)e^{-st}\, dt = -f(0) + s\varphi(s)$$

Thus, the desired equality is proven.

1.16 This relationship follows from the chain of simple transformations:

$$\int_0^\infty \left[\int_0^t f(x)\, dx\right]e^{-st}\, dt = -\frac{1}{s}\int_0^\infty \left[\int_0^t f(x)\, dx\right] de^{-st}$$

$$= \frac{1}{s}\left\{e^{-st}\int_0^t f(x)\, dx\Big|_0^\infty - \int_0^\infty f(t)e^{-st}\, dt\right\} = \frac{1}{s}\phi(s)$$

Thus, the validity of the equation is proven.

CHAPTER 2

RELIABILITY INDEXES

Reliability indexes are basically needed for the quantitative characterization of a system's ability to perform its operations. These indexes must reflect the most essential operating properties of the system, be understandable from a physical viewpoint, be simple to calculate at the design stage, and be simple to check at the test and/or usage stage.

Sometimes it is practically impossible to characterize a system with only one reliability index. But, at the same time, the number of reliability indexes has to be as small as possible. Psychologists say that more than three numerical characterizations of the quality of some object can only lead to confusion and misinterpretation of a situation. Those who deal with multicriteria optimization also know that the Pareto set should be of a small dimension. (One might recall the classical example from medieval French literature: the Buridan donkey died trying to solve a two-dimensional problem when he could not choose one bunch of hay from two!)

Simultaneously, one has to avoid the use of different "integrated" or "weighted" indexes: Such indexes generally have no clear physical sense and may mask an unacceptable level of one index by uselessly high levels of the others.

Reliability indexes may not only be used for the characterization of a system as a whole, but also some of the indexes may have an intermediate character. For example, the system, considered as an independent object, might be characterized by an availability coefficient. If the same system is part of a more complex structure, it may be more reasonable to characterize it separately with the mean time to failure (MTTF) index and the mean repair time index because they might be used to more accurately express the complex system's availability index. Moreover, the system as a whole can be characterized with indexes in the form of nondimensional real numbers. But for the system's subsystems, we sometimes need to know special functions (distribution functions, failure rates, etc.).

Almost all reliability indexes are of a statistical nature and depend on time.

We now make several points about unrepairable and repairable (renewal) units and systems. The distinction between repairable and unrepairable items is relative. The same system may be considered as repairable in one circumstance and as unrepairable in another. The main indicator is a system's ability to continue its operation after repair. For example, a computer used for routine calculations with no special time restrictions may be considered as repairable. The same computer used for a noninterruptible technological process or in a military action (which is almost always dangerous if interrupted!) may be considered as unrepairable. But if in the latter case the computer is used in an on-duty regime, it might be considered as repairable during the idle period.

Of course, some technical objects are essentially unrepairable. Some of them, such as a light bulb, cannot be repaired at all. As another example, a missile cannot be repaired during its mission. For convenience, in these cases we speak of a "renewal socket" in which unrepairable units are installed one after another in the case of a failure. Thus, after a first unsuccessfully launched missile, another one may be launched; after a first bulb has failed, another one may replace it.

2.1 UNREPAIRABLE SYSTEMS

2.1.1 Mean Time to Failure

If the criterion of a system's failure is chosen and perfectly well defined, we can determine its reliability indexes, in particular, the mean time to failure (MTTF).

After observing N failures of N unrepairable systems, there are records of nonnegative values: $t_1, t_2, \ldots, t_N$. One of the most natural characteristics of this set of observations is the sample mean, or the mean time to failure (MTTF):

$$T = (1/N)(t_1 + t_2 + \cdots + t_N) \tag{2.1}$$

This reliability index means that the system, on average, works T time units before a failure.

Consider these values in increasing order, that is, present the observations as

$$t'_1 < t'_2 < \cdots < t'_N$$

In this new notation the following equivalent equation can be written:

$$T = Nt'_1 + (N - 1)(t'_2 - t'_1) + \cdots + (t'_N - t'_{N-1})$$

The equivalency of the formulas follows from consideration of Figure 2.1 where a histogram of the values t_k is presented.

If a prior distribution $F(t)$ of a system's TTF is known, the expected value of T can be calculated in the standard way:

$$T = \int_0^\infty t \, dF(t) \tag{2.2}$$

For nonnegative random variables, the following equivalent expression can be written (see Exercise 2.1)

$$T = \int_0^\infty P(t) \, dt \tag{2.3}$$

where $P(t) = 1 - F(t)$.

The equivalency of (2.2) and (2.3) follows from the fact that we are only using different means of integrating the same function (see Figure 2.1). On a heuristic level this result may be explained by comparing this case with the analogous discrete case depicted by Figure 2.1.

The reliability index MTTF is very convenient if the system's outcome linearly depends on the time of its successful performance. For example, consider a form for producing some plastic item. After a failure the form is replaced by a new one. Thus, this form can be considered as a socket in the above-mentioned sense. In this case the effectiveness of using the form depends only on the average time to failure. But in other cases this reliability index may prove to be inconvenient.

2.1.2   Probability of a Failure-Free Operation

Consider a system performing an operation with a fixed duration t_0. In this case each $t_k < t_0$ corresponds to a system failure. A natural reliability index in this case is the probability of a failure-free operation, which reflects the frequency of appearance of the condition $t_k > t_0$. We introduce the so-called indicator function

$$d(x, y) = \begin{cases} 1 & \text{if } x > y \\ 0 & \text{otherwise} \end{cases}$$

In other words, we define a system failure in a new form: the system fails when $d = 0$.

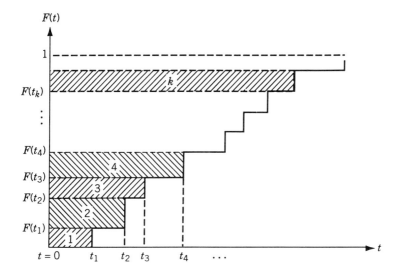

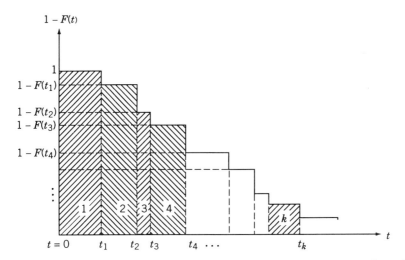

Figure 2.1. Explanation of two types of summation on a histogram of random variables.

For the same data we have to calculate the new reliability index as

$$P(t_0) = (1/N)(d_1 + d_2 + \cdots + d_N)$$

where $d_k = d(t_k, t_0)$. If we know the distribution $F(t)$ of the system TTF, the probability of a successful operation can be expressed as

$$P(t_0) = 1 - F(t_0)$$

Sometimes the duration of a task is a random variable itself, with a distribution $H(t)$. We may then speak about the expected probability of success for the random performance time. The general expression for this index is

$$P = \int_0^\infty P(t)\, dH(t) \tag{2.4}$$

Several particular cases are considered in Exercise 2.2.

2.1.3 Failure Rate

As we mentioned above, we sometimes have to know some special functions in order to calculate the reliability indexes of a complex system. One such important function is the failure rate $\lambda(t)$. In strict probabilistic terms this is the instant conditional density function at moment t under the condition that the random variable under consideration is not less than t, that is,

$$\lambda(t) = \frac{f(t)}{P(t)} \tag{2.5}$$

At first this function, called the hazard rate, appeared in demography connected to the insurance business. The physical sense of this function can be easily explained in the following simple terms. If we know the prior distribution $F(t)$ with density $f(t)$, then an element of the conditional probability

$$\Pr(dt|t) = \lambda(t)\, dt$$

is the probability of the death of an individual of age t during the forthcoming time interval $[t, t + dt]$.

This function has exactly the same sense in reliability theory when one substitutes the corresponding terms. We refer to this function as the failure rate. To explain it, consider the uniform distribution $F(t)$ on the interval $[0, 10]$. In this case $\lambda(0) = f(0) = 0.10$ because $P(0) = 1$ for a nonnegative r.v. Next, consider the moment $t = 1$. The area of the domain for the r.v. under the condition that it is larger than 1 become smaller: now it is $[1, 10]$. So $\lambda(1) = 1/9$. Of course, the same result can be obtained directly from (2.5) if we substitute $f(1) = 0.10$ and $P(1) = 0.90$. Then for the next moment, say $t = 5$, we have $\lambda(t) = 0.20$; for $t = 9$ we have $\lambda(t) = 1.0$; for $t = 9.9$ we have $\lambda(t) = 10.0$; for $t = 9.99$ we have $\lambda(t) = 100.0$; and so on. The function $\lambda(t)$ approaches ∞ at $t = 10$.

For a normal distribution with mean $a = 10$ and standard deviation $\sigma = 1$, we can calculate $\lambda(t)$ using a standard table of the normal distribution.

In both cases we observe that the function $\lambda(t)$ is increasing and unbounded. Thus, the unit's reliability for such a TTF distribution becomes worse in time. Such an aging process is very natural for most real objects. But this type of increasing function is not the only one. As we considered in Chapter 1 for the exponential distribution, the failure rate is constant in time. Moreover, the so-called "mixture of exponential distributions" has a monotonically decreasing function $\lambda(t)$.

For the mixture of two exponential functions, we can write

$$F(t) = 1 - p\left[\exp(-\alpha_1 t)\right] - (1 - p)\left[\exp(-\alpha_2 t)\right] \qquad (2.6)$$

The expression for $\lambda(t)$ can easily be obtained

$$\lambda(t) = \frac{p\alpha_1\left[\exp(-\alpha_1 t)\right] + (1 - p)\alpha_2\left[\exp(-\alpha_2 t)\right]}{p\left[\exp(-\alpha_1 t)\right] + (1 - p)\left[\exp(-\alpha_2 t)\right]}$$

We will analyze this equation "on a verbal level" using only simple explanations. For $t = 0$ we have

$$\lambda(0) = p\alpha_1 + (1 - p)\alpha_2$$

that is, $\lambda(0)$ is a weighted hazard rate at this moment. Then note that the function $\lambda(t)$ is a monotonically decreasing function. If $\alpha_1 > \alpha_2$, then $\lim_{t \to \infty} \lambda(t) = \alpha_2$.

From (2.5) it follows that

$$\lambda(t) = \frac{\dfrac{dF(t)}{dt}}{P(t)} = -\frac{\dfrac{dP(t)}{dt}}{P(t)} = -\frac{d\left[\ln P(t)\right]}{dt}$$

This immediately yields

$$P(t) = \exp\left[-\int_0^t \lambda(x)\, dx\right] \qquad (2.7)$$

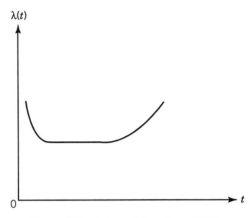

Figure 2.2. U-shaped function of $\lambda(t)$.

From the condition $0 \le P(t) \le 1$, it follows that, for any t,

$$0 \le \int_0^t \lambda(t)\, dt < \infty \tag{2.8}$$

and

$$\lim_{t \to \infty} \int_0^t \lambda(t)\, dt \to \infty \tag{2.9}$$

Thus, the function is such that $\lambda(t) \ge 0$ and possesses properties (2.7) and (2.8).

In most practical cases we observe a so-called "U-shaped form" of the function $\lambda(t)$, as depicted in Figure 2.2. During the first period of time, we observe a "burning-out" process. This process consists in the early failing of weak or defective items. Then follows a period of "normal" operation during which the failure rate is constant. During this period a failure only occurs "completely incidentally," or, as one sometimes says, "as a brick fallen from the roof." It is a period of wearing-out, fatigue, and other normal phenomena of aging.

We will show below that qualitative knowledge about the failure rate is very important for reliability analysis.

2.2 REPAIRABLE SYSTEM

2.2.1 Description of the Process

During the observation of a repairable system, we can record the sequence of periods, each of which consists of a successful performance time plus an idle

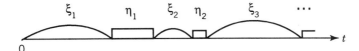

Figure 2.3. Graphic description of an alternating process.

time. Such a process is illustrated in Figure 2.3. Let ξ_k denote the random time from the completion of the $(k-1)$th repair to the kth failure, and let η_k denote the duration of the kth repair (renewal).

In the simplest case with a socket (a one-unit system), we suppose that a repair is equivalent to the replacing of a failed unit. This corresponds to a complete renewal. In this case we consider an alternating stochastic process with i.i.d. r.v.'s ξ and η, having distributions $F(t)$ and $G(t)$, respectively. We denote this alternating stochastic process by $\{\xi, \eta\}$.

Of course, the corresponding process for a system consisting of several renewal units may be much more complicated. Almost all the following explanations will be—for simplicity—presented for a renewal unit, or a socket.

All indexes used for an unrepairable system can also be used in this case for the appropriate purpose. But for repairable units and systems we have to consider several special indexes. They are more complicated and need more explanation.

2.2.2 Availability Coefficient

Consider a system which has to work in a "waiting" regime and, at the same time, the duration of the task performance is negligibly small. In this case a natural reliability index is the so-called *availability coefficient* $K(t)$. This index is the probability that the system will be in an operating state at a specified moment t in the future.

The numerical value of $K(t)$ depends on the specified moment of time t. For example, if we know that at $t=0$ the system is new and, consequently, is in an operating state, then at moment ε, where ε is small, the probability that the system is in an operating state is close to 1 and approximately equals $K(\varepsilon) \approx P(\varepsilon)$.

The behavior of $K(t)$ in time can be periodically attenuating or strictly attenuating. This depends on the types of d.f.'s $F(t)$ and $G(t)$. For illustrative purposes, consider the case where $F(t)$ is a normal d.f. with a small coefficient of variation k and $G(t)$ is a degenerate function (i.e., η is constant). $K(t)$ for this case is presented in Figure 2.4.

It is clear that the first time to failure has a normal d.f. with some mean T and a relatively small σ. The renewal completion time has the same distribution biased on the time axis. If $\eta > 3\sigma$, there may be some interval between T and $T + \eta$ where $K(t) \approx 0$. The second time to failure also has a normal

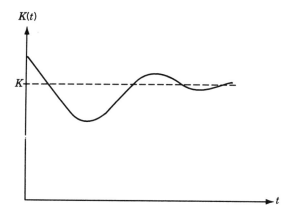

Figure 2.4. Example of the oscillating behavior of $K(t)$ in time.

d.f. but with the standard deviation larger by $\sqrt{2}$ times. Thus, the zone between $2T$ and $2(T + \eta)$, where $K(t) \approx 0$, will be smaller. Finally, for $t \gg T + \eta$, $K(t)$ will be almost constant.

If both d.f.'s $F(t)$ and $G(t)$ are exponential, the function $K(t)$ is strictly decreasing with exponential speed. We will consider this case later.

For large t the initial state has practically no influence on the behavior of $K(t)$. In this case the probability that the system is in an operating state equals the proportion of the total up time to the total operating time. Later we will show that for this case

$$K = \frac{E(\xi)}{E(\xi) + E(\eta)} \qquad (2.10)$$

The index K is called the *stationary availability coefficient*, or simply, the *availability coefficient*.

Sometimes we are interested not in a "point" characteristic $K(t)$ but in the average time spent in an operating state during some period of time, say t. We introduce the index

$$K^*(t) = \frac{1}{t} \int_0^t K(x)\, dx$$

If $t \to \infty$, both $K(t)$ and $K^*(t)$ have the same limit, namely, K defined in (2.10).

2.2.3 Coefficient of Interval Availability

If the duration of the system's task is not negligibly small, we speak about the coefficient of interval availability, that is, the probability that at a time t the system is found in an up state and will not fail during the performance of a task of length, say t_0. Denote this index $R(t, t_0)$:

$$R(t, t_0) = K(t) P(t_0 | t)$$

We will consider this index later in more detail but now we note that $P(t_0 | t) = P(t_0)$ only when $F(t)$ is exponential.

If the system is not a socket with renewal unit, the situation is more complicated. In Chapter 7 we will illustrate this statement on a duplicated system consisting of two identical renewal units.

2.2.4 Mean Time Between Failures and Related Indexes

Mean Time Between Neighboring Failures In general, the mean time to a first failure $T_{(1)}$ differs from the mean time from the first repair termination to the second failure $T_{(2)}$, and so on. In other words, all intervals of failure-free operations $T_{(k)}$, $k = 1, 2, \ldots$, may be different. We consider several typical situations.

For a socket, or a one-unit system, the MTBF coincides with the MTTF because a new unit, put into the socket, is supposed to be identical to the failed one. But this equivalence of the MTTF and MTBF cannot be extended, even for the simplest two-unit system.

Consider a Markov model of a redundant system of independent and identical units [in other words, both $F(t)$ and $G(t)$ are exponential]. Assume that we know how to compute the mean time to a forthcoming failure for this system for the following two cases: starting with the state when two units are up ($T^{[2]}$), and starting with the state when only one unit is up ($T^{[1]}$) (see Figure 2.5).

On average, the time to failure from state 2 is larger than the TTF from state 1. Indeed, in order to fail starting from state 2, the system must first enter state 1 and then from this state transit to the failure state. (Of course, from state 1 the system might even return to state 2 again.) In other words, $T^{[2]} = T^* + T^{[1]}$, where T^* is the time of the system staying in state 2 until entering state 1.

In this particular case all of the remaining intervals of failure-free periods are i.i.d. r.v.'s, because for this particular Markov process all of the initial conditions are the same.

Now consider the behavior of a series system of two independent units. For simplicity, suppose that the repair time is negligible. Let each unit have a normal d.f. of TTFs with a mean T and a very small variation coefficient. If at the moment $t = 0$ both units are new, then the first and second failures of the system are expected to appear close to each other and are around $t = T$. If the random TTFs of the units are denoted by ξ_1 and ξ_2, respectively, then $T_{[1]} = \min\{\xi_1, \xi_2\}$ and $T_{[2]} = \max\{\xi_1, \xi_2\} - \min\{\xi_1, \xi_2\}$. If the r.v.'s ξ's are as

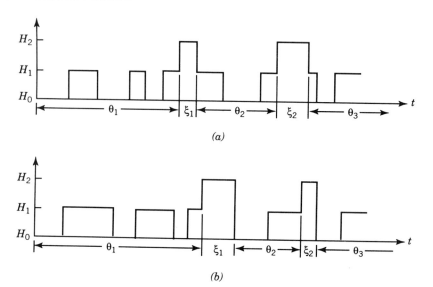

Figure 2.5. Examples of two time diagrams for a two-unit system: (*a*) a system of independent units; (*b*) a system working until both units have failed.

described, $T_{[1]} \approx E\{\xi\}$, and $T_{[2]}$ has the order of σ. By assumption, $\sigma \ll T$ (see Figure 2.6).

The next couple of failures are expected to appear close to $t = 2T$, but the expected deviation is $\sqrt{2}$ times larger than the initial deviation. For the case $\sigma \ll T$,

$$T_{[3]} \approx \min\left(\xi_1^{(1)} + \xi_1^{(2)}, \xi_2^{(1)} + \xi_2^{(2)}\right) \tag{2.11}$$

and

$$T_{[4]} \approx \max\left(\xi_1^{(1)} + \xi_1^{(2)}, \xi_2^{(1)} + \xi_2^{(2)}\right) \tag{2.12}$$

[Notice that strictly speaking (2.11) and (2.12) should be expressed in a more complicated way. We must take into account the mixture of the failure flows of the two sockets. This is explained by the appearance of extremely small and extremely large r.v.'s.]

Thus, if $T_{[1]} > T_{[2]}$, then $T_{[2]} < T_{[3]}$ and $T_{[4]} < T_{[3]}$. At the same time, $T_{[1]} > T_{[3]}$ and $T_{[2]} < T_{[4]}$. The process continues in the same manner for larger numbers of interfailure intervals.

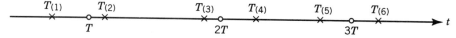

Figure 2.6. Example of the failure flows of a series system of two units.

With $n \gg 1$, for any variance σ^2, the value $\sigma\sqrt{n}$ begins to be larger than T. This leads to a strong mixture of moments of failure of both sockets of the system. In theoretical and practical terms, this means that $T_{[n]} \approx T_{[n+1]}$, and, moreover, $T_{[n]} \to \infty$ with $n \to \infty$. Thus, even for the simplest two-unit system, all MTBFs are different (though they may have the same asymptotical value). More complex cases appear when we consider a series system of more than two units. Notice that in reliability theory the term *MTBF* is usually used for the stationary regime, that is, as $t \to \infty$.

There are other indexes used in reliability theory which are connected with the time to failure. One of them is the instantaneous MTTF at time t. This is the mean time to failure from a specified moment t under the condition that a failure has happened just at this moment. From qualitative arguments it is clear that for t comparable with T this new index will differ from the MTTF. We remark that for the stationary regime, the values of both of these indexes coincide.

To conclude the discussion about the MTBF, we must emphasize that each time one should understand what kind of a particular TTF is under consideration. If we again regard a renewal series system of n units, each of them with a normally distributed TTF with a very small coefficient of variation, then we have the following:

1. For the MTTF

$$T \approx \min_{1 \le i \le n} \xi_i$$

2. The next $n - 1$ MTTFs might be extremely small depending on the number of units and the smallness of the variation coefficient:

$$T^{(i)} = E\{\xi^{(i+1)} - \xi^{(i)}\}$$

where $\xi^{(i)}$ is the ith-ordered statistic, $1 \le i \le n - 1$. A possible behavior of $t^{(k)}$ for a series system is presented in Figure 2.7.

3. The stationary MTTF for any recurrent point process with continuous distributions of TTFs is

$$T = \frac{1}{\displaystyle\sum_{1 \le i \le n} \frac{1}{T_i}}$$

This value is the limit for $T^{(k)}$ when $k \to \infty$.

In practice, we are often interested in the mean time of a failure-free operation starting from some specified moment t. In the theory of renewal

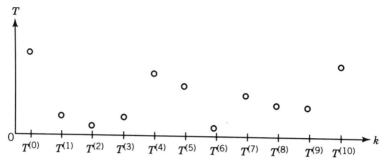

Figure 2.7. Example of changing the mean time between failures depending on the current number of failure for which this value is evaluated.

processes, this value is called the mean residual time. In general, this index differs from the MTTF and any version of the MTBF.

If $t \to \infty$, for a recurrent process this index differs from both of those mentioned above. The exception is a Poisson process for which all three indexes coincide.

2.3   SPECIAL INDEXES

Now we consider some special reliability indexes for repairable systems. These indexes are nontraditional: they describe not a failure-free operation but rather a successful operation during a specified time. In some sense, there is no "local" failure criterion. The determination of a successful or unsuccessful operation is made, not at the moment of a current failure, but only after the completion of the entire system's performance during an acceptable operational time. This means that some interruptions of the system operation might be considered as not being destructive.

2.3.1   Extra Time Resource for Performance

Sometimes a system has some reserve time to perform its task; that is, the interval of time θ_0 given for the performance of the system operation is more than the time t_0 required for a successful operation.

Examples of such situations can be taken from different areas of applications: conveyer production lines, electronic equipment with special power supplies, a computer performing routine calculations not in real time, and so forth. (Other detailed examples will be provided below.) In all of these cases not all failures of the system lead to the failure of the overall system's performance.

Consider a computer performing a computational task whose duration is t_0. The computer has a resource of time θ_0 for its performance which is larger than the required time t_0. Random negligibly short interruptions (errors) may appear, each of which will destroy the results of all of the performed calculations.

In this case the probability of success can be written as

$$\Pr\{\text{at least one } t_k > t_0 | k : t_k \in \theta_0\} \qquad (2.13)$$

Let the computer's task be segmented into phases and assume that the computer works in a restarting regime. After the completion of each phase, all intermediate results are put into the computer's memory. Each short failure has destroyed only the very last phase of the overall solving task. After the error has been found, the calculations for this particular phase are repeated. We do not give the formal definition of the corresponding reliability index but it is understandable that this index may be defined with the help of (2.13). We will consider this case in the special section dedicated to time redundancy.

2.3.2 Collecting Total Failure-Free Time

Suppose a system is required to accumulate some given amount of successful operating time during some given period θ_0 for the successful performance of the task. The probability of success can be written as

$$\Pr\left\{\sum_k t_k > \theta_0 | k : t_k \in \theta_0\right\} \qquad (2.14)$$

As an example, we may again consider a computer system with a restarting regime for which each failure takes some time to repair but the operation of the system can be continued without loss of the previously obtained results. This situation may be observed if the restarting phases are very short, so there is practically no loss of the intermediate results but the system needs some time for restoration.

A close phenomenon occurs when one considers the transportation of some load. Failures and consequent repairs may only delay the termination of the task, but will not lead to a total failure. Of course, if the total idle time exceeds some limit, the task should be considered as not fulfilled (e.g., in the transportation of fresh food).

This index can obviously be written as

$$\Pr\left\{\sum_k x_k < \Psi_0 | k : x_k \in \Psi_0\right\} \qquad (2.15)$$

where Ψ_0 is the specified allowable total down time during the period θ_0.

2.3.3 Acceptable Idle Intervals

Some systems possess the property of *time inertia*: they are insensitive to short breakdowns. As an example, consider a responsible computer system which has an independent power supply to prevent the system from occasional short failures of the common power system. In this case, if a failure of the main power system occurs, the computer system can operate with the help of this special power supply. Another example can be represented by a multistage conveyer system with an intermediate storage of spare subproducts in the case of a breakdown in the previous stages.

Thus, roughly speaking, an operational interruption of any such system can be noticed only if the duration of the down time x_k exceeds some specified value x_0. In this case the reliability index is

$$\Pr\{\text{all } x_k < x_0 | x_k \in \theta_0\} \tag{2.16}$$

In real life we meet more complicated situations. For example, a redundant power supply may demand substantial time to recharge, and this fact must be taken into account.

Of course, some combinations of the listed criteria for a system's failure may be considered. Some of them will be presented later.

2.4   CHOICE OF INDEXES AND THEIR QUANTITATIVE NORM

2.4.1 Choice of Indexes

The problem of choosing a reliability index arises before an operations research analysis. The solution of this problem depends on the nature of the object to be analyzed, its operations, and its expected results.

Depending on the operational level of the system, reliability indexes can be divided into two groups: operational and technical. If we deal with a system performing its individual and independent operation with a concrete final output, the reliability indexes should characterize the system's ability to perform its operation successfully. Such indexes are called *operational*.

If we deal with an object that is a subsystem and only performs some functions that are necessary to fulfill the operation of the system as a whole, the reliability indexes may be auxiliary. We can express the operational indexes of the system as a whole through these indexes. Such indexes are called *technical*. They are used to describe the reliability of a system's components and parts.

Starting with operational indexes, consider a computer that can be used to perform several quite different operations. The computer which is used for routine calculating tasks may be characterized with the help of an average percentage of useful operational time. The availability coefficient is the appropriate reliability index in this case.

The same computer used for supporting a long and noninterrupted techno-logical process can naturally be characterized by the *probability of a failure-free operation* (PFFO).

If the computer is used for an automated landing system in an airport, and the duration of each operation is negligibly small in comparison with the computer's mean time to failure (MTTF), the reliability index should reflect the number of successfully served airplanes. In this case the availability coefficient is also the most appropriate reliability index.

If the same computer is unrepairable, for instance, its task consists of collecting and processing information in a spy satellite, the best characteriza-tion of it is the MTTF.

In all of these cases the reliability index corresponds to the system predestination and to the nature of its use.

Now consider an example when a reliability index is used to characterize "inner" technical abilities. Consider two identical computers connected in parallel. The natural reliability index for this duplicated system is the PFFO. In this case each computer is only a subsystem taking part in the performance of a system's operation. What should we know about each separate computer to characterize this system? To compute the complex PFFO, one needs to know the probability distributions of both the time to failure and the repair time of the computer, as well as the parameters of these distributions. The distribution itself is *not* an index; it is a function. But parameters of the distribution can be considered as *technical* reliability indexes of the com-puter. These parameters have no relation to a system's operation, they only reflect an ability to work in general.

Note that the type of reliability index chosen does not depend on the responsibility of the performed operation. The responsibility of the system's operation defines the level of requirements but is not part of the nomencla-ture of reliability indexes.

When choosing reliability indexes we should take into account the follow-ing simple recommendations based on common sense:

- The chosen indexes must allow one to analyze them with the help of analytic or computer models at the stage of system design.
- The total number of reliability indexes chosen for a system's characteri-zation should be as small as possible.
- The chosen indexes should be easily interpreted.
- The indexes should allow one to formulate clear requirements on relia-bility.
- The indexes must allow one to estimate the achieved reliability level after field tests or experimental exploitation.
- Complex "integrated" indexes must be avoided: various "convolutions" and "weightings" of different indexes usually have no real meaning.

2.4.2   Required Reliability Level

The problem of choosing the level of reliability is a very complex one. In practice, this problem is usually solved on the basis of engineering experience. For the purposes of determining reliability requirements, equipment may be divided into three groups by its "level": systems, subsystems, and units (components).

A system is considered as an object with its own goals of performance. A system performance is evaluated with the help of operational reliability indexes which are a measure of its success.

A subsystem is a more or less independent part of the system. It is considered to be an assembly of objects within a system. Each subsystem performs functions that are necessary for the operation of the system as a whole. The system's subsystems can be characterized by operational indexes if their functions can be measured with independent indexes or by technical indexes if these indices are used to express the system's performance effectiveness index.

A unit, or a component, is the smallest indivisible part of an object. The term *unit* is sometimes also used as a generic term for one physically separate item. In general, the term *component* is usually used for the smallest technological part of an object: electronic components, mechanical details, and so on.

The only problem which can be formulated as a mathematical problem is the assignment of reliability requirements among subsystems (parts of the system) when the requested level of reliability is known for the system as a whole. In this case the problem is reduced to the problem of the optimal allocation of some resources used for the improvement of reliability. The technical aspects of the problem will be considered in Chapter 10. Here we explain the nature of the problem.

Consider a system consisting of N independent subsystems. Assume that the probability of successful operation of the system as a whole must not be less than R_0. Each subsystem can be designed with different levels of reliability. Such levels depend on the expenditure of some kind of resource, for example, money. Suppose that we know all functions $P_k(c)$ which reflect how the reliability index of the kth subsystem increases as a function of the expenditure of the resource c.

If the system's reliability index can be represented as

$$P(C_0) = \prod_{1 \le k \le N} P_k(C_k)$$

and the value of C_0 is specified, then the problem is to find the optimal allocation of the total resource C_0 in such a way that the resulting system index is maximal, that is, find $\mathbf{C}^* = \{C_1, C_2, \ldots, C_N\}$ such that

$$P(\mathbf{C}^*) = \max\left\{ \prod_{1 \le k \le N} P_k(C_k) \,\middle|\, \sum_{1 \le k \le N} C_k \le C_0 \right\}$$

Of course, we might also formulate the inverse problem: to design a system with a required level of reliability, for example, R_0. Then the problem may be reformulated as

$$P(\mathbf{C}^*) = \min\left\{\sum_{1 \le k \le N} C_k \,\middle|\, \prod_{1 \le k \le N} P_k(C_k) \ge R_0\right\}$$

A solution to both of these problems is presented in Chapter 10.

Unfortunately, even such simple problems cannot usually be solved in practice because the functions $P_k(C_k)$ are often unknown. In such cases we usually use heuristic methods based on a proportional distribution of reliability "quotas" among the system's units.

What must one do in the general case when it is necessary to assign a reliability level to the system as a whole? In our opinion, there is only one thing to do: perform an evaluation based on engineering experience. Prototypes can be used for comparison with the designed system and, on the basis of this, the decision about a possible or desirable reliability level might be made.

Naturally, if one fixes the amount of available resources for the production of some type of technical system, then we not only have to solve the problem of an optimal reliability level, but we must also answer the question of how many such systems we intend to produce? In turn, the number of systems of some chosen type depends on the number of other "competing" systems in the same area of use or utility. Assume that we are considering the design of a new type of jet. First of all, too high a level of reliability will demand a high level of expenses for the production of each jet and, as a consequence, will lead to a decrease in the total number of jets produced. It is clear that it is useless to have only one extremely highly reliable jet and it is equally unreasonable to have a large number of jets, each of which has a very low reliability. To choose "a golden middle" is a problem which lies outside the scope of mathematics and even outside the scope of engineering. The only way to solve this problem is to rely on expert's opinions and traditions.

But the experts' opinions are also not isolated. Taking into account all considerations concerning this particular type of a jet, experts have to think about the number and reliability of other jets owned by the airline. But this total number depends on a specific situation, considering the transportation system of the country as a whole. In turn, it depends on the level of the national economy. The level of the national economy depends on a number of unformulated and nonformulated factors: the political stability of the country, the external situation in the world, and so forth. Thus, we are convinced that any attempt to try to solve this problem in some "precise" sense is doomed.

But then one may ask: Why use mathematical methods at all? Why not rely on experts' opinions to solve all problems of this kind? The answer is that mathematical methods of analysis of situations help one to make logically strong decisions; mathematical models of technical systems help one to understand the nature of systems being designed. We begin to make local solutions in optimal ways. This leads to a kind of process of "natural selection." As in nature this process allows for the survival of only those who have best adapted to existing environments. And, in this situation, those technical systems which are "locally optimally designed" have a better chance to "survive" under currently existing circumstances.

We now consider possible methods of establishing reliability.

System Level   Consider two principal cases. One of them consists in the use of practical experience and engineering intuition. Mostly it is based on an analysis of prototypes of the new technical system to be investigated. This method needs no special comments.

Practically, the only time a system's reliability requirement appears is if:

- The system's outcome can be measured in cost units, that is, in the same units as the costs of the system's design, production, and maintenance.
- The system's structure and operational modes are well known in advance.
- Necessary statistical data for all components are determined with a satisfactory degree of confidence.

In this case the designer has an opportunity to compare M different variants of the system's design and to choose the most profitable one. The objective function of the system's performance for the kth variant can be written in the form

$$F_k(R) = E_k(R) - \gamma C_k(R)$$

where R is the system's reliability index, $E_k(R)$ is the outcome of the kth variant of the designed system, and $C_k(R)$ is the expenditure needed to design, produce, and maintain the system with index R, $1 \le k \le M$ and γ is a dimensional coefficient analogous to a Lagrange multiplier. The value of R depends on the structure of the kth variant, S_k, and on the reliability indexes of the subsystems used, $r_i^{(k)}$, $1 \le i \le n_k$, where n_k is the number of subsystems in the kth variant of the system. Thus, R itself can be written in a general form as

$$R = R\left(S_k, r_i^{(k)}; 1 \le k \le M, 1 \le i \le n_k\right)$$

For simplicity, suppose that all functions are differentiable. Then the optimal level of the reliability index R can usually be determined by solving the equation

$$\frac{d}{dR}F_k(R) = 0$$

or, equivalently,

$$\frac{dE_k(R)}{dR} = \frac{dC_k(R)}{dR}$$

Each optimum R_k^{opt} can be evaluated and then the variant k with the highest value of $F_k(R_k^{\text{opt}})$ is selected. Unfortunately, such an ideal situation appears extremely rare in engineering practice.

Subsystem Level Suppose that the system's reliability requirement is specified. Then the problem is to distribute the given value of the index over the subsystems. We consider several cases, each of them representing different information concerning the system's structure and the availability of statistical input data.

Uniform Allocation of Requirements This method is usually used when one can imagine only the approximate size of a subsystem of the main system. A reliability index R of a probabilistic nature (e.g., the probability of success or the availability coefficient) is specified for the system as a whole. The simplest assumption is that the system has a series structure and consist of n subsystems. The reliability requirement for each subsystem is then given by

$$R_i = \sqrt[n]{R} \qquad 1 \leq i \leq n$$

Clearly, if subsystem indices are chosen in such a way, the system reliability index equals R.

If requirements can be specified as the system's MTTF T, we can choose for each ith subsystem

$$T_i = nT$$

This means that we additionally assume that the TTF of any subsystem has an exponential distribution.

Allocation in Proportion to the Number of Units Assume that the same conditions exist as before, but in addition subsystem i consists of n_i units which are essentially similar in their complexity. In this case the requirement (in terms of the probability of success) should be chosen to be

$$R_i = \sqrt[a_i]{R} \qquad 1 \leq i \leq n \tag{2.17}$$

where

$$a_i = \frac{n_i}{\displaystyle\sum_{1 \leq i \leq N} n_i}$$

When all distributions are assumed to be exponential, the requirements can be formulated in terms of the MTTF as

$$T_i = \frac{T}{a_i}$$

This method can be useful if different subsystems are designed by different subcontractors. It is reasonable to specify "softer" demands for more complex subsystems.

Allocation in Proportion to the Expected Failure Rate   Suppose a designer has more complete information about the system: the unit failure rates are known (perhaps from previous experience), and the hypothesis about the exponential distribution can be considered valid. In this case the previous method can be improved. We can use (2.17) but we substitute a_i defined by

$$a_i = \frac{\displaystyle\sum_{1 \le j \le M} \lambda_j n_{ji}}{\displaystyle\sum_{1 \le i \le n} \sum_{1 \le j \le M} \lambda_j n_{ji}}$$

where M is the number of types of units and n_{ji} is the number of units of the jth type in subsystem i.

Optimal Allocation of Reliability Requirements   This method is applied if we know the system's structure S and can predict the cost–reliability trade-off for each subsystem. The problem is to find the values of R_i, $1 \le i \le n$, that yield the required reliability index R_0 at the lowest cost.

This problem can be written in mathematical terms as

$$\min\left\{ \sum_{1 \le i \le n} C_i(R_i) \middle| R(R_i, 1 \le i \le n|S) \ge R_0 \right\}$$

where $C_i(R_i)$ represents the subsystem's costs as a function of its reliability and S is the conditional notation of the system structure. For instance, if we consider a series system, the reliability function can be represented as

$$R(R_i(C_i), 1 \le i \le n) = \prod_{1 \le i \le n} R_i(C_i)$$

In other words, the optimal allocation of reliability requirements between subsystems is a type of optimal redundancy problem (see Chapter 10).

Reliability Requirements for a Component Almost all equipment components in engineering are of general usage. The only method in this case is based on the "natural selection" principle. In other words, the better and cheaper components among existing ones survive the competition in a technical and economic environment. And, at the same time, new components appear and replace technical "dinosaurs."

CONCLUSION

This chapter does not need any special comments. In one form or another, reliability parameters are discussed in any book on reliability engineering or theory. As examples, we refer the reader to the wide list of general references at the end of this book.

The nomenclature of reliability indexes in a systematic and structured form can be found in Kozlov and Ushakov (1970) and Ushakov (1985, 1994). The methodological problems of choosing indexes and quantitative requirements in reliability engineering are discussed in Gnedenko, Kozlov, and Ushakov (1969).

REFERENCES

Gnedenko, B. V., B. A. Kozlov, and I. A. Ushakov (1969). The role and place of reliability theory in engineering. In *Theory of Reliability and Queuing Systems* (in Russian). Moscow: Nauka.

Kozlov, B. A., and I. A. Ushakov (1970). *Reliability Handbook*. New York: Holt, Rinehart, and Winston.

Kozlov, B. A., and I. A. Ushakov (1975). *Handbook on Reliability of Radio and Automation Systems* (in Russian). Moscow: Soviet Radio.

Ushakov, I. A., ed. (1985). *Reliability of Technical Systems: Handbook* (in Russian). Moscow: Radio i Sviaz.

Ushakov, I. A., ed. (1994). *Handbook of Reliability Engineering*. New York: Wiley.

EXERCISES

2.1 Prove that the mean value of a nonnegative r.v. ν with distribution $F(t)$ can be expressed in the form of (2.4) which is equivalent to (2.3).

2.2 A system has an exponentially distributed TTF with parameter λ. The operation to be performed also has a random duration. Find the

probability that the system successfully performs its operation if
(a) the operation duration is distributed exponentially with parameter α;
(b) the operation duration is distributed normally

$$f_N(x|a, \sigma) = \frac{1}{\sigma\sqrt{2\pi}} e^{-(x-a)^2/2\sigma}$$

with mean equal to a and variance equal to σ^2. We also assume that $\lambda\sigma \ll 1$.

2.3 Build the graph of the failure rate for the mixture of two exponential distributions (2.6) with the following parameters, respectively,
(a) $\lambda_1 = 1$ [1/hour], $\lambda_2 = 1$ [1/hour];
(b) $\lambda_1 = 0.5$ [1/hour], $\lambda_2 = 1$ [1/hour];
(c) $\lambda_1 = 2$ [1/hour], $\lambda_2 = 1$ [1/hour].

SOLUTIONS

1.1

$$T = \int_0^\infty t\,dF(t) = \int_0^\infty t\,d[1 - P(t)] = -\int_0^\infty t\,dP(t)$$

and, after integrating by parts,

$$-\int_0^\infty t\,dP(t) = -tP(t)\Big|_0^\infty + \int_0^\infty P(t)\,dt = \int_0^\infty P(t)\,dt = \int_0^\infty [1 - F(t)]\,dt$$

1.2
(a) Using (2.4), one writes

$$P = \int_0^\infty e^{-\lambda x}\alpha e^{-\alpha x}\,dx = \alpha\int_0^\infty e^{-(\alpha+\lambda)x}\,dx = \frac{\alpha}{\alpha + \lambda}$$

(b) First of all, consider the given conditions. Almost all "probabilistic mass" is concentrated in a relatively very compact area related to the MTTF of the system. This means that in this area the exponential function can be successfully approximated by the set with at most two terms:

$$e^{-\lambda x} \approx 1 - \lambda x + \frac{(\lambda x)^2}{2}$$

Thus, one has

$$P = \int_0^\infty e^{-\lambda x} f_N(x|a,\sigma)\, dx \approx \int_0^\infty \left[1 - \lambda x + \frac{(\lambda x)^2}{2} \right] f_N(x|a,\sigma)\, dx$$

$$= \int_0^\infty f_N(x|a,\sigma)\, dx - \int_0^\infty \lambda x_N(x|a,\sigma)\, dx + \int_0^\infty \frac{(\lambda x)^2}{2} f_N(x|a,\sigma)\, dx$$

$$= 1 - a\lambda + \frac{(\sigma\lambda)^2}{2}$$

2.3 In the first case there is no mixture at all; the second and third cases differ only by the scale. In general, one can write

$$f(t) = p_1 \lambda_1 e^{-\lambda_1 t} + p_2 \lambda_2 e^{-\lambda_2 t}$$

and

$$\lambda(t) = \frac{p_1 \lambda_1 e^{-\lambda_1 t} + p_2 \lambda_2 e^{-\lambda_2 t}}{p_1 e^{-\lambda_1 t} + p_2 e^{-\lambda_2 t}}$$

(The numerical solution is left to the reader.)

CHAPTER 3

UNREPAIRABLE SYSTEMS

In this chapter we will consider the main types of unrepairable systems. The only type that we will not address is a general network structure, which will be considered in a later chapter.

3.1 STRUCTURE FUNCTION

For convenience in future mathematical explanations, let us introduce the so-called *indicator function* x_i for unit i:

$$x_i = \begin{cases} 1 & \text{if the } i\text{th unit is operating} \\ 0 & \text{otherwise} \end{cases} \tag{3.1}$$

Let us introduce a similar function for the system as a whole. This new function depends on all of the x_i's, the system's structure, and the criterion of system failure that has been chosen:

$$f(x_1, x_2, \ldots, x_n) = \begin{cases} 1 & \text{if the system is operating} \\ 0 & \text{otherwise} \end{cases} \tag{3.2}$$

In reliability theory this function is called the *structure function* of a system. If each unit has two states—up and down—then a system of n units may have 2^n different states determined by states of the individual system's units. The function (3.2) is determined by the system failure criterion.

Of course, system states may differ from each other by their level of operational effectiveness. This case will be considered in Chapter 8. Here we

110

Figure 3.1. System with a series structure.

restrict ourselves to the case where a system has only two possible states: up and down.

From the definition of (3.2) it is clear that the x_i's are Boolean variables and $f(x_1, x_2, \ldots, x_n)$ is a Boolean function. We also denote this function by $f(\mathbf{X})$ where $\mathbf{X} = (x_1, x_2, \ldots, x_n)$. System reliability structures are often displayed as a two-pole network. One of the simplest examples of such a network is presented in Figure 3.1. The connectedness of this two-pole network is equivalent to the equality of the Boolean function (3.2) to 1.

Each unit i may be in state $x_i = 1$ or $x_i = 0$ in random. If each Boolean variable x_i is considered as a Bernoulli r.v., then $E\{x_i\}$ is interpreted as the probability that unit i is in an up state, and $E\{f(\mathbf{X})\}$ is defined as the probability of the system's successful operation:

$$p_i = E\{x_i\} \qquad \text{and} \qquad P_{\text{syst}} = E\{f(\mathbf{X})\}$$

We consider only monotone functions $f(\mathbf{X})$ for which $f(\mathbf{X}) \geq f(\mathbf{X}')$ if $\mathbf{X} > \mathbf{X}'$. Here the inequality $\mathbf{X} > \mathbf{X}'$ means that $x_i \geq x_i'$ for all i and there is a strict inequality at least for one i. This assumption is very natural. Indeed, a unit failure generally will not improve a system's operational state. Therefore, if a system is down in state $\mathbf{X}$, it cannot be up in state $\mathbf{X}'$ with some additionally failed units. (Of course, it is correct under the assumption that the system was correctly designed.) We emphasize that it relates only to systems whose operation can be described in terms of Boolean functions.

3.2 SERIES SYSTEMS

The series structure is one of the most common structures considered in engineering practice. A system with such a structure consists of units which are absolutely necessary to perform the system's operation: a failure of any of one of them leads to a system failure. Schematically, this structure is represented in Figure 3.1.

Of course, the series system in a reliability sense does not always correspond to a real physical series connection of the system units. For example, the parallel connection of capacities (Figure 3.2) subjected to failures of a shortage type corresponds to a series structure in reliability terms.

Let us denote the structure function of a series system as $\alpha(x_1, x_2, \ldots, x_n)$. This function is

$$\alpha(\mathbf{X}) = \alpha(x_1, x_2, \ldots, x_n) = \bigcap_{1 \leq i \leq n} x_i \qquad (3.3)$$

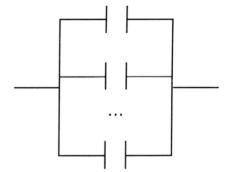

Figure 3.2. A parallel connection of capacitors which represents a series structure in a reliability sense.

where the symbol ∩ denotes the Boolean product (disjunction). The same
expression can be written in an equivalent form

$$\alpha(\mathbf{X}) = \min_{1 \le i \le n} x_i$$

In reliability theory systems consisting of independent units are usually
considered. In this case the computation of the probability of a successful
system operation is easy. We are interested in the probability

$$\Pr\{\alpha(x_1, x_2, \ldots, x_n) = 1\} = E\{\alpha(x_1, x_2, \ldots, x_n)\} \qquad (3.4)$$

For independent units (3.4) might be rewritten in two equivalent ways

$$\Pr\left\{\bigcap_{1 \le i \le n} x_i = 1\right\} = \prod_{1 \le i \le n} \Pr\{x_i = 1\} = \prod_{1 \le i \le n} p_i \qquad (3.5)$$

or

$$E\left\{\bigcap_{1 \le i \le n} x_i\right\} = \prod_{1 \le i \le n} E\{x_i = 1\} = \prod_{1 \le i \le n} p_i \qquad (3.6)$$

Expressions (3.5) and (3.6) make the following statements true:

1. A series system's reliability decreases (increases) if the reliability of any
   unit decreases (increases).
2. A series system's reliability decreases (increases) if the number of units
   increases (decreases).
3. A series system's reliability is worse than the reliability of any of its
   units.

The first two statements reflect the monotonicity property.

Above we have considered the static case where probabilities are specified as constant. But the process of a system's operation develops over time, so it is reasonable to consider a random function $x_i(t)$:

$$x_i(t) = \begin{cases} 1 & \text{if the } i\text{th unit is operating at moment } t \\ 0 & \text{otherwise} \end{cases} \tag{3.7}$$

. This function is monotone and nonincreasing over time for unrepairable units; that is, after a failure the unit cannot return to state 1. In other words, $x_i(t + \Delta) \leq x_i(t)$ for any $\Delta > 0$. Thus, for the system as a whole, it follows that $f(\mathbf{X}(t + \Delta)) \leq f(\mathbf{X}(t))$.

From (3.5) it follows that

$$P_{\text{syst}}(t) = \prod_{1 \leq i \leq n} p_i(t) \tag{3.8}$$

Obviously, (3.5) and (3.8) can be written in a direct way from the verbal definition of a series system's successful operation:

Pr{a series system operates successfully}

$\quad$ = Pr{all system's units are up}

$\quad$ = Pr{unit 1 is up, AND unit 2 is up, ..., AND unit n is up}

$\quad$ = Pr{unit 1 is up} Pr{unit 2 is up} ... Pr{unit n is up}

In a more general case, direct calculations must be used for obtaining the function $P(t)$ for the system. But in one important particular case, when each $p_i(t)$ is an exponential function, we can write a very simple expression

$$P_{\text{syst}}(t) = \prod_{1 \leq i \leq n} e^{-\lambda_i t} = \exp\left(-t \sum_{1 \leq i \leq n} \lambda_i\right) = e^{-\Lambda t} \tag{3.9}$$

where

$$\Lambda = \sum_{1 \leq i \leq n} \lambda_i$$

Suppose a system consists of highly reliable units: $p_i(t) = 1 - \varepsilon_i(t)$ where $\varepsilon_i(t)$ is very small; for example,

$$\max_{1 \leq i \leq n} \varepsilon(t) \ll \frac{1}{n}$$

[It is clear that the smallness of $\varepsilon_i(t)$ for each particular case depends on the number of units.] Then, for a system with units having arbitrary distributions of time to failure,

$$P_{\text{syst}}(t) = \prod_{1 \le i \le n} [1 - \varepsilon_i(t)] \approx 1 - \sum_{1 \le i \le n} \varepsilon_i(t) \qquad (3.10)$$

The error of the calculation in this case will not exceed the value

$$P_{\text{syst}}(t) - \prod_{1 \le i \le n} [1 - \varepsilon_i(t)] < \sum_{1 \le i < j \le n} \varepsilon_i(t)\varepsilon_j(t) < \binom{n}{2} \left[\max_{1 \le i \le n} \varepsilon_i(t) \right]^2$$

$$(3.11)$$

Let us consider a particular case. Suppose a series system consists of n units, each of which has a continuous failure distribution with a nonzero first derivative at $t = 0$. Suppose the system is operating during a small period of time t_0. The Taylor series restricted to the first term is

$$\varepsilon_i(t_0) \approx \left[\frac{d}{dt} F(t) \Big|_{t=0} \right] t_0 = f(0)t_0$$

Note that, at $t = 0$,

$$\lambda(0) = \frac{f(0)}{P(0)} = f(0)$$

Then, for a series system consisting of a large number of highly reliable identical units with an arbitrary d.f. $F(t)$,

$$P_{\text{syst}}(t_0) \approx [1 - \lambda(0)t_0]^n \approx \exp[-n\lambda(0)t_0] \qquad (3.12)$$

If the units are different but some of them have distribution functions $F_i(t)$, $i \in \alpha$, with nonzero first derivatives at $t = 0$ equal to $\lambda_i(0)$, then for small t_0

$$P_{\text{syst}}(t_0) \approx \prod_{i \in \alpha} (1 - \lambda_i(0)t_0) \approx \exp\left[-t_0 \sum_{i \in \alpha} \lambda_i(0) \right] \qquad (3.13)$$

Of course, we assumed that $|\alpha| \gg 1$; that is, the number of distributions with a nonzero derivative is large. Therefore, we see one more example of an exponential distribution in reliability theory.

If the distribution of a unit's TTF is such that $d^i F(0)/dt^i = 0$ for $i < k$ and $d^k F(0)/dt^k = \alpha$, then

$$P_{\text{syst}}(t_0) = \left[1 - \alpha(t_0)^k\right]^n \tag{3.14}$$

For large n one can write the approximation

$$P_{\text{syst}}(t) \approx e^{-At^k}$$

where $A = n\alpha$. Thus, this series system has a Weibull–Gnedenko distribution of time to failure. One practical example of such a system concerns a set of bearings in a machine. Another example will be presented in Section 3.4.

In the ideal case, if all of the series system's units have a constant TTF, that is, a degenerate distribution of the type

$$p_i(t) = \begin{cases} 1 & \text{if} \quad t \le T_i \\ 0 & \text{otherwise} \end{cases} \tag{3.15}$$

then $P(t)$ coincides with the $p_i(t)$ of the worst unit, that is,

$$P_{\text{syst}}(t) = \begin{cases} 1 & \text{if} \quad t \le \min T_i \\ 0 & \text{otherwise} \end{cases} \tag{3.16}$$

Of course, such a distribution does not exist in a real life. (Mathematics always deals with ideal objects!) But normally distributed r.v.'s with very small coefficients of variation can be considered as "almost constant" or "almost nonrandom."

Now consider the MTTF of a series system. For any series system the random TTF, say Y, can be expressed through the random TTFs of its units (y_i) in the following way:

$$Y = \min y_i \tag{3.17}$$

The MTTF can be found in a standard way as

$$T_{\text{syst}} = E\{Y\} = \int_0^\infty P_{\text{syst}}(t)\, dt \tag{3.18}$$

where $P_{\text{syst}}(t)$ is determined above.

For an exponential distribution, $p_i(t) = \exp(-\lambda_i t)$,

$$T_{\text{syst}} = \int_0^\infty \exp\left(-t \sum_{1 \le i \le n} \lambda_i\right) dt = \frac{1}{\displaystyle\sum_{1 \le i \le n} \lambda_i} = \frac{1}{\displaystyle\sum_{1 \le i \le n} \frac{1}{T_i}} \tag{3.19}$$

where T_i is the MTTF of the ith unit.

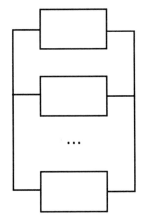

Figure 3.3. System with a parallel structure.

For units with a degenerate distribution

$$T_{\text{syst}} = \min T_i \qquad (3.20)$$

that is, the MTTF of the system equals the MTTF of the worst unit.

3.3 PARALLEL STRUCTURE

3.3.1 Simple Redundant Group

Another principal structure in reliability theory is a parallel connection of units (Figure 3.3). This system consists of one main unit and $m - 1$ redundant units. We call such a system a *simple redundant group*. A system failure occurs if and only if all of the system's units have failed. In other words, the system is operating as long as at least one of its units is operating. Sometimes parallel systems are called *systems with an active (or loaded) redundancy*. Thus, the redundant units are in a working regime during the entire time of the system's operation. A main feature of active redundancy is that all of the reliability characteristics of the redundant units are assumed to be the same as the system's operational units.

The structure function of a parallel system, $\beta(\mathbf{X})$, is

$$\beta(\mathbf{X}) = \beta(x_1, x_2, \ldots, x_m) = \bigcup_{1 \leq i \leq m} x_i \qquad (3.21)$$

where the symbol $\bigcup$ denotes Boolean summation (conjunction). The same expression can be written in an equivalent form:

$$\beta(X) = \max_{1 \le i \le m} x_i$$

For further discussion we need to acknowledge the following result.

DeMorgan's Rule For two Boolean variables x and y, the following equivalences are true (see the exercises):

$$x \lor y = \overline{x} \land \overline{y} \qquad (3.22a)$$

$$x \land y = \overline{x} \lor \overline{y} \qquad (3.22b)$$

$$\overline{x} \lor \overline{y} = \overline{x \land y} \qquad (3.22c)$$

$$\overline{x} \land \overline{y} = \overline{x \lor y} \qquad (3.22d)$$

All of these equivalences express the same property but in slightly different form. The most important one for us is $(3.22a)$. If one considers a series system of two units x and y, and "1" means an up state, then $x \lor y = 1$ means unit x and/or unit y are in an up state; that is, the system is in an up state. At the same time, $x \land y = 0$ means unit x and unit y are in a down state; that is, the system is in a down state. It is clear that these two events are complementary. To prove (3.22), one may use a *Venn diagram*. This diagram graphically depicts random events, their sum and intersection, complementary events, and so on. A simple case with two events A and B is presented in Figure 3.4. The proof of $(3.22c)$ one can find in Figure 3.5.

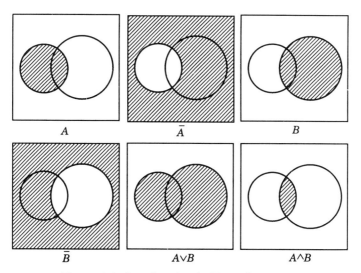

Figure 3.4. Samples of main Venn diagrams.

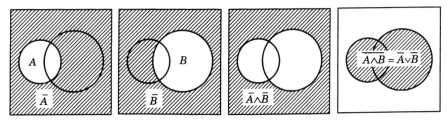

Figure 3.5. The proof of the DeMorgan Rule with the help of the Venn Diagram.

From the above-given particular forms of DeMorgan's rule, the following generalizations can be easily obtained:

$$\bigcup_{1 \le i \le n} x_i = \overline{\bigcap_{1 \le i \le n} \bar{x}_i} \qquad (3.23a)$$

$$\bigcap_{1 \le i \le n} x_i = \overline{\bigcup_{1 \le i \le n} \bar{x}_i} \qquad (3.23b)$$

$$\bigcup_{1 \le i \le n} \bar{x}_i = \overline{\bigcap_{1 \le i \le n} x_i} \qquad (3.23c)$$

$$\bigcap_{1 \le i \le n} \bar{x}_i = \overline{\bigcup_{1 \le i \le n} x_i} \qquad (3.23d)$$

These latter statements can be proved by induction and we leave their proofs to the exercises.

Another (almost purely verbal) explanation of (3.23) follows from the definition of a parallel system's failure which was given at the very beginning of this section:

Pr{a parallel system operates successfully}

$= $ Pr{at least one unit operates successfully}

$= $ Pr{unit x_1 is up, OR unit x_2 is up, ..., OR unit x_m is up}

$$= \Pr\left\{ \bigcup_{1 \le i \le m} x_i = 1 \right\}$$

At the same time,

Pr{a parallel system has failed}

$= $ Pr{all of its units have failed}

$= $ Pr{unit x_1 is down, AND unit x_2 is down, ...,

AND unit x_m is down}

$$= \Pr\left\{ \bigcap_{1 \le i \le m} \bar{x}_i = 1 \right\} = E\left\{ \bigcap_{1 \le i \le m} \bar{x}_i \right\} = \prod_{1 \le i \le m} E\{\bar{x}_i\} = \prod_{1 \le i \le m} q_i$$

We note that if two events, say z and $\bar{z}$, are complementary, then

$$\Pr\{z = 1\} + \Pr\{\bar{z} = 1\} = 1$$

Consequently,

$$\Pr\left\{\bigcup_{1 \le i \le m} x_i = 1\right\} = 1 - \Pr\left\{\bigcap_{1 \le i \le m} \bar{x}_i\right\} = 1 - \prod_{1 \le i \le m} q_i \qquad (3.24)$$

Now the equivalence of (3.23) can be confirmed in an inverse way by the equality of the probabilities.

We repeat that a detailed inference was done above only from a methodological viewpoint to provide further discussion. Of course, it was enough to use a verbal definition of a parallel system of independent units and to write the final expression. Sometimes a different form equivalent to (3.24) is used

$$P_{\text{syst}} = p_1 + q_1 p_2 + q_1 q_2 p_3 + \cdots + q_1 q_2 \cdots \cdots q_{m-1} p_m$$

$$= p_1 + q_1(p_2 + q_2(p_3 + \cdots))$$

This expression can be explained as follows:

$\Pr\{$a parallel system operates successfully$\}$

$= \Pr\{$the first unit is up, OR if the first unit has failed, the second one

has not failed; OR if both of these units have failed,

then the third one has not failed, OR $\cdots\}$

If each of the system's units has an exponential TTF distribution, $p_i(t) = \exp(-t\lambda_i)$, for a highly reliable system where max $q_i(t) \ll 1/m$, one can write $q_i(t) \approx t\lambda_i$, and, finally,

$$P_{\text{syst}}(t) = 1 - t^n \prod_{1 \le i \le m} \lambda_i \qquad (3.25)$$

If each unit of a parallel system has a constant time to failure (a degenerate distribution of TTF), then

$$P_{\text{syst}}(t) = \begin{cases} 1 & \text{for} \quad t \le \max T_i \\ 0 & \text{otherwise} \end{cases} \qquad (3.26)$$

Now consider a parallel system's MTTF. For this system the random TTF (ξ_{syst}) is expressed through the random TTFs of its units (ξ_i) as

$$\xi_{syst} = \max \xi_i \qquad (3.27)$$

Thus, this is equivalent to the statement that a parallel system operates successfully until the last failure of its units.

When each unit has an exponential distribution of TTF, an analytic expression can be derived. For this purpose write the probability of failure-free operation in the form

$$P_{syst}(t) = 1 - \prod_{1 \le i \le m} (1 - e^{-\lambda_i t})$$

$$= \sum_{1 \le i \le m} e^{-\lambda_i t} - \sum_{1 \le i < j \le m} e^{-(\lambda_i + \lambda_j)t} + \sum_{1 \le i < j < k \le m} e^{-(\lambda_i + \lambda_j + \lambda_k)t} - \cdots$$

$$+ (-1)^m \exp\left(-t \sum_{1 \le i \le m} \lambda_i\right) \qquad (3.28)$$

Integrating (3.28) gives

$$T_{syst} = \sum_{1 \le i \le m} T_i - \sum_{1 \le i < j \le m} \frac{1}{\lambda_i + \lambda_j}$$

$$+ \sum_{1 \le i < j < k \le m} \frac{1}{\lambda_i + \lambda_j + \lambda_k} - \cdots + (-1)^n \frac{1}{\sum_{1 \le i \le m} \lambda_i} \qquad (3.29)$$

If, at the same time, all units are identical

$$P_{syst} = 1 - \left(1 - e^{-\lambda t}\right)^m$$

In this case the MTTF has the form

$$T_{syst} = T\left(1 + \frac{1}{2} + \frac{1}{3} + \cdots + \frac{1}{m}\right) \qquad (3.30)$$

where T is the MTTF of a single unit. For large m, a well-known approximation for a harmonic set can be applied:

$$T_{syst} \approx T(\ln m + C) \qquad (3.31)$$

where C is Euler's constant: $C = .57712\ldots$.

TABLE 3.1   Dependence of a System's MTTF on Its Units' MTTF

Number of Redundant Units	Relative Growth of the System MTTF
0	1
9	2.88
99	5.18
999	7.48
$\cdots$	$\cdots$
10^{10}	23.6

Formula (3.30) can be explained in a simple and understandable way with the use of the memoryless property of the exponential distribution. At the moment $t = 0$ the system of m active redundant units has a failure rate $\lambda_m = m\lambda$. The first failure occurs in a random time Z_m with an exponential distribution with parameter λ_m. After this failure the system consists of $m - 1$ units, so its failure rate is now $\lambda_{m-1} = (m - 1)\lambda$. The second failure occurs in a random time Z_{m-1} with an exponential distribution with parameter λ_{m-1}. And so on, until the last unit has failed.

The total time of a successful system's operation consists of the sum of all these intervals, that is, $T_{\text{syst}} = E\{Z_1 + Z_2 + \cdots + Z_m\}$. Obviously, this result coincides with (3.30).

From (3.30) and (3.31) it follows that, at least theoretically, the use of active redundancy potentially allows one to construct a system with an arbitrarily large MTTF value. Of course, one needs to understand that such a mathematical model is strongly idealized. First of all, one must take into account the necessity to use a switching device which itself possesses a nonideal reliability. On the other hand, even with absolutely reliable switching devices, the growth of the system's MTTF is very slow. Several examples are shown in Table 3.1.

Hardly anybody would ever use such redundancy (even with absolutely reliable switches!) to improve the MTTF. But this kind of redundancy can be successfully used if one considers other indexes of reliability, for example, the probability of a system's successful performance. In this case if the initial value of $q(t)$ is much less than 1, each new parallel unit decreases the system's unreliability level by the order q.

Note that for a nonexponentially distributed TTF with an increasing failure rate (i.e., for "aging" units), the growth of a system's MTTF is even slower.

3.3.2   "k out of n" Structure

For some technical schemes one sometimes considers a special structure—the so-called "*k out of n*" *structure*, or voting system. In engineering practice such a system almost always consists of identical units. In this case the system operates successfully if at least k out of its total n units are operating. The structure function of the system is illustrated here by a simple example with "2 out of 3":

$$f(x_1, x_2, x_3) = x_1 \wedge x_2 \wedge x_3 \vee \bar{x}_1 \wedge x_2 \wedge x_3 \vee x_1 \wedge \bar{x}_2 \wedge x_3 \vee x_1 \wedge x_2 \wedge \bar{x}_3$$

(This case is most often encountered in engineering practice.)

In general, the structure function of a "k out of n" structure can be written in the form

$$\varphi(\mathbf{X}) = \left[\sum_{1 \le i \le n} x_i - k \right]_+$$

where

$$[z]_+ = \begin{cases} 1 & \text{if} \quad z \ge 0 \\ 0 & \text{otherwise} \end{cases}$$

We will use an explanation based on combinatorial methods, avoiding the structure function. Considering a "k out of n" structure corresponds to the binomial test scheme, so

$$\Pr\{\nu = j\} = \binom{n}{j} p^j q^{n-j}$$

and, consequently, the probability of a system's failure-free operation equals

$$P_{\text{syst}}(t) = \Pr\{\nu \ge k\} = \sum_{k \le j \le n} \binom{n}{j} p^j q^{n-j} \tag{3.32}$$

or

$$P_{\text{syst}}(t) = 1 - \Pr\{\nu < k\} = 1 - \sum_{0 \le j \le k-1} \binom{n}{j} p^j q^{n-j} \tag{3.33}$$

For a highly reliable system where $q \ll 1/n$ from (3.33) one can easily write

$$P_{\text{syst}}(t) \approx 1 - \binom{n}{n-k+1} p^{k-1} q^{n-k+1}$$

The task of finding the MTTF of such a system for arbitrary unit failure distributions is not simple. One may use any numerical method. The only case where it is possible to obtain an analytical result is the case of the exponential distribution $p(t) = \exp(-\lambda t)$.

We will not integrate (3.32), but we will use the method described above. The system starts with n operating units and operates, on the average, $1/n\lambda$

units of time until the first failure. After the first failure, there are $n - 1$ operating units in the system. They work until the next failure, on the average, $1/(n - 1)\lambda$ units of time, and so on until the $(k + 1)$th failure has occurred. Thus, the system's MTTF equals

$$T_{\text{syst}} = \frac{1}{\lambda} \sum_{k \leq j \leq n} \frac{1}{j}$$

For an arbitrary distribution $p(t)$, one should use a direct numerical integration of $P_{\text{syst}}(t)$ or a Monte Carlo simulation.

3.4 MIXED STRUCTURES

Pure series or pure parallel systems are rarely encountered in practice. Indeed, mixed structures with series and parallel fragments are common. For example, a duplicate computer system may be used for monitoring a production line. Each of these two computers, in turn, is represented by a series structure, and so on.

A combination of series and parallel structures can generate various mixed structures. First, let us consider "pure" series–parallel and parallel–series types of structures (see Figures 3.6a and 3.7a) because they will be of interest in further discussions.

For these structures the following expressions can be easily written. For a parallel–series structure, one has

$$P_{\text{PS}}(t) = \text{E}\left\{ \bigcup_{1 \leq i \leq M} A_i(\mathbf{X}_i) \right\}$$

where

$$\mathbf{X}_i = (x_{i1}, x_{i2}, \ldots, x_{iN})$$

or

$$P_{\text{PS}}(t) = 1 - \text{E}\left\{ \bigcap_{1 \leq i \leq M} \overline{x_{i1} \wedge x_{i2} \wedge \cdots \wedge x_{iN}} \right\}$$

$$= 1 - \prod_{1 \leq i \leq M} \left(1 - \prod_{1 \leq j \leq N} p_{ij} \right) \tag{3.34}$$

where N is the number of units in a series subsystem.

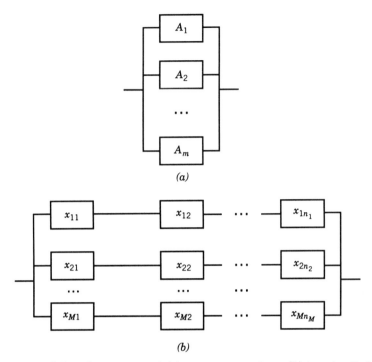

Figure 3.6. Parallel–series structure: (*a*) in an aggregate form; (*b*) in a detailed form.

For a series–parallel structure, one writes

$$P_{\text{SP}}(t) = \text{E}\left\{ \bigcap_{1 \le i \le N} B_i(\mathbf{X}_i) \right\}$$

or

$$P_{\text{SP}}(t) = \text{E}\left\{ \bigcap_{1 \le i \le N} \overline{\bar{x}_{1i} \wedge \bar{x}_{2i} \wedge \cdots \wedge \bar{x}_{Mi}} \right\}$$

$$= \prod_{1 \le j \le N} \left(1 - \prod_{1 \le i \le M} q_{ji} \right) \tag{3.35}$$

where M is the number of units in a parallel subsystem and $q = 1 - p$.

In conclusion, we make the following remark. If we would like to improve the reliability of a series system of N units using redundancy, there are two ways to do so. The first way is to use M redundant systems as a whole. The second way is to use M redundant units for each of the main units (see Figure 3.8).

Comparing (3.34) and (3.35), one can find that it is more effective to use a series–parallel structure rather than a parallel–series structure. In particular,

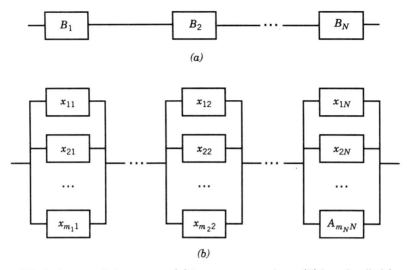

Figure 3.7. Series–parallel structure: (a) in an aggregate form; (b) in a detailed form.

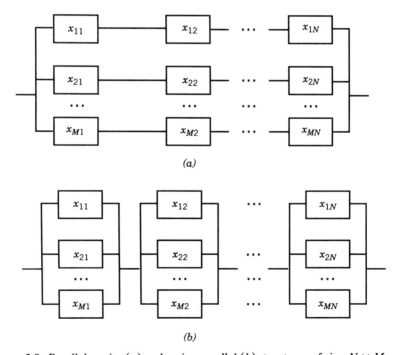

Figure 3.8. Parallel–series (a) and series-parallel (b) structures of size $N \times M$.

for identical units

$$1 - \left(1 - p^N\right)^M \leq \left(1 - q^M\right)^N \tag{3.36}$$

From Figure 3.8 one sees that in a series–parallel system there are more "degrees of freedom," more possibilities to avoid failures. To check this statement, we suggest the extremely simple and clear proof of the statement based on the inequality

$$\max_{1 \leq i \leq M} \min_{1 \leq j \leq N} x_{ij} \leq \min_{1 \leq i \leq N} \max_{1 \leq j \leq M} x_{ij} \tag{3.37}$$

This inequality means that under any splitting of the set of x_{ij}'s by subsets, the minimal value among the maximal values for all these subsets is always larger (not smaller) than the maximal value among the minimal values.

Now, using this fact, one can prove the statement. Notice that if ξ_{ji} is the random TTF of the jth unit in the ith subsystem of series units, then the random TTF of this subsystem is

$$\xi_i = \min_{1 \leq j \leq n_i} \xi_{ji} \tag{3.38}$$

and, consequently,

$$\xi_{PS} = \max_{1 \leq i \leq M} \xi_i \tag{3.39}$$

is the random TTF of the parallel–series system as a whole.

Consider the same set $\{\xi_{ji}\}$ divided in such a way that ξ_{ji} is the random TTF of the jth unit in the ith subsystem of parallel units. Then the random TTF of this subsystem is

$$\xi_i = \max_{1 \leq j \leq m_i} \xi_{ji} \tag{3.40}$$

and, consequently,

$$\xi_{SP} = \min_{1 \leq i \leq N} \xi_i \tag{3.41}$$

is the random TTF of the series–parallel system as a whole.

A substitution of (3.38) to (3.41) in (3.37) gives, for any sample of r.v.'s ξ_{ji}, that $\xi_{SP} \geq \xi_{PS}$. From this it automatically follows that

$$T_{PS} = E\{\xi_{PS}\} \geq E\{\xi_{SP}\} = T_{SP}$$

and

$$P_{PS}(t) = \Pr\{\xi_{PS} \geq t\} \geq \Pr\{\xi_{SP} \geq t\} = P_{SP}(t)$$

For a "long" series–parallel system (when $N \gg 1$), the Weibull–Gnedenko distribution might be applied if the system's reliability is relatively high. Consider a system of independent identical units. The distribution of the TTF of each parallel subsystem is such that M is the first order of the derivative which differs from 0. As we considered in Section 3.1, in this case for small t_0 and relatively large N, the Weibull–Gnedenko distribution can be used for the description of the TTF of the system as a whole.

Thus, any series–parallel or parallel–series system can be understood as a two-pole network of a special type. This network possesses the so-called *reducible structure*. A sequential application of the following procedures—(a) replacement of each series connection by a single equivalent unit and (b) replacement of each parallel connection (or "k out of n" structure) by a single equivalent unit—allows one to transform any reducible structure into a single equivalent unit.

Such a reduction is very convenient for the calculation of the probability of a system's successful operation. For instance, consider the structure shown in Figure 3.9. This figure depicts the sequential steps of the system reduction. We hope that the figure is self-explanatory. Using a similar procedure in an

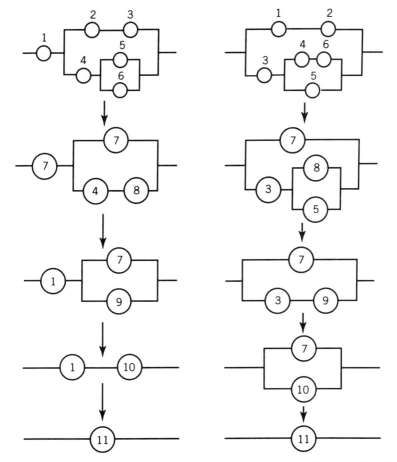

Figure 3.9. Examples of the reduction of a complex structure with parallel and series inner substructures to the simplest kinds of structures.

inverse way, one can construct various reducible structures from a single equivalent unit by a detailed disaggregation at each step. Examples of irreducible structures (different arbitrary networks) are presented in a later chapter.

3.5 STANDBY REDUNDANCY

A number of systems have standby redundant units. These units are not included in an "active" system's structure. Moreover, these redundant units cannot fail before they occupy an active position. A standby unit instantly replaces its corresponding main unit after the latter's failure. Generally, such a replacement does not occur instantly, but most mathematical models assume that the time of the standby unit's switching into a main position is 0.

The spare parts supply problem is usually analyzed in terms of standby redundancy. For this problem the sufficiency of the stock of spare parts for replacement (or repair performance) is considered rather than the system's successful operation. (Usually, in this case, one considers the inventory system itself. But for simplicity of terminology we will use the terms *system failure* and *instant replacement* in further discussion.)

For standby redundancy it is convenient to speak about "a socket" and a set of units which are put into it one by one after each failure. All units for the socket are considered to be identical. In reality, a standby unit of some type must be able to replace the main unit only with a unit of the same type.

3.5.1 Simple Redundant Group

If a system consists of a single main unit and $m - 1$ standby units, we call it a simple redundant group. In this case the random time of the system's successful operation θ_m equals

$$\theta_m = \sum_{1 \le i \le m} \xi_i$$

Thus, a system's MTTF can immediately be written as

$$T_{\text{syst}} = E\{\theta_m\} = E\left\{ \sum_{1 \le i \le m} \xi_i \right\} = \sum_{1 \le i \le m} E\{\xi_i\} = \sum_{1 \le i \le m} T_i \quad (3.42)$$

The MTTF does not depend on the switching order of the steady units. If all units of the system are identical,

$$T_{\text{syst}} = mT$$

where T is the single unit's MTTF.

It is clear that standby redundancy is much more effective than active redundancy: here the growth of T_{syst} is linear, and for the active redundancy case the growth is only logarithmic. But again we would like to emphasize that this mathematical model is a very idealized picture of reality. Remember the well-known property of the mean: formula (3.42) is valid even if the standby units are dependent.

The probability of a system's successful operation $P_s(t)$ can be written as

$$P_{\text{syst}}(t) = \Pr\{\theta_m \geq t\} = \Pr\left\{ \sum_{1 \leq i \leq m} \xi_i \geq t \right\} = \Pr\{\theta_{m-1} + \xi_m \geq t\} \quad (3.43)$$

The system's TTF, $\xi_{\text{syst}} = \theta_m$, represents the sum of independent r.v.'s. As we know from Chapter 1, in this case

$$P_{\text{syst}}^{(m)}(t) = 1 - F^{*m}(t) = \int_0^t P_{\text{syst}}^{(m-1)}(t - x) \, dF(x)$$

where $P_{\text{syst}}^{(k)}(t)$ is the probability of a failure-free operation of the system with $k - 1$ standby units (k units in the redundant group).

As known from Section 1.3.1, only a very restricted number of d.f.'s allow one to find convolutions in convuluted form. The reader can use the above-mentioned results for probability calculations. Of course, if the number of standby units in the redundant group is large, a normal approximation based on the central limit theorem (see Section 1.3.2) can be used.

In engineering practice, especially in electronics, the most frequently used distribution $F(t)$ is exponential. The standby group's random TTF has an Erlang d.f. of the mth order, and the probability of a failure-free operation is

$$P_{\text{syst}}(t) = \Pr\left\{ \sum_{1 \leq i \leq m} \xi_i \geq t \right\} = \sum_{1 \leq i \leq m-1} \frac{(\lambda t)^i}{i!} e^{-\lambda t} = 1 - \sum_{m \leq i < \infty} \frac{(\lambda t)^i}{i!} e^{-\lambda t}$$

$$(3.44)$$

For $\lambda t \ll 1$ the approximation can be written as

$$P_{\text{syst}}(t) \approx 1 - \frac{(\lambda t)^m}{m!} \quad (3.45)$$

If λt is not too small, the following inequality is true:

$$\sum_{m \le i < \infty} \frac{(\lambda t)^i}{i!} e^{-\lambda t} = \frac{(\lambda t)^m}{m!} e^{-\lambda t} \left[1 + \frac{\lambda t}{n+1} + \frac{(\lambda t)^2}{(n+1)(n+2)} + \cdots \right]$$

$$\le \frac{(\lambda t)^m}{m!} e^{-\lambda t} \left[1 + \frac{(\lambda t}{n+1} + \frac{(\lambda t)^2}{(n+1)^2} + \cdots \right]$$

$$= \frac{(\lambda t)^m}{m! \left(1 - \dfrac{\lambda t}{n+1} \right)} e^{-\lambda t} \tag{3.46}$$

This can be used for approximate computations. The substitution of (3.46) into (3.45) produces an approximation of $P_{\text{syst}}(t)$:

$$P_{\text{syst}}(t) \approx 1 - \frac{(\lambda t)^m}{m! \left(1 - \dfrac{\lambda t}{n+1} \right)} e^{-\lambda t}$$

Note that this value is smaller than the exact value; that is, it delivers a "guaranteed result."

In conclusion, note that standby redundancy is more effective than active redundancy (at least in theory!). This follows from the simple fact that

$$\xi \,(\text{standby redundancy}) = \sum_{1 \le i \le m} \xi_i \ge \max_{1 \le i \le m} \xi_i = \xi \,(\text{active redundancy})$$

The equality is never attained because of the strongly positive values of the ξ's. (Of course, we are considering the case where $m > 1$.) Of course, the reader should never forget that standby redundancy, in practice, requires some time to switch a unit into an active regime.

Finally, we would like to point out the relationship between the MTTFs for series and parallel systems of two units. (The result can easily be expanded by induction to an arbitrary number of units.) Suppose that one unit has a random TTF ξ_1 and another has ξ_2. It is clear that

$$\xi_1 + \xi_2 = \min(\xi_1, \xi_2) + \max(\xi_1, \xi_2)$$

because one of these r.v.'s is obviously larger and another is smaller. Taking the mean of both sides of the equality, one gets

$$E\{\xi_1 + \xi_2\} = E\{\min(\xi_1, \xi_2)\} + E\{\max(\xi_1, \xi_2)\}$$

Now we can see that these values are, in order:

- The first is the MTTF of a duplicated system of two standby redundant units which are working sequentially one after another.
- The second is the MTTF of a series connection of these units.
- The third is the MTTF of a parallel connection of these units.

In particular, when both ξ's are exponentially distributed, one can obtain a convenient expression for the MTTF of a parallel system of two different units:

$$\frac{1}{\lambda_1} + \frac{1}{\lambda_2} = \frac{1}{\lambda_1 + \lambda_2} + T_{\text{parallel}}$$

or, in final form,

$$T_{\text{parallel}} = \frac{1}{\lambda_1} + \frac{1}{\lambda_2} - \frac{1}{\lambda_1 + \lambda_2}$$

Of course, the latter expression has such a simple form only because both distributions are exponential.

3.5.2 "k out of n" Redundancy

The use of standby units for several main units is very common. For example, consider a system which includes k main units. To support the system, there are $n - k$ spare units which can replace any main unit of the group. This method of redundancy is very efficient because of the large number of "degrees of freedom" in the usage of standby units. Indeed, no unit is predetermined to replace some specified main unit. (We repeat that this mathematical model is mainly used to describe a spare units supply system.)

For standby redundancy the formulas for $P_{\text{syst}}(t)$ and T_{syst} cannot be written in a convoluted form except for the case of an exponentially distributed random TTF of the units. We may write the result basing our explanation on simple arguments.

Recall again that we assume that the units are independent. The system consists of k identical units and has $n - k$ standby units. The system failure rate equals $k\lambda$. After a first failure the failed unit is replaced by a redundant unit and the system continues its operation. The random TTF equals t_1 and has an exponential distribution with parameter $k\lambda$. The MTTF in this case equals T/k where T is the MTTF of a single unit. The memoryless property of the exponential distribution and the independence of the units ensure the exponentiality of the system's random TTF. Hence, a random period of a system's successful operation consists of the sum of $n - k + 1$ i.i.d. TTFs.

(One period to the first system failure and then $n - k$ replacements of spare units.)

Therefore, the system's MTTF is

$$T_{syst} = E\left\{ \sum_{1 \leq i \leq n-k+1} \min_{1 \leq j \leq k} \xi_{ji} \right\} = (n - k + 1)\frac{T}{k} \qquad (3.47)$$

The probability of a system's successful operation when its units have exponential distributions is

$$P_{syst}(t) = Pr\left\{ \sum_{1 \leq i \leq n-k+1} \min_{1 \leq j \leq k} \xi_{ji} \geq t \right\} = \sum_{0 \leq i \leq n-k+1} \frac{(\lambda kt)^i}{i!} e^{-\lambda kt} \qquad (3.48)$$

In general, the problem is very complicated. The most reasonable way to calculate accurate values of the reliability indexes $P_{syst}(t)$ and T_{syst} is via Monte Carlo simulation.

Below we give a simple method for obtaining lower and upper bounds for these reliability indexes. It is clear that the best use of the standby units would be in a so-called "time-sharing" regime. Here the MTTF of the "k out of n" structure could be calculated as the total operation time of all units divided by k. The upper bound for T_{syst} follows:

$$T_{syst} \leq \frac{1}{k}E\left\{ \sum_{1 \leq i \leq n} \xi_i \right\} = \frac{n}{k}T \qquad (3.49)$$

Comparison of (3.47) and (3.49) shows the difference between the accurate value of T_{syst} for the exponential distribution and its upper bound. An upper bound for $P_{syst}(t)$ can be obtained via the use of similar explanations:

$$P_{syst}(t) \leq Pr\left\{ \frac{1}{k} \sum_{1 \leq i \leq n} \xi_i \geq t \right\} = Pr\left\{ \sum_{1 \leq i \leq n} \xi_i \geq kt \right\} \qquad (3.50)$$

To obtain lower bounds, we use the fact that the joint use of redundant units is more effective than an individual one. Let us equally allocate all redundant units among k initially operational units of the system. Then we have k series subsystems, each with n/k redundant units.

If n/k is not an integer, the procedure will be slightly more difficult. Denote the integer part of n/k by $m^* = [n/k]$. Then $a = n - km^*$ subsystems have $m^* + 1$ redundant units and all of the remaining $b = k - (n - km^*)$ ones have m^* standby units. Thus, a lower bound for $P_{syst}(t)$ is

$$P_{syst}(t) \geq 1 - [1 - F^{*m^*}(t)]^b[1 - F^{*(m^*+1)}(t)]^a \qquad (3.51)$$

where $F(t)$ is the d.f. of the random TTF of a single unit. A lower bound of the MTTF can be found by integrating (3.51). If the coefficient of variation of $F(t)$ is small, (3.51) can be reduced to

$$P_{\text{syst}}(t) \geq 1 - \left[1 - F^{*m^*}(t)\right]^b$$

where the previous notation is preserved.

3.5.3 On-Duty Redundancy

The use of standby redundant units in an operation requires a special regime on duty. This regime is intermediate between the two previously considered types: active and standby.

We illustrate the subject with several examples. An electronic monitor needs at least a portion of a second to be ready to display information. A redundant computer in a control system must be supplied with current information before it is switched to an operational regime. Usually, the unit on duty has a regime which is lighter than the working unit but harder than a total standby one. Practically, there is no realistic input data for this on-duty regime and, moreover, even a confident knowledge about the process is absent. Even with the appropriate input data, the problem of a reliability evaluation in this case is hardly solvable analytically under general assumptions. As a rule, Monte Carlo simulation allows one to obtain numerical results. But even in this case a lack of input data makes the result very problematic.

The only mathematically acceptable model arises when all units have an exponential distribution of their TTFs: the main ones with parameter λ, and the redundant ones with parameter $\alpha\lambda$ where $\alpha < 1$. In general, a system may have on-duty units which are used for the replacement of failed main units and standby units which are switched into an on-duty position.

Consider a system of k main (operational) units, l on-duty units, and m standby units. Let $N = k + l + m$. For this case one can build the transition graph in Figure 3.10. On the basis of this graph, the following system of

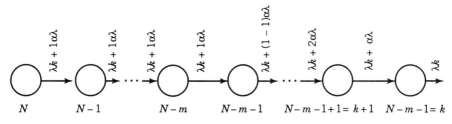

Figure 3.10. Transition graph for an unrepairable system of k main, l on-duty, and m standby units.

differential equations can be written:

$$p'_N(t) = -p_N(t)[k\lambda + l\alpha\lambda]$$

$$p'_{N-1}(t) = -p_{N-1}(t)[k\lambda + l\alpha\lambda]$$

$$\vdots$$

$$p'_{N-m}(t) = -p_{N-m}(t)[k\lambda + l\alpha\lambda]$$

$$p'_{N-m-1}(t) = -p_{N-m-1}(t)[k\lambda + (l-1)\alpha\lambda]$$

$$\vdots$$

$$p'_{N-m-l+1}(t) = p'_{k+1}(t) = -p_{k+1}(t)[k\lambda + \alpha\lambda]$$

$$p'_{N-m-l}(t) = p'_k(t) = -p_k(t)k\lambda$$

The initial state for this process is the system state with all operating units, so $p_N^{(0)} = 1$. The system MTTF can be found immediately from the transition graph in Figure 3.10:

$$T_{\text{syst}} = m\frac{1}{\lambda(k+l\alpha)} + \sum_{1\leq j\leq l}\frac{1}{\lambda(k+(j-1)\alpha)}$$

The method described in Chapter 2 can be used to find the probability of a failure-free operation. But we avoid writing bulky and boring expressions. The obvious upper- and lower-bound models can be written with common sense: one for an upper bound (see the transition graph in Figure 3.11), and another for a lower bound (see Figure 3.12). The first graph contains all transition intensities equal to the maximal one, $\lambda(k + \alpha l)$. The second one is constructed under the observation that as soon as at least one on-duty redundant unit has been spent, all of the remaining units become standby units.

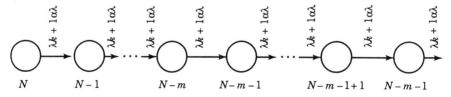

Figure 3.11. Transition graph for obtaining the lower bound.

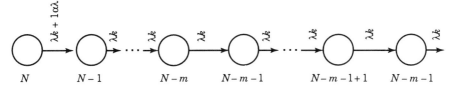

Figure 3.12. Transition graph for obtaining the upper bound evaluation.

For the graph of Figure 3.11, we can write the result immediately without solving the system of differential equations. Indeed, we have the sum of $m + l$ i.i.d. exponential r.v.'s with means equal to $1/\lambda(k + \alpha l)$. Therefore,

$$P_{\text{syst}}(t) = \sum_{1 \leq j \leq k+l} \frac{\left[\lambda(k + \alpha l)t\right]^j}{j!} e^{-\lambda(k + \alpha l)t}$$

For the second case we have the sum of m exponential r.v.'s with means $1/\lambda(k + \alpha l)$ and of one exponential r.v. with mean $1/\lambda k$:

$$P_{\text{syst}}(t) = F_1 * F_2(t)$$

where

$$F_1(t) = \sum_{0 \leq j \leq m} \frac{\left[\lambda(k + \alpha l)t\right]^j}{j!} e^{-\lambda(k + \alpha l)t}$$

and

$$F_2(t) = \sum_{0 \leq j \leq m+l} \frac{(\lambda kt)^j}{j!} e^{-\lambda kt}$$

These boundary estimates are given not as essential results but rather as examples of the possible thinking involved in finding simple solutions to complex problems.

We repeat that on-duty redundancy, in general, is a real problem in practice, not because of the solution difficulties but because of the lack of information. Usually, nobody has the slightest idea of the kind of distribution parameters (or even the distributions of the TTFs themselves!) which a unit in an on-duty regime has.

3.6 SWITCHING AND MONITORING

Traditional mathematical models of redundant systems reflect an idealized redundancy: no switching device, no operational monitoring, and no maintenance are investigated. These idealized mathematical models help a researcher to understand the redundancy phenomenon but they could also lead to very harmful mistakes if these models are used without corrections.

For example, the mathematical formulas show that it is possible to reach any specified level of system reliability with the use of redundant units. But engineering practice convinces us that real system reliability improvement depends on the reliability and quality of the monitoring and switching devices. Both of them are usually far from ideal themselves.

Of course, taking into account all of these various factors will lead to more complex and less elegant mathematical results. But in engineering, the results must be not only elegant but also useful!

3.6.1 Unreliable Common Switching Device

In practice, active redundant units are not really operating in parallel. In the case of electronic equipment, the simultaneous presence of several output signals from all redundant units could lead to a real mess. In the case of an information system, the superposition of output information from several computers can produce false signals. Usually, active redundant units are operating in an on-duty regime although their reliability characteristics may not be distinguished from the main unit. All functions of monitoring, switching, and special interface duties are performed by some special device which we call a switching device (SD). Of course, a model of such a group of redundant units almost coincides with the model of a group of active redundant units. But at the same time one needs to take into account the presence of the SD.

Consider a group of m redundant units which uses a common SD for switching from a failed unit to an active redundant one. First of all, note that the SD itself might be one of two main types:

1. The SD is always necessary for the normal operation of the redundant group as a whole.
2. The SD is necessary only at the moment of switching performance.

In the first case, the SD can be, for example, an interface between the redundant group and the remaining equipment. It can be of a various physical nature (electrical, mechanical, hydraulic, etc.). The successful operation of the redundant group depends directly on a failure-free operation of the SD.

In the second case, the SD becomes necessary only at the moment of switching. Even if the SD has failed, the system can successfully operate until the main unit fails. But then the system will have failed even if there are available redundant units.

Necessary Switching Device   Denote the random TTF of the ith unit of the redundant group of m units by ξ_i and the random TTF of the SD by θ. For the first case, we can write that the random TTF of the redundant group ζ is $\min(\theta, \max \xi_i)$. For the redundant group as a whole,

$$P_{\text{syst}}(t) = P_{\text{SD}}(t)P_m(t) \qquad (3.52)$$

where $P_m(t)$ is the probability of a failure-free operation of the redundant group. In other words, for such an SD, the redundant group can be investigated as a simple series–parallel system (see Figure 3.13). The MTTF for this case can be generally found by integrating $P_{\text{syst}}(t)$ as determined in (3.52).

Figure 3.13. Approximate representation of the duplicated system with the switch as a series connection of the two redundant units and a non-ideal switch.

Switching Device Using Only for Switching   The second case should be considered in more detail. There are two possibilities for an SD failure:

1. The SD may fail during the time of the system's operation.
2. The SD may fail with probability Q only at the moment of switching, independent of the time when it has occurred.

Switching Device Failure Depending on Time   Consider a redundant group of identical and independent units with an exponential distribution of their random TTFs. The probability of a successful operation of the redundant group is denoted by $P_{\text{RG}}(t)$. The distribution of the switching device TTF $F_{\text{SD}}(t)$ is arbitrary. There are two possibilities for the system's operation:

1. Both the redundant group and the SD have not failed during a specified interval of time t.
2. The SD fails at some moment $x < t$, the redundant group has not failed until x, and, after this moment, the current main unit does not fail during the remaining time $t - x$.

Notice that both cases—active and standby redundancy—are equivalent, in the sense of the general formula (3.53), under the assumption of exponentiality of a unit's TTF. These conditions can be written as

$$P_{\text{syst}}(t) = P_{\text{SD}}(t) P_{\text{RG}}(t) + \int_0^t P_{\text{RG}}(x) e^{-\lambda(t-x)} \, dF_{\text{SD}}(x) \qquad (3.53)$$

where $P_{\text{SD}}(t) = 1 - F_{\text{SD}}(t)$.

The exponentiality of a unit's TTF permits us to use (3.53) for both active and standby redundancy. The expressions for the redundant group, $P_{\text{RG}}(t)$, are different, but the residual time of the remaining unit after an SD failure is exponential in both cases.

From (3.53) one sees how the reliability of an SD influences the reliability of the system as a whole.

A system's MTTF T_{syst} can be obtained only by integrating the corresponding $P_{\text{syst}}(t)$. Now we would like to consider some limiting cases. If $T_{\text{RG}} \gg T_{\text{SD}}$, then T_{syst} equals T_{RG} for all practical purposes. If $T_{\text{RG}} \ll T_{\text{SD}}$, then T_{syst} approximately equals $T_{\text{SD}} + T$, where T is the MTTF of a single unit.

Example 3.1 Consider a duplicate system with an active redundant unit. The distributions of the random TTFs for the unit and for the SD are exponential with parameters λ and λ_{SD}, respectively. Find the probability of a failure-free operation during a time interval t:

$$P_{\text{syst}}(t) = e^{-\lambda_{\text{SD}} t} \left[1 - \left(1 - e^{-\lambda t} \right)^2 \right]$$

$$+ \int_0^t \left[1 - \left(1 - e^{-\lambda x} \right)^2 \right] e^{-\lambda(t-x)} \lambda_{\text{SD}} e^{-\lambda_{\text{SD}} x} \, dx$$

This solution can be easily obtained in closed form.

For a unit with an arbitrary TTF distribution, the solution is not so simple, though its general form does not seem especially awkward. (Very often—including this case—a "simple" form of a formula hides numerical difficulties which arise during computation!)

Consider active redundancy. Assume that the current operating unit of the redundant group is chosen randomly. The residual unit's TTF begins from the moment x (the SD failure). Equation (3.53) transforms into

$$P_{\text{syst}}(t) = P_{\text{SD}}(t) \left(1 - [q(t)]^n \right) + p(t) \int_0^t \left(1 - [q(x)]^{n-1} \right) \frac{1}{p(x)} \, dF_{\text{SD}}(x)$$

$$(3.54)$$

where $p(t)$ is the probability of a unit's failure-free operation and $q(t) = 1 - p(t)$.

For a standby redundant group, the expression is slightly bulky: the conditional distribution of the residual time of a unit which appears in the operational position at the moment x depends on the number of failures which have occurred before that moment. We obtain an approximation by considering the process of failures before x as renewal.

Then we can write

$$P_{\text{syst}}(t) \approx P_{\text{SD}}(t) P_{\text{standby RG}}(t)$$

$$+ \int_0^t P_{\text{standby RG}}(x) \left[\int_0^x p(x - y) p(t - x | x - y) \, dH(y) \right] dx$$

$$= P_{\text{SD}}(t) P_{\text{standby RG}}(t)$$

$$+ \int_0^t \left[\int_0^x p(t - y) \, dH(y) \right] dF_{\text{SD}}(x) \tag{3.55}$$

where $H(t)$ is the renewal function. In other words, $H(t) \, dt$ is the probability that some failure has occurred in the time interval $[t, t + dt]$. We use $H(t)$ though we observe a finite sequence of r.v.'s but not the point renewal process. We should remark that, for highly reliable systems, this approximation is quite good.

Both cases also allow the following approximation:

$$P_{\text{syst}}^{(m)} = p(t) + q(t) P_{\text{SD}}(t) P_{\text{syst}}^{(m-1)}(t) \tag{3.56}$$

where $P_{\text{syst}}^{(k)}$ is the system with an SD and a redundant group of size k.

This approximation gives a lower bound on the probability of interest because we assume that the SD operates successfully during the entire period t. As a matter of fact, an SD failure may not lead to a system failure.

3.6.2 Common Switching Device with Unreliable Switching

Consider an active redundant group of n independent and identical units. The system's successful operation is possible in two situations:

1. The first unit chosen at random operates without a failure during period t.
2. The first unit chosen at random fails at some moment $x < t$, the SD performs a successful switch to one of the operating units of the remaining redundant group of $m - 1$ units, and the new system performs successfully up to time t.

This permits one to write a recurrent relationship in the form

$$P_{syst}^{(m)} = p(t) + R \int_0^t P_{syst}^{(m-1)}(t - x|x) \, dq(x) \qquad (3.57)$$

where R is the probability of a successful switching.

For a standby redundant group of m independent and identical units, the system's successful operation is possible in two situations:

1. The first unit chosen at random operates without a failure during period t.
2. The first unit chosen at random fails at some moment $x < t$, the SD performs a successful switch, and from this moment on the new system of $m - 1$ redundant units and SD perform successfully until time t.

This description permits one to write the recurrent relationship

$$P_{syst}^{(m)} = p(t) + R \int_0^t P_{syst}^{(m-1)}(t - x) \, dq(x) \qquad (3.58)$$

where $P_{syst}^{(k)}(t)$ is the probability of a successful operation of the active redundant group of k units during a time interval t.

3.6.3 Individual Switching Devices

We assume that each unit of a redundant group may be chosen to replace a failed main unit at random. After the main unit's failure, an individual SD associated with the next unit in the redundant group may successfully perform the next connection, or it may fail. Assume also that a unit's failure leads to a corresponding SD's failure. The absence of an operating unit with an operating SD leads to the system's failure. As follows from the above description, the SD is necessary for the unit to operate.

Switching Devices That Fail with Time   For active redundancy a system operates successfully in the following situations:

1. The first unit operates successfully.
2. After its failure there is a group of randomly chosen redundant units with operating SDs; this new system operates successfully during the remaining time.

The probability of a system's successful operation can be written as

$P_{\text{syst}}^{(m)}$

$$= p(t) + \int_0^t \left[\sum_{1 \le k \le m-1} P_{\text{syst}}^{(k)}(t-x|x) \binom{m-1}{k} \tilde{p}^k(x) \tilde{q}^{m-k-1}(x) \right] dq(x) \quad (3.59)$$

where $\tilde{p}(t) = p_{\text{SD}}(t)p(t)$.

For a standby redundant group, the expression is simpler. The following are situations wherein a system operates successfully:

1. The first unit operates successfully.
2. After its failure there is a group of $m-1$ redundant units with a random number of operating SDs; this new system operates successfully during the remaining time. The random number of operating SDs appears because of SD failures during the on-duty regime.

The probability of a system's successful operation can be written as

$P_{\text{syst}}^{(m)}(t) = p(t)$

$$+ \int_0^t \left\{ \sum_{1 \le k \le m-1} P_{\text{syst}}^{(k)}(t-x) \binom{m-1}{k} [p_{\text{SD}}(x)]^k [q_{\text{SD}}(x)]^{m-k-1} \right\} dq(x) \quad (3.60)$$

Of course, (3.59) and (3.60) can be practically utilized only with the aid of a computer.

Switching Devices That Fail at the Time of Switching   First, consider an active redundant group. There are again two situations where the system can successfully perform its operation:

1. The first unit operates successfully.
2. After its failure at some moment x, there is a group of redundant units. The size of this group is random because some of them might have failed before the moment x. Let the number of operating redundant units at the moment x equal j. In some order we try to switch each of these j operating units to the main position until a first successful switching occurs. The number of attempts before a success is distributed geometrically with parameter R. After k SDs have failed during switching, a successful attempt occurs (k is random). This means

that the new system of $j - k$ redundant units must successfully operate during the remaining time period $t - x$.

This verbal description corresponds to the expression

$$P_{\text{syst}}^{(m)}(t) = p(t) + R \int_0^t \sum_{1 \le j \le m-1} \binom{m-1}{j} [p(x)]^j [q(x)]^{m-j-1}$$

$$\cdot \left\{ \sum_{1 \le k \le j} Q^k P_{\text{syst}}^{(k)}(t - x | x) \right\} dq(x) \tag{3.61}$$

where $Q = 1 - R$.

When we consider standby redundancy, the system can successfully operate if:

1. The first unit operates successfully.
2. After its failure at some moment x, there is a group of $m - 1$ standby units. In some order we try to switch each of these units to the main position until a first successful switching occurs. The number of attempts before success is distributed geometrically with parameter R. After k SDs have failed during switching, a successful attempt occurs (k is random). The new system of $m - k - 1$ redundant units must successfully operate during the remaining time interval $t - x$. The appropriate probability is

$$P_{\text{syst}}^{(m)} = p(t) + R \int_0^t \sum_{1 \le k \le m-1} Q^k P_{\text{syst}}^{(m-k-1)}(t - x) \, dq(x) \tag{3.62}$$

The MTTF for both systems can only be found numerically. Note that for the exponential distribution and large m, (3.62) can be approximated by

$$P_{\text{syst}}^{(m)} \approx e^{-Q\lambda t} \tag{3.63}$$

and the MTTF can be approximated by

$$T_{\text{syst}} = \frac{T}{Q} \tag{3.64}$$

Both (3.63) and (3.64) are obtained under the assumption of the correctness of the application of the result of the random summation to exponentially distributed r.v.'s. (Note that in our case we consider a fixed number of Bernoulli trials.)

This brief review does not, of course, cover the entire theme. There are various cases in practice which lie outside the scope of this material. Our main purpose was to display some inferences in this area and not to give the reader a "cook book."

3.6.4 Periodic Monitoring

In this section we do not try to examine the monitoring problem but rather give a simple example of the possible influence of monitoring on a system's reliability. Above we considered a redundant system with the possibility of an instant replacement of a failed main unit by a redundant one. In many practical cases such a situation is unrealistic. In many cases the state of the units, main or redundant, can be checked only at some prespecified moments, usually at periodic intervals.

Consider a simple system consisting of two parallel units and one standby redundant unit which cannot be switched immediately to either of the parallel units. This redundant unit can replace either failed parallel unit only at some predetermined moments $t_s = s\Delta$, $s = 1, 2, \ldots$. At these moments the state of the two parallel units is checked and a failure may be detected. (In other words, the monitoring of the units is not continuous.) All units are assumed identical and independent, and their TTF distributions are assumed exponential with parameter λ.

The system is considered to have failed if:

1. Both parallel units have failed inside a period between two neighboring check points, even if there is a standby unit.
2. There are no units operating at some moment.

Consider the probability of a system's failure-free operation during N cycles. For this case the following discrete recurrent equation can be written:

$$P_{syst}(N) = p^2 P_{syst}(N-1) + 2pq[1 - Q(N-1)] \qquad (3.65)$$

with $P_{syst}(0) = 1$. Here $Q(K)$ is the probability of a failure-free operation of two parallel units during K cycles, $Q(K) = 1 - [1 - p^K]^2$, and $p = 1 - e^{-\lambda\Delta}$. Equation (3.65) can be solved numerically.

For the system's MTTF, one can write

$$T_{syst} = p^2[\Delta + T_{syst}] + 2pq[\Delta + T_2] + q^2\Delta^* \qquad (3.66)$$

where after a successfully operating cycle of length Δ the system starts its failure-free operations from the beginning. A cycle with two failures contains a portion of useful time which is denoted by Δ^*. Setting $\Delta^* = \Delta$, we can write the approximation

$$T_{\text{syst}} = \Delta + p^2 T_{\text{syst}} + 2pqT_2 \tag{3.67}$$

where T_2 is the average number of successful cycles of the two parallel units.

We essentially use here the Markov nature of the model. Even in this simple case we have no strong results in a simple form. But we see that monitoring essentially changes the operation process and, consequently, changes the reliability indexes of the system.

Equations (3.65) and (3.66) are complicated enough to make some quantitative conclusions, but we consider two simple limiting cases. It is clear that $\Delta \to 0$ leads to the continuous monitoring model, and, hence, the system reverts to a system composed of two active redundant units and one standby redundant unit. Incidentally, the MTTF of such a system equals

$$T_{\text{syst}} = \frac{1}{2\lambda} + \frac{1}{2\lambda} + \frac{1}{\lambda} = \frac{2}{\lambda}$$

If, on the contrary, we assume that $\Delta \to \infty$, it means that factually the system has no redundant units at all because they will never be used. In this case $T_{\text{syst}} = 3/2\lambda$. Thus, for intermediate Δ's, the value of T_{syst} lies somewhere between the mentioned values.

It is clear that for a series system of units with an exponentially distributed TTF, it is totally unreasonable to have any redundant group which can be switched only after periodically checking the system's state. (We consider reliability indexes such as the probability of failure-free operation or the MTTF.)

3.7   DYNAMIC REDUNDANCY

This interesting redundancy class is very close to the classical problems of inventory control. Consider a redundant group of n units. A part of them might be used as active and another part as standby. There is a possiblility of checking the current state of the active units only at some predetermined moments. Thus, there is no feedback information within the interval between two neighboring check points. The system can be found to be failed at a moment of checking even if there are some standby units available to be used. The following questions arise: How many units should be reasonably switched between two checking moments? How does one refill the active redundant group?

From a qualitative point of view, it is clear that it is not reasonable to switch all redundant units to an active state: the reliability at the first stage of operation will be high but the redundant units will be rapidly spent. To switch a small number into the active redundant group is unreasonable because a system failure can, with a high probability, appear before a current check.

This kind of problem may appear in connection with nonmonitoring technical systems, for example, space vehicles designated for an investigation of the solar system. The time of response can be excessively long, and it becomes impossible to control the situation, so that one needs to find some prior rule of autonomous switching of the redundant units over time without external signals.

A possible solution to the problem is to choose moments of switching spare units into acting positions. We will discuss the problem of finding the optimal moments of switching in Chapter 11. Now we only consider how to calculate the reliability indices for such a system.

We call such a kind of redundancy a *dynamic redundancy*. We will only investigate dynamic redundancy with exponentially distributed unit ITFs. All units are also supposed to be identical and independent.

3.7.1 Independent Stages

The system possesses n identical and independent units to perform its function up to some time t_0. Then an initial group of units, n_0, is installed as an active redundant group and all remaining units are placed in a standby regime. The duration of the system's operation is divided into k stages. There are moments $0 < \tau_1 < \tau_2 < \cdots < \tau_k < t_0$ when the new group of standby redundant units are to be switched into an active regime. When we consider independent stages, such switching is performed at some predetermined moments. Such a procedure is called a *programmed controlled switching*. The previous group of units is expelled from the operation with no consideration of their real state. (As a matter of fact, no active units may fail before the beginning of the next stage.) In this case all stages are independent. Such a situation arises if the deployment of previously used units for use at the next stage is a difficult or even impossible engineering task.

In this simple case the probability of a system's successful operation during a time interval t_0 equals

$$P_{\text{syst}}(t) = \prod_{0 \leq j \leq k} \left\{ 1 - \left[q_j(t_j) \right]^{n_j} \right\} \qquad (3.68)$$

where $q_j(t_j)$ is the probability of a failure-free operation of a single unit.

The calculation of a system's MTTF is not very simple in this case. Assume that a failure of the system occurs at some stage k. This means that the system operates successfully for $k - 1$ stages and during some random time within the last stage. The kth stage duration equals $\Delta_k = \tau_k - \tau_{k-1}$. The conditional value of a failure-free operational time during stage k (denote this by ζ_k) is

$$\zeta_k = \left\{ \max_{1 \leq i \leq n_k} \xi_i | \Delta_k \right\}$$

This conditional mean time can be found in the standard way,

$$E\{\zeta_k\} = \frac{\int_0^\Delta p(x)\,dx}{p(\Delta)}$$

and only the group of units at the last stage operates until complete exhaustion of all redundant units.

Finally, for the system we have

$$T_{\text{syst}} = \sum_{1 \le j \le k-1} q_j(\Delta_j) \prod_{1 \le i \le j-1} p(\Delta_i)\left[\sum_{1 \le i \le j-1} \Delta_i + E\{\zeta_j\}\right] + E\left\{\max_{1 \le i \le n_k} \xi_i\right\}$$

$$(3.69)$$

Once can use an approximation by replacing ζ_k, with Δ_k or one can obtain lower and upper bounds by substituting $\xi_k = 0$ and $\zeta_k = \Delta_k$, respectively.

3.7.2 Possibility of Transferring Units

A more interesting and more complicated case arises if one considers the possibility of using all nonfailed units at some stage for the next stage. Of course, in this case it is possible to analyze only the systems whose units have an exponentially distributed TTF. If stage j has a duration Δ_j and there are m units in the active redundant group (including those from the previous stage), then the probability of a failure-free operation is given by

$$P_j(\Delta_j) = 1 - \left[1 - \exp(-\lambda\Delta_j)\right]^m$$

After the first stage of a successful operation, the system has a random number, say y, $1 \le j \le n_1$, of operating units. These units can be used at the second stage, starting at the moment τ_1 with n_2 new units switched in by the prior rule. Thus, the total number of units acting at this stage equals $n_2 + j$.

The probability of exactly j units $\{j > 0\}$ being transferred to the second stage is

$$\binom{n_1}{j} p_1^j q_1^{n_1-j}$$

where $p = 1 - q$. If the system performed successfully during the first stage, j units $(j > 0)$ leave to operate at the second stage. At the same time, at moment τ_1, new n_2 units are switched into the system. So, for a two-stage process with $n = n_1 + n_2$, one can write the probability of interest as

$$P_{\text{syst}}(\tau_2) = \sum_{1 \le j \le n_1} \binom{n_2 + j}{j} p_1^j q_1^{n_1-j}\left(1 - q_2^{n_2+j}\right) \qquad (3.70)$$

Similarly, the expression for a system with three stages can be written as

$$P_{syst}(\tau_3) = \sum_{1 \le j \le n_1} \binom{n_1}{j} p_1^j q_1^{n_1-j} \sum_{1 \le s \le n_2+j} \binom{n_2+j}{s} p_2^s q_2^{n_2+j-s}(1 - q_3^{n_3+s})$$

$$(3.71)$$

Of course, equations such as (3.71) might be considered as the basis for computational algorithms, not for hand calculations. At the same time, it is possible to write a recurrent equation which could be used for computer calculations:

$$P_{syst}(t_0|n_1; n) = \sum_{1 \le j \le n_1} \binom{n_1}{j} r_1^j q_1^{n_1-k} P_{syst}(t_0 - \Delta_1|n_2 + j; n - n_1) \quad (3.72)$$

Notice that in such systems the most important thing is to define the optimal intervals Δ_k and the number of units n_j that should be switched each time. A simple heuristic solution of this optimization problem is presented in Chapter 13.

3.8 SYSTEMS WITH DEPENDENT UNITS

In the real world different random events and random variables are often "associated" and not independent. For instance, units of a system can be dependent through the system's structure, the functioning environment, the inner state changing, and so on. In all of these situations, reliability usually tends to change in the same way for all units: all of a unit's parameters increase or all decrease.

Two r.v.'s X and Y are *associated* if their covariance is positive:

$$Cov(X, Y) = Cov(Y, X) = E\{(X - E\{x\})(Y - E\{y\})\} \ge 0$$

A stronger requirement for the association of two r.v.'s demands that the inequality

$$Cov[f_1(X, Y), f_2(X, Y)] \ge 0$$

holds, where both f_1 and f_2 are increasing or both are decreasing functions. The vector $\mathbf{X} = (X_1, X_2, \ldots, X_n)$ consists of associated components if

$$Cov[f_1(\mathbf{X}), f_2(\mathbf{X})] \ge 0$$

A more formal discussion of associated r.v.'s can be found in Barlow and Proschan (1975).

3.8.1 Series Systems

Consider a series system of two units. Let x_i, $i = 1, 2$, be the indicator function of the ith unit. Suppose the x_i's are associated. For instance, both of them are dependent on the same environmental factor Ω; that is, notationally, they are $x_1(t|\Omega)$ and $x_2(t|\Omega)$ for some specified Ω. Then for these two units one can write

$$\Pr\{x_1 \wedge x_2 = 1\} = \Pr\{x_1 = 1\}\Pr\{x_2 = 1\} + \rho(x_1, x_2) \qquad (3.73)$$

where $\rho(x_1, x_2)$ is the correlation coefficient

$$\rho(x_1, x_2) = \frac{\text{Cov}(x_1, x_2)}{\text{Var}\{x_1\}\,\text{Var}\{x_2\}}$$

This normalized value satisfies the condition $-1 \leq \rho \leq 1$. For n associated r.v.'s it is possible to consider only the case $\rho \geq 1$. From this condition and (3.73), for a series system of two associated units, it follows that

$$\Pr\{x_1 \wedge x_2 = 1\} \geq \Pr\{x_1 = 1\}\Pr\{x_2 = 1\} \qquad (3.74)$$

This result can be immediately generalized for a series system of n associated units:

$$P_{\text{syst}} = \Pr\{\alpha(\mathbf{X}) = 1\} \geq \prod_{1 \leq i \leq n} p_i \qquad (3.75)$$

From (3.75) it automatically follows that for a series system of n associated units:

$$T_{\text{syst}} \geq \int_0^\infty \prod_{1 \leq i \leq n} p_i(t) \qquad (3.76)$$

Consider the example when each unit of a series system depends on the same factor ρ, for instance, the temperature. The system is designed for use at two different temperatures, ρ_1 and ρ_2. The designer decides to check the probability of a failure-free operation of the series system of n units. For this purpose, the designer arranges for a unit testing under these two conditions.

The probabilities of the unit's failure-free operation under these two conditions are p_1 and p_2, respectively. The average unit failure-free operation probability equals $p = (1/2)(p_1 + p_2)$. At a first glance, it is very attractive to try to compute the system reliability index as $P^*_{\text{syst}} = p^n$ if we know nothing about the real conditions of the system's use.

But let us assume that the first condition appears in practice with frequency R, and the second condition appears with frequency $Q = 1 - R$. (Of course, the frequency R can be considered to be a probability.) Then a realistic value of the index is $P_{syst} = Rp_1^n + Qp_2^2$. It is easy to check that $P_{syst} \geq P_{syst}^*$. (To convince yourself in a particular case, do it with $n = 2$ and $R = 1.2$.) Of course, the same phenomenon will be observed if one considers more than two different environmental conditions.

Another example of a system with associated units is a system operating in a changing regime. Assume that a system operates with probability p_k at the kth regime. Under this regime the system's units have a failure rate equal to λ_k. It may happen if the system switches from regime to regime periodically (or, perhaps, randomly). In this case

$$P_{syst}(t) = \sum_{\forall k} p_k e^{-\lambda_k nt}$$

This is larger than

$$P_{syst}^*(t) = e^{-E\{\lambda\}nt} = \exp\left(-t \sum_{\forall k} p_k \lambda_k n\right)$$

So, for a series system we can use the hypothesis of the unit's independence to obtain a conservative bound on the reliability index of types $P(t)$ or T.

3.8.2 Parallel Systems

Now consider a parallel system of two associated units. For this system we have

$$\Pr\{x_1 \vee x_2 = 1\}$$

$$= 1 - \Pr\{\overline{\bar{x}_1 \wedge \bar{x}_2}\} = 1 - \left[\Pr\{\bar{x}_1 = 1\} \Pr\{\bar{x}_1 = 1\} + \rho(\bar{x}_1, \bar{x}_2)\right] \quad (3.77)$$

$$\leq 1 - \Pr\{\bar{x}_1 = 1\} \Pr\{\bar{x}_1 = 1\} = 1 - q_1(t)q_2(t)$$

where $\rho(\bar{x}_1, \bar{x}_2)$ is the correlation coefficient for the indicator functions. It is easy to show that

$$\rho(\bar{x}_1, \bar{x}_2) = \frac{\text{Cov}(\bar{x}_1, \bar{x}_2)}{\text{Var}\{\bar{x}_1\} \text{Var}\{\bar{x}_2\}} \quad (3.78)$$

But $\text{Cov}(\bar{x}_1, \bar{x}_2) = \text{Cov}(x_1, x_2)$ and $\text{Var}\{\bar{x}_i\} = \text{Var}\{x_i\}$, $i = 1, 2$.

Equation (3.78) can immediately be generalized for a parallel system of m associated units:

$$P_{syst} = \Pr\{\beta(\mathbf{X}) = 1\} \leq 1 - \prod_{1 \leq i \leq m} q_i \quad (3.79)$$

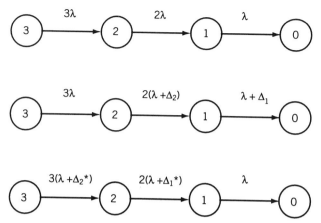

Figure 3.14. Transition graph for obtaining the upper and lower bounds for a parallel system with dependent units.

The reader can consider the previous examples applied to a parallel system. We consider several examples connected with the death process which, as we mentioned above, can successfully be used for describing unrepairable redundant units. We consider a special type of dependence.

For simplicity, consider a parallel system of three units. All units operate in the system in a nominal regime. For such a regime, each unit has a failure rate λ. If the units are independent, the transition graph is presented in Figure 3.14a. Assume that, after the failure of the first active redundant unit, the two remaining units are forced to operate in a harsher regime. For example, in an electrical parallel circuit, as the flow through each resistor becomes larger, the resistors produce more heat, the surrounding temperature increases, and, consequently, the failure rate increases. In a hydraulic circuit, after one of the parallel pipes is closed, the remaining are under a higher pressure and, consequently, can fail with higher probability. Thus, a unit's failure rate often depends on the state of the other units.

Assume that λ of each unit is an increasing function of the flow through the unit. In this case $\lambda_3 = 3\lambda$, but $\lambda_2 = 2(\lambda + \Delta_2)$ and $\lambda_1 \text{-} \lambda + \Delta_1$ where $\Delta_1 \geq \Delta_2$. The transition graph for this case is presented in Figure 3.14b. It is clear that this system of associated units is less reliable than the initial system of independent units.

Now let us consider a case which is, in some sense, the inverse of the previous one. All parallel units are operating in a restricted room. A single unit operating in this room has a nominal failure rate λ. Each working unit generates a heat which accumulates in the room and influences all of the remaining units. Thus, the more units that are operating, the higher the temperature, which leads to the decreasing reliability of each unit. (At the same time, remember that the system is a parallel system!) We begin the analysis of the transition graph for this case with state 1.

A single unit operates with $\lambda_1 = \lambda$. When two units are operating, the temperature in the room is higher and, consequently, $\lambda_2 = 2(\lambda + \Delta_1^*)$. When all three units are operating, $\lambda_3 = 3(\lambda + \Delta_2^*)$. Under the conditions of the example, $\Delta_1^* \leq \Delta_2^*$. The transition graph for this case is shown in Figure 3.14c.

A comparison of these two cases with the initial system of independent units shows that both of them possess a less favorable reliability index than the initial system: in each case the transition intensities are higher than in the initial case. Thus, on average, systems with associated units reach a failure state more quickly. We finish this comparison with a comparison of the MTTFs:

$$T_{\text{syst}} = \frac{1}{3\lambda} + \frac{1}{2\lambda} + \frac{1}{\lambda} \geq \frac{1}{3\lambda} + \frac{1}{2(\lambda + \Delta_2)} + \frac{1}{\lambda + \Delta_1}$$

and

$$T_{\text{syst}} = \frac{1}{3\lambda} + \frac{1}{2\lambda} + \frac{1}{\lambda} \geq \frac{1}{3(\lambda + \Delta_2^*)} + \frac{1}{2(\lambda + \Delta_1^*)} + \frac{1}{\lambda}$$

3.8.3 Mixed Structures

We consider this concept in more detail in Chapter 9 when we address two-pole network bounds. Here we only illustrate some kind of dependence between the system's units when the system has a mixed structure. Consider the simplest series–parallel and parallel–series systems with different forms of unit dependence. To perform its operation, the system should have at least one unit of type A and at least one unit of type B. Assume that we analyze a system whose two units (say functional blocks) have their own power supply (PS). The power supply is not absolutely reliable. Of course, a power supply failure leads to an immediate failure of both units which are supplied by this PS.

First, consider a series–parallel system. There are two possibilities of switching the power supply (see Figure 3.15). Denote the probabilities of a successful operation by p_A, p_B, and p_{PS}. Then for structure (a) we can write

$$P_a = p_{\text{PS}}^2 \left[1 - (1 - p_A p_B)^2 \right] + 2 p_{\text{PS}} q_{\text{PS}} P_A P_B$$

and for structure (b) we can write

$$P_b = p_{\text{PS}}^2 \left[1 - (1 - P_A P_B)^2 \right]$$

It is obvious that structure (a) is better than structure (b).

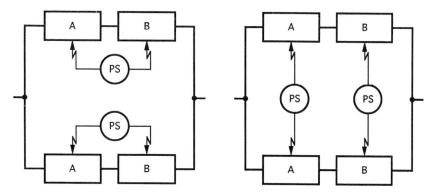

Figure 3.15. Two variants of the power supply of a series–parallel structure.

Now consider two variants of the parallel–series structure (see Figure 3.16). For structure (c) we have

$$P_c = p_{PS}^2(1 - q_A^2)(1 - q_B^2) + 2p_{PS}q_{PS}p_A p_B$$

and for structure (d) we have

$$P_d = p_{PS}^2(1 - q_A^2)(1 - q_B^2)$$

Again, we can deduce that $P_c \geq P_d$ without calculation, based only on our previous knowledge about the reliability of a series system with associated units.

A consideration of these examples shows us that the reliability of some auxiliary units may have an influence on other system units in such a way that the reliability of a parallel–series structure might be worse than the reliability

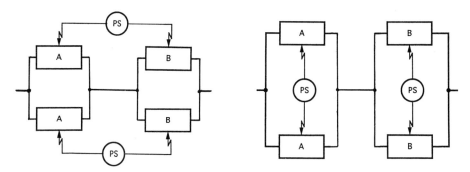

Figure 3.16. Two variants of the power supply of a parallel–series structure.

of a series–parallel structure. Indeed, it is possible that for some fixed p_A, p_B, and p_{PS}, for instance, the inequality $p_a > p_d$ is true:

$$p_{PS}^2\left[1 - (1\text{-}p_A p_B)^2\right] + 2p_{PS}q_{PS}p_A p_B > p_{PS}^2(1 - q_A^2)(1 - q_B^2)$$

Avoiding general deductions, take $p_A = p_B = p$. Denote for simplicity $p_{PS} = R$ and $Q = 1 - R$. Then the condition

$$R^2\left[1 - (1 - p^2)^2\right] + 2RQp^2 > R^2(1 - q^2)^2$$

is equivalent to

$$\frac{Q}{R^2} > \frac{(1 - q^2)^2 - \left[1 - (1 - p^2)^2\right]}{2p^2}$$

The right part of the inequality is restricted by 1 for any p. Thus if $Q > R^2$ this inequality holds for any p. The solution of the corresponding equality gives

$$Q = \frac{3 \pm \sqrt{5}}{2} \approx 0.382$$

In other words, for an unreliable common unit (in our case the power supply), with a reliability index lower than approximately 0.6, one should choose a parallel–series structure rather than a series–parallel one.

For some additional examples of the analysis of systems consisting of dependent units, see Gnedenko, Belyaev, and Solovyev (1969).

3.9 TWO TYPES OF FAILURES

Some units have two types of failures. For instance, a resistor may be disconnected (leaving an open circuit) in an electric circuit, and in this case no flow goes through the unit. Or it may burn out, and so will not provide any resistance at all (a short circuit). One observes an analogous effect with capacitors: no capacity at all (a short circuit) or infinite capacity (disconnection). In pipelines, holes allow the leaking of pumped liquid, which decreases the user's consumption and, simultaneously, decreases the hydraulic resistance. Rubbish in the pipe results in the same decrease in user consumption, but, at the same time, increases the hydraulic resistance.

In a most essential way, this phenomenon appears in relay circuits. These circuits are assigned for connection and disconnection and the nature of their failure can be one of two kinds: they may fail to connect or they may fail to disconnect. Each unit (relay) itself is subjected to two similar failures. It makes the problem of redundancy of such systems more difficult: a parallel structure of relays fails if at least one unit makes a false connection when it should be disconnected, and a series structure fails if at least one unit causes a false disconnection when it should be connected. As a matter of fact, mixed series–parallel (parallel–series) structures are more effective in this case. Moreover, for a relay with known probabilities of failure of both types, there is an optimal mixed structure. We consider this problem separately in Chapter 11.

Consider a parallel–series relay system. This system can be considered as a two-pole network with an input on the left and an output on the right. Each unit of the system at any moment of time can be in one of three jointly exclusive possible states: failure-free with probability p, failed in a "connected" state with probability c, or failed in a "disconnected" state with probability d. First, we consider the case where the system must provide a connection between the input and output. For each series circuit of n units, the probability of a successful connection R_{con} can be written as

$$R_{con} = (p + c)^n \tag{3.80}$$

and, for the system as a whole,

$$P_{con} = 1 - (1 - R_{con})^m = 1 - \left[1 - (p + c)^n\right]^m \tag{3.81}$$

If the system must provide a disconnection, the corresponding probabilities are

$$R_{discon} = 1 - (1 - p - d)^n = 1 - c^n \tag{3.82}$$

and

$$P_{discon} = \left[1 - c^n\right]^m \tag{3.83}$$

A relay system operation consists of alternating cycles of connections and disconnections. It seems that for this system it is reasonable to choose a reliability index in the form

$$P_{syst} = \min(P_{con}, P_{discon}) \tag{3.84}$$

It is clear that a single relay with the same parameters can perform successfully in both cases only with probability p. (Any kind of failure makes one of the operations totally impossible.)

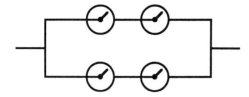

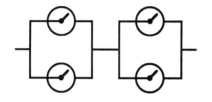

Figure 3.17. Parallel–series and series–parallel relay schemes for Examples 3.2 and 3.3.

Example 3.2   Consider a parallel–series system with $n = m = 2$, $p = 0.8$, and $c = d = 0.1$ (see Figure 3.17a). For this system,

$$P_{con} = 1 - \left[1 - 0.9^2\right]^2 \approx 0.96$$

and

$$P_{discon} = \left[1 - 0.1^2\right]^2 \approx 0.98$$

Thus, the system operates successfully with a probability of not less than 0.96 under either type of operation: connection or disconnection. Both probabilities P_{con} and P_{discon} are larger than the corresponding initial probability of a single unit (under the condition that it equal 0.9).

Now let us consider a general series–parallel relay system. Again consider first the case when the system must provide a connection between the input and output. For each parallel circuit of m units, the probability of a successful connection R_{con} is

$$R_{con} = 1 - (1 - p - c)^m = 1 - d^m \tag{3.85}$$

and for the system as a whole

$$P_{con} = R_{con}^n = (1 - d^m)^n \tag{3.86}$$

If the system must provide a disconnection, the corresponding probabilities are

$$R_{discon} = (p + d)^m = (1 - c)^m \tag{3.87}$$

and

$$P_{\text{discon}} = 1 - \left[1 - (p + d)^m\right]^n \tag{3.88}$$

Again we can use (3.84) to characterize the system as a whole.

Again in this case a single relay with the same parameters can perform successfully in both cases only with probability p.

Example 3.3   Consider a system with $n = m = 2$, $p = 0.8$, and $c = d = 0.1$ (see Figure 3.17b). For this system,

$$P_{\text{con}} = 1 - \left[1 - 0.1^2\right]^2 \approx 0.98$$

and

$$P_{\text{discon}} = \left[1 - 0.9^2\right]^2 \approx 0.96$$

Thus, the system operates successfully with a probability of not less than 0.96 under either type of operation: connection or disconnection. Again both probabilities P_{con} and P_{discon} are larger than the corresponding initial probability of a single unit.

One can notice that the structures of Figures 3.17a and b are "mirror images" with respect to the probabilities c and d. Thus, both structures are equivalent for the relay with $c = d$.

3.10   MIXED STRUCTURES WITH PHYSICAL PARAMETERS

A unit presented with an indicator function x_i reflects a "dichotomic" object which can only be one of two states: for reliability problems they are termed "success" and "failure." But sometimes we need to analyze systems consisting of units with physical parameters whose particular value plays an essential role.

In Chapter 1 we introduced the *generalized generating sequence* (GGS). Here we use it and make some concrete additions to the general method. These additions are helpful for the designing of appropriate computer algorithms.

We present the discussion via simple examples.

Series System   This case has been considered in Chapter 1. Thus, we consider only simple examples.

Example 3.4   Consider an oil pipeline consisting of n pipes (units) connected in series. Each unit has a random capacity which decreases for different reasons: the accumulation of so-called "heavy" fractions on the pipe walls, a deformation of pipes, and so forth. The distribution of the capacity for each pipe is supposed to be known. We also assume that the distributions are determined by a finite number of values, say v_i, for the ith pipe. The GGS for the ith pipe is presented in the form of the legion

$$L_i = \{(c_{i1}, p_{i1}), \ldots, (c_{iv_i}, p_{iv_i})\}$$

Here we try to avoid the complexity of a general notation and so denote $M_{ik1} = c_{kk}$ and $M_{ik2} = p_{ik}$, which corresponds to their natural notation as a capacity and a probability.

The interaction of n legions produces $N = \prod v_i$ different cohorts C_k, $1 \leq k \leq N$,

$$C_k = (c_k, p_k)$$

The capacity and the probability are determined by the rules of the cohort interactions: the "cells" with the values of capacities and the "cells" with the values of probabilities are considered separately. The capacity is determined by

$$c_k = \Omega_c^M \underset{\substack{1 \leq i \leq n \\ j_i \in k}}{c_{ij_i}} = \underset{\substack{1 \leq i \leq n \\ j_i \in k}}{\min} c_{ij_i}$$

and the probability is determined by

$$p_k = \Omega_p^M \underset{\substack{1 \leq i \leq n \\ j_i \in k}}{p_{ij_i}} = \underset{\substack{i \leq i \leq n \\ j_i \in k}}{\prod} p_{ij_i}$$

The operator Ω^L in this particular case possesses the following property. If for two terms of the final GGS there are C_k and C_{k+1} with $c_k = c_{k+1}$, then these two terms form a new term with parameters c_k^* and p_k^* determined by

$$c_k^* = c_k = c_{k+1} \quad \text{and} \quad p_k^* = p_k = p_{k+1}$$

Let us call this the *absorption property*.

Now assume that there is a known failure criterion for this pipeline, for example, suppose it is considered to be failed when $c_k < c^0$. In this case, to obtain the resulting reliability index, one has to revise the operators Ω^L and Ω^C in an appropriate way.

If c_k must be larger than c^0, the actual capacity does not play any role. The operator Ω^C must be determined in such a way that any $c_k \geq c^0$ might be considered as some c_{accept} and the remaining c_k's are set equal to 0. In this case one has cohorts of two types: the ones with c_{accept} and the other with 0. Incidentally, a computer procedure for finding the minimal value may be solved in a sequential way:

$$c_2^* = \min(c_1, c_2)$$
$$c_3^* = \min(c_2^*, c_3)$$
$$\vdots$$
$$c_n^* = \min(c_{n-1}^*, c_n) = \min(c_1, \ldots, c_n)$$

One may stop the procedure as soon as the value $c_k^* < c^0$ appears at some intermediate step of the calculation.

Now let Ω^L possess the above-mentioned absorption property and, additionally, the *preference property*. In our case the latter means that if two cohorts have different sets of maniples, then under some specified conditions the one which possesses the "better" maniple is kept for further consideration and the one with the "worse" maniple is excluded.

In the case of cohort interaction we, at first, use the absorption property and obtain the final legion in an intermediate form

$$L = \big((c_{\text{accept}}, p), (0, p^0) \big)$$

where p is the sum of all p's of the cohorts with c_{accept}. The resulting legion, after applying the preference operation, will have the form

$$L = (c_{\text{accept}}, p)$$

It is clear that $R_{\text{syst}} = p$.

A Parallel System   The formal technique used for parallel systems completely coincides with the above-described method. But, for convenience, we will use the corresponding operators $\mho^L$, $\mho^C$, and $\mho^M$. We use these new symbols to distinguish the operations over the maniples. Indeed, for instance, for resistance

$$\Omega^M_{1 \le i \le n} = \sum_{1 \le i \le n} r_i$$

and

$$\mho^M_{1 \le i \le n} = \sum_{1 \le i \le n} \frac{1}{r_i}$$

for the time to failure caused by a short connection

$$\Omega^M_{1 \le i \le n} \xi_i = \max_{1 \le i \le n} \xi_i$$

TABLE 3.2   Main Kinds of Maniple Interactions

Physical Nature of Maniple	Series Structure $\Omega^{(M)}$	Parallel Structure $\Omega^{(M)}$
Probability of success (a cutoff failure type)	$w_1 w_2$	$1 - (1 - w_1)(1 - w_2)$
Probability of success (a short-connection failure type)	$1 - (1 - w_1)(1 - w_2)$	$w_1 w_2$
Probability of failure (a cutoff failure type)	$1 - (1 - w_1)(1 - w_2)$	$w_1 w_2$
Probability of failure (a short-connection failure type)	$w_1 w_2$	$1 - (1 - w_1)(1 - w_2)$
Random TTF for a cutoff	$\min\{w_1, w_2\}$	$\max\{w_1, w_2\}$
Random TTF for a short connection	$\max\{w_1, w_2\}$	$\min\{w_1, w_2\}$
Electrical capacity	$[w_1^{-1} + w_2^{-1}]^{-1}$	$w_1 + w_2$
Ohmic resistance	$w_1 + w_2$	$[w_1^{-1} + w_2^{-1}]^{-1}$
Ohmic conductivity	$[w_1^{-1} + w_2^{-1}]^{-1}$	$w_1 + w_2$
Capacity of communication channel	$\min\{w_1, w_2\}$	$w_1 + w_2$
Cost of transportation through network	$w_1 + w_2$	$\min\{w_1, w_2\}$

and

$$\mathbf{\mho}_{1 \leq i \leq n}^{M} \xi_i = \min_{1 \leq i \leq n} \xi_i$$

and so on.

In Table 3.2 we present the principal kinds of maniple interaction operators for systems with series and parallel structures (the considered parameters are denoted in the table by w).

Of course, the functions listed in Table 3.2 do not exhaust all possibilities.

After this brief review of the possible interactions between maniples, we might begin with a consideration of the system with a reducible structure.

Mixed Structures   Here we illustrate how to use the GGS for the analysis of a mixed structure with a simple example. We will not produce detailed transformations and calculations because for us and for the reader it is more reasonable to leave it to a computer. We ignore the physical nature of the system and its units for the moment.

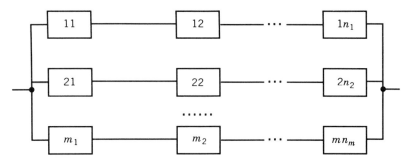

Figure 3.18. Structure of the system considered in Example 3.5.

Example 3.5 A parallel–series system is presented in Figure 3.18. For the system

$$L = \bigcup_{1 \le i \le m}^{L} \bigcap_{1 \le j \le n_i}^{L} L_{ji}$$

The remaining interactions depend on the concrete nature of the system.

Example 3.6 The series–parallel system is presented in Figure 3.19. For the system

$$L = \bigcap_{1 \le i \le n}^{L} \bigcup_{1 \le j \le m_i}^{L} L_{ji}$$

The remaining interactions again depend on the concrete nature of the system.

Example 3.7 The system with a mixed structure is presented in Figure 3.20. For this system the following chain of operations for obtaining the resulting

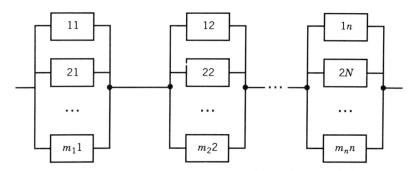

Figure 3.19. Structure of the system considered in Example 3.6.

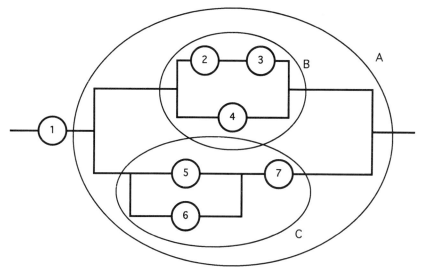

Figure 3.20. Structure of the system considered in Example 3.7.

legion can be written: For the system as a whole, Figure 3.21a, we obtain

$$L = \Omega^L(L_1, L^{(A)})$$

Subsystem (A) can be presented itself as Figure 3.21b. Then

$$L^{(A)} = \mho_L(L^{(B)}, L^{(C)})$$

We will write expressions for $L^{(B)}$ and $L^{(C)}$, again with no explanations (use the ancient rule: "see the sketch"):

$$L^{(B)} = \mho^L(L_4, \Omega^L(L_2, L_3))$$

and

$$L^{(C)} = \Omega^L(L_7, \mho^L(L_5, L_6))$$

Therefore, the final macroalgorithm for the system GGS computation can now be written in the final form

$$L = \Omega^L(L_1, \mho_L(\mho^L(L_4, \Omega^L(L_2, L_3)), \Omega^L(L_7, \mho^L(L_5, L_6))))$$

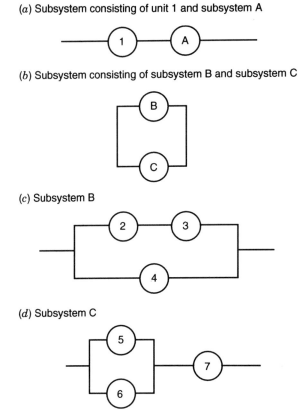

(*a*) Subsystem consisting of unit 1 and subsystem A

(*b*) Subsystem consisting of subsystem B and subsystem C

(*c*) Subsystem B

(*d*) Subsystem C

Figure 3.21. Sequence of the system structure transformation for Example 3.7.

CONCLUSION

In general, the investigation of unrepairable systems—series and parallel—can be reduced to combinatorial problems. It is almost impossible to find a track to the first works in this area. We suspect that if one finds such a work it would be (in terms of the terminology) a work of one of the three Bernoullis—Jacob, Daniel, or Nicholas! Seriously speaking, almost all of the first works and reports on reliability contained such types of analysis. The methods of analysis of unloaded redundancy have the same long history.

Therefore, we restrict ourselves to the following comments. We would only like to mention that some special problems (aging systems, systems with an irreducible structure) will be considered in the following chapters. The reader can find material dedicated to this problem in almost any book on reliability theory or engineering (see the list of general references at the end of this book). We find that for general purposes it is enough to refer to handbooks.

REFERENCES

Barlow R. E.,and F. Proschan (1975). *Statistical Theory of Reliability and Life Testing*. New York: Holt, Rinehart and Winston.

Gnedenko, B. V., Yu. K. Belyaev, and A. D. Solovyev (1969). *Mathematical Methods in Reliability Theory*. San Diego: Academic Press.

Kozlov, B. A., and I. A. Ushakov (1970). *Reliability Handbook*. New York: Holt, Rinehart, and Winston.

Ushakov, I. A., ed. (1985). *Reliability of Technical Systems: Handbook* (in Russian). Moscow: Radio i Sviaz.

Ushakov, I. A., ed. (1994). *Handbook of Reliability Engineering*. New York: Wiley.

EXERCISES

3.1 Prove $(3.22a)$ using the Venn diagram.

3.2 Prove $(3.22b)$, $(3.22c)$, and $(3.22d)$ on the basis of the result of Exercise 3.1. (*Hint:* Use the "double rejection" rule of Boolean algebra: $\bar{\bar{x}} = x$.)

3.3 Prove identities from $(3.23a)$ to $(3.23d)$.

3.4 Write the Boolean function $\varphi(X_i)$ for the scheme depicted in Figure E.3.2.

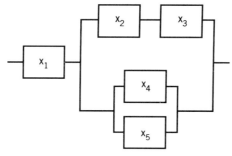

Figure E3.2.

3.5 Write the Boolean function $\varphi(X_i)$ for the scheme depicted in Figure E3.3.

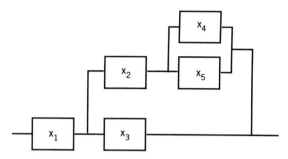

Figure E3.3.

3.6   A system consists of 10 identical and independent units connected in series. The requirement of the probability of a failure-free operation equals 0.99. What reliability level must a system unit have to satisfy the system requirements?

3.7   A system consists of three identical units connected in parallel. The requirement of the probability of a failure-free operation equals 0.999. What reliability level must a system unit have to satisfy the system requirements?

SOLUTIONS

3.1   For given sets X and Y (see the shadowed areas in Figures E3.1a and b, the union is a set of elements belonging to at least one of them (see the shadowed area in Figure E3.1c). Then $\overline{X}$ and $\overline{Y}$ are depicted in Figures E3.1d and e (see the shadowed areas). In Figure E3.1f one finds the area $\overline{\overline{X} \wedge \overline{Y}}$ shadowed. Consequently, the complementary area is $\overline{X} \wedge \overline{Y}$. Obviously, the latter area coincides with the shadowed area in Figure E3.1c. Thus, the desired result is obtained.

3.2   For example, let us prove identity (3.22b)

$$X \wedge Y = \overline{\overline{X} \vee \overline{Y}}$$

Take a rejection operation from both sides of the identity which does not violate it

$$\overline{X \wedge Y} = \overline{\overline{\overline{X} \vee \overline{Y}}} = \overline{X} \vee \overline{Y}$$

Now use a rejection operation to all arguments which also does not

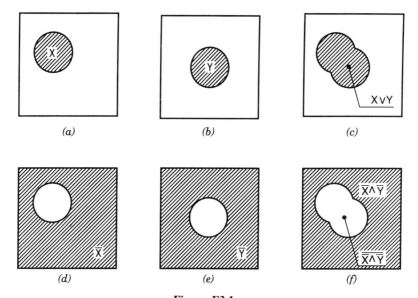

Figure E3.1.

violate the identity

$$\overline{X \wedge \overline{Y}} = \overline{\overline{X}} \vee \overline{\overline{Y}} = X \vee Y$$

Thus, this identity is reduced to the first one, (3.22 a), which was proven in the previous exercise.

3.3 Using (3.22 a), one has for the three arguments

$$X_1 \vee X_2 \vee X_3 = (X_1 \vee X_2) \vee X_3 = \overline{\overline{(X_1 \vee X_2)} \wedge \overline{X_3}}$$

Using the identity

$$\overline{X_1 \vee X_2} = \overline{X_1} \wedge \overline{X_2}$$

one finally obtains

$$X_1 \vee X_2 \vee X_3 = \overline{\overline{X_1} \wedge \overline{X_2} \wedge \overline{X_3}}$$

Now if for $n - 1$ arguments we have

$$\bigcup_{1 \le i \le n-1} X_i = \overline{\bigcap_{1 \le i \le n-1} \overline{X_i}}$$

then, using the previous rule, one finally obtains

$$\bigcup_{1 \le i \le n-1} X_i = \left(\bigcup_{1 < i \le n-1} X_i \right) \vee X_n$$

$$= \overline{\left(\overline{\bigcup_{1 \le i \le n-1} X_i} \right) \wedge \overline{X}_n} = \overline{\left(\overline{\overline{\bigcap_{1 \le i \le n-1} \overline{X}_i}} \right) \wedge \overline{X}_n}$$

$$= \overline{\left(\bigcap_{1 \le i \le n-1} \overline{X}_i \right) \wedge \overline{X}_n} = \bigcap_{1 < i \le n} \overline{X}_i$$

This completes the proof.

3.4 Denote $Y_1 = X_1 \wedge X_2$ and $Y_2 = X_4 \vee X_5$. In this notation

$$\varphi(X_i) = X_1 \wedge [(X_1 \wedge X_2) \vee (X_4 \vee X_5)]$$

In reliability computational practice, one usually uses such expressions without "ORs"; that is, one reduces an initial form in a special way using DeMorgan's rule. In the example under consideration, one has

$$\varphi(X_i) = X_1 \wedge \left[\overline{\overline{(X_1 \wedge X_2)} \wedge \overline{(X_4 \wedge X_5)}} \right]$$

3.5 Denote $X_4 \vee X_5 = Y_1$, $X_2 \wedge Y_1 = Y_2$, $Y_2 \vee X_3 = Y_4$. In this notation $\varphi(X_i) = X_1 \wedge Y_4$ or in open form

$$\varphi(\mathbf{X}) = X_1 \wedge \{X_3 \vee [X_2 \wedge (X_4 \vee X_5)]\}$$

The final expression using only logic AND and rejection operators has the form

$$X_1 \wedge \left\{ \overline{\overline{X}_3 \wedge \left[\overline{X_2 \wedge (\overline{\overline{X}_4 \wedge \overline{X}_5})} \right]} \right\}$$

3.6 A unit has to have $p = \sqrt[10]{0.99} \approx 0.999$.

3.7 Let an unknown probability of failure of a unit be denoted by q. For the system under consideration, one can write

$$1 - 0.999 = q^3$$

Thus, $q = \sqrt[3]{0.001} = 0.1$, that is, $p = 0.9$.

CHAPTER 4

LOAD – STRENGTH
RELIABILITY MODELS

For many reliability engineering applications, one needs to investigate the ability of a structure or a piece of equipment to survive under extreme conditions. For a mechanical construction, one speaks of the probability that it can withstand a specified external load (a shock, vibration, etc.) or internal tension. For electronic equipment, one is concerned with the probability that it is able to withstand a specified voltage jump in its power supply or a significant change in its input signals.

Both an external load and a construction strength might be considered as random. The first is random in a very natural way, as it depends on environmental factors. The second is random because of the inherent instability of any technological process.

4.1 STATIC RELIABILITY PROBLEMS OF "LOAD – STRENGTH" TYPE

4.1.1 General Expressions

Generally, the construction strength X and the applied load Y are random. The problem is to find R, the probability of the system's successful operation, that is, the probability that the applied load does not exceed the actual level of construction strength:

$$R = \Pr\{X \geq Y\} \tag{4.1}$$

167

Let $\Pr\{X \le x\} = F(x)$ and $\Pr\{Y \le x\} = G(x)$. Then the probability of a successful operation of the system can be calculated as

$$R = \int_{-\infty}^{\infty} \Pr(X \ge x)\, dG(x) = \int_{-\infty}^{\infty} \Pr(Y \le x)\, dF(x)$$

$$= \int_{-\infty}^{\infty} [1 - F(x)]\, dG(x) = \int_{-\infty}^{\infty} G(x)\, dF(x) \tag{4.2}$$

If both distributions are continuous, then

$$R = \int_{-\infty}^{\infty} \left[\int_{-\infty}^{y} f(x)\, dx \right] g(y)\, dy = \int_{-\infty}^{\infty} \left[\int_{x}^{\infty} g(x)\, dx \right] f(x)\, dx$$

where $f(x)$ is the density of $F(x)$ and $g(x)$ is the density of $G(x)$.

If X and Y are considered to be independent r.v.'s, it is convenient to introduce a new random variable, $Z = X - Y$, with distribution $H(x)$. Then (4.1) can be rewritten in the form

$$R = \Pr(Z \ge 0) = \int_{0}^{\infty} dH(x)\, dx \tag{4.3}$$

4.1.2 Several Particular Cases

F(x) and G(x) Are Normal   In this case

$$f(x) = \frac{1}{\sigma_f \sqrt{2\pi}} e^{-(x-S)^2/2\sigma_f^2} \tag{4.4}$$

where S and σ_f are, respectively, the mean and the standard deviation of the strength's distribution $F(t)$, and

$$g(x) = \frac{1}{\sigma_g \sqrt{2\pi}} e^{-(x-L)^2/2\sigma_g^2} \tag{4.5}$$

where L and σ_g are, respectively, the mean and the standard deviation of the load's distribution $G(t)$.

Notice that we consider the area of domain of both distributions to range from $-\infty$ to ∞. Of course, one should consider truncated distributions such that their r.v.'s cannot be negative. But, in practice, $S > 3\sigma_f$ and $L > 3\sigma_g$, so that such a truncation does not lead to any crucial numerical errors.

Now introduce the new r.v., $Z = X - Y$. The mean of this new r.v. equals $E\{Z\} = S - L$ and

$$\sigma_h = \sqrt{\sigma_f^2 + \sigma_g^2}$$

which immediately gives the required result

$$R = \Pr(Z \geq 0) = \int_0^\infty \frac{1}{\sigma_h \sqrt{2\pi}} \exp\left[\frac{-(x - E\{Z\})^2}{2\sigma_h^2}\right] dx = \Phi\left(\frac{S - L}{\sqrt{\sigma_f^2 + \sigma_g^2}}\right)$$

(4.6)

Numerical results can be found from a standard table of the normal distribution.

From (4.6) one can see that the reliability of the construction decreases if the variances of X and/or Y increase. Roughly speaking, the more uncertain the conditions of use and the more unstable the quality of the construction, the lower is the reliability of the construction.

Example 4.1 The span of a bridge has a safety coefficient c_s equal to 5. The safety coefficients is determined as $c_s = S/L$. The coefficient of variation of the strength K_s equals 0.05 and that of the land K_l equals 0.2. (a) What is the probability of a successful operation of the construction? (b) What is the probability of a successful operation of the construction if the coefficient of variation of the strength is twice as large?

Solution. (a) Assume that $L = 1$. (By an appropriate normalizing this is always possible.) Then, taking into account the value c_s, we obtain $S = 5$. By definition, the coefficient of variation of the r.v. Z is the ratio $\text{Var}\{Z\}/E\{Z\}^2$. Therefore, $\sigma_g = 0.2$ and $\sigma_f = (0.05)(25) = 1.25$. The probability of a successful operation equals

$$\Phi\left(\frac{5 - 1}{\sqrt{1.25 + 0.2}}\right) = \Phi\left(\frac{4}{1.2}\right) = \Phi(3.33) = 0.999517$$

(b) In this case the value of the variance is 2.50 and the probability of a successful operation equals

$$\Phi\left(\frac{5 - 1}{\sqrt{2.5 + 0.2}}\right) = \Phi\left(\frac{4}{1.6}\right) = \Phi(2.50) = 0.99379$$

F(x) and G(x) Are Exponential   In this case

$$f(x) = \frac{1}{S} e^{-x/S} \tag{4.7}$$

where S is the main strength, and

$$g(x) = \frac{1}{L} e^{-x/L} \tag{4.8}$$

where L is the mean load. Using (4.2), we obtain

$$R = \int_0^\infty \frac{1}{L} e^{-x/L} e^{-x/S} \, dx = \frac{1}{L} \int_0^\infty \exp[-(1/L + 1/S)x] \, dx = \frac{S}{S+L} \tag{4.9}$$

In this case the variances do not influence the resulting probability. Of course, it should be mentioned that exponential distributions in problems such as this are very seldom encountered in practice (especially for a distribution of strength). One can find an example of this in Exercise 4.2.

F(x) Is Normal and G(x) Is Exponential   Let us use the expression

$$R = \int_0^\infty f(x) \left[\int_0^x g(y) \, dy \right] dx$$

Notice that

$$\int_0^x g(y) \, dy = \int_0^x \lambda e^{-\lambda y} \, dy = 1 - e^{-\lambda x}$$

Therefore, we can write

$$R = \int_0^\infty \frac{1}{\sigma_f \sqrt{2\pi}} \exp\left[-\frac{1}{2} \left(\frac{x-S}{\sigma_f} \right)^2 \right] \left[1 - \exp\left(-\frac{x}{L} \right) \right] dx$$

$$= \frac{1}{\sigma_f \sqrt{2\pi}} \int_0^\infty \exp\left[-\frac{1}{2} \left(\frac{x-S}{\sigma_f} \right)^2 \right] dx$$

$$- \frac{1}{\sigma_f \sqrt{2\pi}} \int_0^\infty \exp\left[-\frac{1}{2} \left(\frac{x-S}{\sigma_f} \right)^2 \right] \exp\left(-\frac{x}{L} \right) dx$$

Combine the powers of the exponential functions of the second terms to get the complete square form and the free term:

$$-\frac{1}{2}\left(\frac{x-S}{\sigma_f}\right)^2 - \frac{x}{L} = -\frac{1}{2\sigma_f^2}\left[\left(x-S+\frac{\sigma_f^2}{L}\right)^2 + 2S\frac{\sigma_f^2}{L} - \left(\frac{\sigma_f^2}{L}\right)^2\right]$$

Then

$$R = 1 - \Phi\left(-\frac{S}{\sigma_f}\right) - \frac{1}{\sigma_f\sqrt{2\pi}}$$

$$\times \int_0^\infty \exp\left\{-\frac{1}{2\sigma_f^2}\left[\left(x-S+\frac{\sigma_f^2}{L}\right)^2 + 2S\frac{\sigma_f^2}{L} - 2\frac{\sigma_f^4}{L^2}\right]\right\} dx$$

Change the variables as

$$t = \frac{1}{\sigma_f}\left(x - S + \lambda\sigma_f^2\right) \qquad \text{and} \qquad \sigma_f\, dt = dx$$

Now the final expression becomes

$$R = 1 - \Phi\left(-\frac{S}{\sigma_f}\right) - \frac{1}{\sqrt{2\pi}}\int_{[-S-(\sigma_f^2/L)]/\sigma_f}^\infty \exp\left(-\frac{t^2}{2}\right)\exp\left[-\frac{1}{2}\left(2\frac{S}{L} - \frac{\sigma_f^2}{L^2}\right)\right] dt$$

$$= 1 - \Phi\left(-\frac{S}{\sigma_f}\right) - \exp\left[-\frac{1}{2}\left(2\frac{S}{L} - \frac{\sigma_f^2}{L^2}\right)\right]\left[1 - \Phi\left(-\frac{S - \dfrac{\sigma_f^2}{L}}{\sigma_f}\right)\right] \qquad (4.10)$$

Notice that for most practical problems such as this, the strength S should be located "far" from the point 0. This means that the value $S/L \gg 1$. Incidentally, this corresponds well to the assumption that we do not take into account the truncation of the normal distribution in $t = 0$. In this case, of course,

$$R \approx 1 - \exp\left[-\frac{1}{2}\left(2\frac{S}{L} - \frac{\sigma_f^2}{L^2}\right)\right]\left[1 - \Phi\left(-\frac{S - \dfrac{\sigma_f^2}{L}}{\sigma_f}\right)\right] \qquad (4.11)$$

If, in addition, $\lambda \sigma_f$ is small, say of order 1, then it is possible to write the next approximation

$$R \approx 1 - \exp\left[-\frac{1}{2}\left(2\frac{S}{L} - \frac{\sigma_f^2}{L^2}\right)\right] \qquad (4.12)$$

Therefore, if one takes into account that the mean load equals $L = 1/\lambda$, (4.12) can be rewritten as

$$R \approx 1 - \exp\left(-S\lambda + \frac{1}{2}(\lambda\sigma_f)^2\right) \qquad (4.13)$$

F(x) Is Normal and G(x) Is Biased Exponential   The biased exponential distribution with parameter $\lambda = 1/L$ and bias l^* is presented in Figure 4.1. In this case

$$R = \int_0^\infty \frac{1}{\sigma_f\sqrt{2\pi}}\exp\left[-\frac{1}{2}\left(\frac{x-S}{\sigma_f}\right)^2\right]G^*(x)\,dx$$

where

$$G^*(x) = \begin{cases} 0 & \text{for} \quad x \le l^* \\ 1 - e^{-\lambda(x-l^*)} & \text{for} \quad x > l^* \end{cases} \qquad (4.14)$$

After changing variables, (4.14) becomes

$$R = \int_{l^*}^\infty \frac{1}{\sigma_f\sqrt{2\pi}}\exp\left[-\frac{1}{2}\left(\frac{x-(S-l^*)}{\sigma_f}\right)^2\right]\left[1 - e^{-\lambda x}\right]dx \qquad (4.15)$$

Omitting the transformations which are quite similar to the above, we

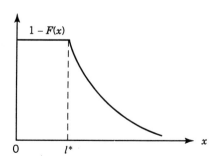

Figure 4.1. Sample of a biased exponential distribution.

present the final result

$$R = 1 - \Phi\left(-\frac{S - l^*}{\sigma_f}\right) - \exp\left[-\frac{1}{2}\left[2(S - l^*)\lambda + \lambda^2\sigma_f^2\right]\right]$$

$$\times\left[1 - \Phi\left(-\frac{(S - l^*) - \lambda\sigma_f^2}{\sigma_f}\right)\right] \tag{4.16}$$

In this case the similar approximate expressions become

$$R \approx 1 - \exp\left[-\frac{1}{2}\left[2(S - l^*)\lambda + \lambda^2\sigma_f^2\right]\right]\left[1 - \Phi\left(-\frac{(S - l^*) - \lambda\sigma_f^2}{\sigma_f}\right)\right] \tag{4.17}$$

for $(S - l^*)/\sigma_f \gg 1$ and

$$R \approx 1 - \exp\left[-\frac{1}{2}\left[2(S - l^*)\lambda + \lambda^2\sigma_f^2\right]\right] \tag{4.18}$$

for small values of $\lambda\sigma_f$.

F(x) Is Biased Exponential and G(x) Is Normal Consider the biased distribution of the strength. We do this because it is unreasonable to consider any construction with a strength equal to 0. By assumption, the strength might not be less than s^*, so in this case

$$R = \int_0^\infty g(x)\left[\int_{x \geq s^*}^\infty f(y)\,dy\right]dx = \int_0^{s^*} g(x)\,dx + \int_{s^*}^\infty g(x)\left[\int_x^\infty f((y)\,dy\right]dx \tag{4.19}$$

Notice that we again use the lower limit of 0 in the integral. As we pointed out above, for numerical calculations the truncation of the normal distribution at $t = 0$ to the left can be neglected.

A simple transformation leads to

$$R = \int_{-\infty}^{s^*} dG(x) + \int_{s^*}^\infty \varphi(x)e^{-(x - s^*)\mu}\,dx$$

$$= \Phi\left(\frac{s^* - L}{\sigma_g}\right) + \int_0^\infty \varphi(x + s^*)e^{-\mu x}\,dx \tag{4.20}$$

or, in detailed form,

$$R = \Phi\left(\frac{s^* - L}{\sigma_g}\right) + \int_0^\infty \frac{1}{\sigma_g\sqrt{2\pi}} \exp\left[-\frac{(x + (s^* - L))^2}{2\sigma_g^2}\right] e^{-\mu x} \, dx \quad (4.21)$$

Avoiding repetition of the transformation which completely coincide with those above, we write the final result directly as

$$R = \Phi\left(\frac{s^* - L}{\sigma_g}\right) + \exp\left[-(s^* - L)\mu + \frac{\mu^2\sigma_g^2}{2}\right]\Phi\left(\frac{s^* - L + \mu\sigma_g^2}{\sigma_g}\right) \quad (4.22)$$

Additional results for some other important particular cases can be found in Kapur and Lamberson(1977). We would like to mention that this reference contains useful formulas for the Weibull–Gnedenko distribution which is important for description of the strength of mechanical construction.

4.1.3 Numerical Method

In general, it is reasonable to use an approximate numerical method. This method is good for calculations using histograms as well as standard statistical tables. In the first case, the approximation is defined by restricted statistical data and their inevitably discrete nature. In the second case, the approximate nature of the solution is explained by a discrete representation of continuous distributions. Because of the approximate nature of these calculations, it is sometimes reasonable to consider the upper and lower bounds of the calculated values.

First, assume that a set of statistical input data is given. The set of observed values of the material strength is $X_1, \ldots, X_n$ and the set of observed values of the load is $Y_1, \ldots, Y_m$. Arrange the ordered set $W_1 \leq \cdots \leq W_{n+m}$ where each W_s is one of the X_i's or one of the Y_j's.

For each $W_s = X_i$ calculate the number of W_r's where $W_r = Y_j$ and $r < s$. Denote this value by k_s. This value means that, on the average, in k_s cases of m possible observations of the r.v. Y, the load will be smaller than the given strength X. In other words, we might say that with conditional probability k_s/m, the investigated system with fixed strength X_i will operate successfully if the load will take on one of the possible values of Y. Thus, the complete probability of success is

$$R = \frac{1}{n} \sum_{1 \leq s \leq n} k_s$$

Obviously, the same numerical result can be obtained if we consider $W_s = Y_j$ and calculate the number of W_r's where $W_r = X_i$ for each $W_s = Y_j$, $r < s$. Denote this value by k_s^*. This value means that in k_s^* cases of m

possible observations of the r.v. X, the strength will be smaller than the specified load Y. It means that with conditional probability k_s^*/m, the investigated system will fail under the load Y; that is, the complete probability of success is

$$R = 1 - \frac{1}{m} \sum_{1 \le s \le m} k_s^*$$

Example 4.2 The following data are available: $X_1 = 98.1$, $X_2 = 98.2$, $X_3 = 99.4$, $X_4 = 100.3$, $X_5 = 101.2$, $X_6 = 103.5$, $X_7 = 103.9$, $X_8 = 104.1, \ldots$, $X_{16} = 110.2$; $Y_1 = 79.1$, $Y_2 = 82.4, \ldots, Y_{18} = 98.0$, $Y_{19} = 98.3$, $Y_{20} = 98.5$, $Y_{21} = 99.5$. Calculate the probability that the construction will operate successfully.

Solution. We find from the data that $k_1 = 18/21$, $k_2 = 18/21$, $k_3 = 20/21$, $k_4 = \cdots = k_{16} = 1$. Thus, the result taken by the first expression is

$$R = \frac{1}{16} \cdot \left[\frac{18}{21} + \frac{18}{21} + \frac{20}{21} + 13 \right] = \frac{329}{336} \approx 0.9792$$

From the same data, one finds that $k_1^* = \cdots = k_{18}^* = 0$, $k_{19}^* = 2/16$, $k_{20}^* = 2/16$, $k_{21}^* = 3/16$. Using the second expression, one has the following result:

$$R = 1 - \frac{1}{21} \cdot \left[\frac{2}{16} + \frac{2}{16} + \frac{3}{16} \right] = 1 - \frac{7}{336} \approx 0.9792$$

If tables of the distributions $F(x)$ and $G(x)$ are available, the numerical calculation of the index R can be performed using the following formulas:

$$R = \sum_{1 \le m \le M} \left[1 - F\left(\left(m + \frac{1}{2} \right) \Delta \right) \right] \left[G((m + 1)\Delta) - G(m\Delta) \right] \quad (4.23)$$

or

$$R = \sum_{1 \le m \le M} G\left(\left(m + \frac{1}{2} \right) \Delta \right) \left[F((m + 1)\Delta) - F(m\Delta) \right] \quad (4.24)$$

where Δ is the chosen increment and M is the number of increments.

It is clear that the summation can only be performed in the area of the distribution's domain where the corresponding values of the product terms are significant. For practical purposes, the increments may be chosen to have a value ranging from 0.5 to 0.05 of the smallest standard deviations of the distributions $F(x)$ and $G(x)$. Obviously, the more accurate the result that is needed, the smaller the increments must be. For practical calculations, the left bound m of the summation must begin with the value $k = \{m: F(-m\Delta) < \varepsilon\}$ where ε is chosen in correspondence with the needed accuracy.

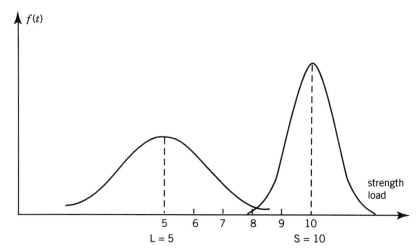

Figure 4.2. Explanation of the solution of Example 4.3.

Example 4.3 The strength has a normal distribution with $S = 10$ and $\sigma_f = 1$, and the load also has a normal distribution with $L = 5$ and $\sigma_g = 2$ (all values are measured in some conditional scales). Calculate the probability of success R using a standard table of the normal distribution.

Solution. We present Figure 4.2 to illustrate the solution. This figure helps us to see that, for example, the point 7 corresponds to $L + \sigma_g$ and, at the same time, corresponds to $S - 3\sigma_f$, and the point 9 corresponds to $L + 2\sigma_g$ and to $S - \sigma_f$, and so on. Use a standard table of the normal distribution and arrange (only for illustrative purposes) the new Table 4.1 with the input data for numerical calculation. Thus, the probability of failure equals 0.98385. A calculation with the use of the strong formula gives

$$\Phi\left(\frac{10 - 5}{\sqrt{1^2 + 2^2}}\right) = \Phi\left(\sqrt{5}\right) = 0.9878$$

TABLE 4.1

Interval $[k, k + 1]$	Value of $G(k + 1) - G(k)$	Argument $k + 1/2$	Value of $F(k + 1/2)$	Intermediate Product
$[7, 8]$	0.0920	7.5	0.00621	0.00057
$[8, 9]$	0.0441	8.5	0.0668	0.00295
$[9, 10]$	0.0164	9.5	0.692	0.01114
$[10, 11]$	0.0049	10.5	0.308	0.00149
				0.01615

The relatively large error is explained by the use of excessively large increments.

REMARK. The unreasonably high level of accuracy of the calculations is presented only to compare the obtained solution with the exact solution. Once more we would like to emphasize that for practical purposes the use of "too accurate" a solution can be considered almost incorrect because of the very rough statistical data which we usually have in practice.

Sometimes it might be more useful to obtain lower and upper bounds on the value R because this allows one to evaluate the accuracy of the result.

Lower bounds can be written as

$$R = \sum_{1 \le m \le M} [1 - F(m + 1)\Delta][G((m + 1)\Delta) - G(m\Delta)] \quad (4.25)$$

$$R = \sum_{1 \le m \le M} G(m)\Delta[F((m + 1)\Delta) - F(m\Delta)] \quad (4.26)$$

and upper bounds as

$$R = \sum_{1 \le m \le M} [1 - F(m)\Delta][G((m + 1)\Delta) - G(m\Delta)] \quad (4.27)$$

or

$$R = \sum_{1 \le m \le M} G((m + 1)\Delta)[F((m + 1)\Delta) - F(m\Delta)] \quad (4.28)$$

Example 4.4   Suppose the construction has a truncated exponential distribution of the strength with parameters $\mu = 1/S = 0.5$ and $s^* = 10$, and a normal distribution of the load with parameters $L = 6$ and $\sigma_g = 2$. Find the probability of success R for this construction.

Solution.   Find upper and lower bounds on the probability R. For the purpose of numerical calculation, construct a special table (see Table 4.2) based on standard tables of the normal and exponential distributions. Table 4.2 contains the meaning of the corresponding distribution $G(x)$ and the

TABLE 4.2

m	x	z_1	$G(m)$	z_2	$F(m)$	$\Delta(m)$
1	10.0	2.00	0.9773	0.00	0.000	0.221
2	10.5	2.25	0.9878	0.25	0.221	0.173
3	11.0	2.50	0.9938	0.50	0.394	0.134
5	11.5	2.75	0.9970	0.75	0.528	0.104
7	12.0	3.00	0.9987	1.00	0.632	—

increments of $F(x)$ in the area of interest. The calculation of lower and upper bounds is performed using formulas (4.26) and (4.28).

In Table 4.2, m is the number of the term in the sum, x is the absolute value, z_1 is the argument of the standard normal distribution, and z_2 is the argument of the standard exponential function $\Delta^*(m) = \Delta(m + 1) - \Delta(m)$.

Using (4.26), we obtain

$$R = (0.9773)(0.221) + (0.9878)(0.173)$$
$$+ (0.9938)(0.134) + (0.9970)(0.104) + \tau = 0.9917$$

where τ is the probability of the "tail" of the strength's distribution with an insignificant influence of the load (all this area must be considered as the area of the "practically absolute" reliability). Using (4.28), we obtain

$$\tilde{R} = (0.09878)(0.221) + (0.9938)(0.173)$$
$$+ (0.9970)(0.134) + (0.9987)(0.104) + \tau = 0.9956$$

The difference between the two values is significant—about 100%. (Notice that if the probabilities are close to 1, one should consider the complementary probabilities, i.e., 0.0083 and 0.0044 in the investigated case.) This means that the values of Δ are chosen too large.

4.2 MODELS OF CYCLE LOADING

The static models of the "strength–load" type which we considered in the previous section may be referred to as one-cycle loading models. Moreover, this single cycle is assumed to be short enough (but not a shock!), so the strength of the material is assumed to be constant in time. In other words, there is no time for any deterioration or fatigue effects to appear. For practical tasks, such a consideration is important even if the cycles are considered to be independent and identical. Anyway, this more accurately reflects the physical process than a totally static situation. A consideration of the cycle loading is supported by the results of the previous section: the probability determined there is considered as a characteristic of the ability of the chosen construction to withstand a specified fixed load during one cycle.

In real life, the strength of a mechanical construct might monotonically change in time, due to deterioration, fatigue and aging processes, environmental influences and so on. (For electronic equipment, the "strength" can fluctuate: the actual tolerance limits can change in time depending on the temperature, humidity, and other environmental influences. Below we often refer to mechanical systems.) The load can also change in time for various obvious reasons. We consider only a simple case: a sequence of shock-type (practically instantaneous) loading. Notice that an investigation of continuous loading with a simultaneous changing of the load and strength is a very sophisticated physical problem which an only be solved for some particular cases.

We will discuss a very particular case of cycle loading when the strength X and the load Y are independent random variables with known distributions $F(x)$ and $G(x)$, respectively. For compactness, the mean value of the strength $E\{X\}$ will be denoted by S and the mean value of the load $E\{Y\}$ by L. The strength is assumed to be fixed (known or unknown) or monotonically changing and the load can be represented by a sequence of independent r.v.'s from cycle to cycle.

4.2.1 Fixed Level of Strength

Known Fixed Level Suppose that the level of strength is known and equals some value s^0, while the load is random with distribution function $G(x)$. The values of the load at each cycle are mutually independent r.v.'s. Denote the random number of failure-free cycles by ν.

The probability that exactly k cycles will be successful equals $\Pr\{\nu = k|s^0\} = p^k q$; that is, the r.v. ν has a geometrical d.f. with $p = G(s^0)$ and $q = 1 - p$. The probability of success during K or more cycles equals

$$\Pr\{\nu \geq K|s_0\} = p^K \qquad \text{where} \quad p = \Pr\{Y \leq s^0\} = G(s^0)$$

The mean number of cycles before failure equals $E\{\nu\} = 1/q$. If $q \ll 1$, an approximation in exponential form can be written as

$$\Pr\{\nu \geq K|s^0\} \approx e^{-qK}$$

Unknown Fixed Level Now assume that s^0 is unknown but constant during the total period of the system's operation. The only thing we know is the prior distribution $F(x)$. In this case

$$\Pr\{\nu \leq K\} = \int_C [G(x)]^K \, dF(x)$$

where C is the domain of the distribution $F(x)$.

Let the probability $q(x) = 1 - G(x)$ be small "on average." In practice, this corresponds to the condition

$$\frac{S - L}{\sqrt{\sigma_S^2 + \sigma_L^2}} \gg 1$$

where σ_S is the standard deviation of the d.f. $F(x)$. We can conclude that the right "tail" of the distribution $G(x)$ is concave in the essential area of the domain of the distribution $F(x)$. Then the following simple bound is true:

$$\Pr\{\nu \ge K\} \le [G(S)]^K$$

The mean number of cycles is

$$E\{\nu\} = \int_C [1 - G(x)]^{-1} dF(x)$$

The corresponding approximation for small q is

$$E\{\nu\} \le [1 - G(S)]^{-1}$$

4.2.2 Deteriorating Strength

If the expected number of successful cycles is very large, that is, the operational time is sufficiently large, we might assume that the level of the system's strength decreases in time. Indeed, most materials deteriorate with time and, consequently, the system strength becomes weaker and weaker. We will consider several simple models.

1. Assume that the material strength decreases from cycle to cycle in such a way that $p_{k+1} = \alpha p_k$ where p_k is the probability of success at the kth cycle and α is constant, $0 < \alpha < 1$. Then

$$\Pr\{\nu \ge K|s^0, \alpha\} = p(p\alpha)(p\alpha^2) \cdots (p\alpha^{K-1})$$
$$= p^K \prod_{0 \le k \le K-1} \alpha^k = p^K \alpha^{(K-1)^2/2}$$

and the mean number of successful cycles is

$$E\{\nu\} = \sum_{0 \le k < \infty} p^k \alpha^{(k-1)^2/2}$$

2. Again consider the case where the level of the strength is known and the deterioration is described by an exponential decrease of this level: at the kth cycle, the level of the strength is $x_k = s^0 \alpha^k$ where $0 < \alpha < 1$. The probability of success over at least K cycles equals

$$\Pr\{\nu \ge K|s^0, \alpha\} = \prod_{1 \le k \le K} G(s^0 \alpha^k)$$

Note that $G(s^0 \alpha^k) > G(s^0 \alpha^{k+1})$ and for the most commonly used distributions, this discrete function is concave. Then

$$\Pr\{\nu \ge K|s_0, \alpha\} \le G(s^0 \alpha^{(K/2)})$$

The mean number of cycles until the system fails equals

$$E\{\nu\} = 1p_1 + 2p_1p_2 + 3p_1p_2p_3 + \cdots$$
$$= p_1(1 + p_2(1 + p_3(1 + \cdots)))$$

3. For a known distribution $F(x)$ of the initial value of the strength, the probability of success equals

$$\Pr\{\nu \geq K|F(x), \alpha\} = \int_C \prod_{1 \leq k \leq K} G(\alpha^k) \, dF(x)$$

We do not have a simple approximation for this case.

4.3 DYNAMIC MODELS OF "STRENGTH – LOAD" TYPE

In the previous sections we considered a simple version of the dynamic loading process, that is, the cycling process. That scheme is sufficiently good to describe some specific mechanical systems. But for most electronic systems the process of "loading" should be described as a continuous stochastic process. Indeed, in this case one considers a process of randomly changing the system parameters inside the tolerance zone. We will consider only a simple case where one-dimensional stochastic process crosses a specified level.

4.3.1 General Case Consider a differentiable stochastic process $x(t)$. We are interested in the distribution of intervals between neighboring intersections of a specified level a by the process. At first, we find the probability that the process will intersect the level a at moment t. This event happens if the two following events have occurred:

$$\{x(t) < a\} \quad \text{and} \quad \{x(t + dt) > a\}$$

In other words, the probability of the event equals

$$\Pr\{(x(t) < a), (x(t + dt) > a)\} \tag{4.29}$$

Let $v(t)$ be the speed of the process, that is, $v(t) = dx(t)/dt$. Now we can rewrite (4.29) in the new form

$$\Pr\{a - v(t) \, dt < x(t) < a\} \tag{4.30}$$

To find this probability, we need to know the density function of the joint distribution $f(x, v|t)$ of the ordinate x of the process $x(t)$ and its derivative for the same moment of time t. Using these terms, we can write

$$\Pr\{a - v(t) \, dt < x(t) < a\} = \int_0^\infty \int_{a-v\,dt}^a f(x, v|t) \, dx \, dt \tag{4.31}$$

The internal integral can be computed instantly because of its special limits

$$\int_{a-v\,dt}^{a} f(x, v|t)\, dx = dt\, v f(a, v|t)$$

Substitution of (4.32) into (4.31) gives us

$$\Pr\{a - v(t)\, dt < x(t) a\} = dt \int_{0}^{\infty} f(a, v|t) v\, dv \tag{4.33}$$

This formula shows that the probability of the intersection of the specified level by the stochastic process during the infinitesimally small time interval dt is proportional to the length of the interval. This allows one to introduce the time density for this probability $p(a|t)$. Using (4.33) gives

$$\Pr\{a - v(t)\, dt < x(t) < a\} = p(a|t)\, dt \tag{4.34}$$

and, consequently,

$$p(a|t) = \int_{0}^{\infty} f(a, v|t) v\, dv \tag{4.35}$$

Analogously, one can find the derivative of the probability $p(a|t)$:

$$\frac{d}{dt} p(a|t) = -\int_{t} f(a, v|t) v\, dv \tag{4.36}$$

Adding and subtracting (4.35) and (4.36), one can easily obtain the two following equations:

$$p(a|t) + \frac{d}{dt} p(a|t) = \int_{-\infty}^{\infty} f(a, v|t) |v|\, dv \tag{4.37}$$

and

$$p(a|t) - \frac{d}{dt} p(a|t) = \int_{-\infty}^{\infty} f(a, v|t) v\, dv \tag{4.38}$$

It is clear that

$$f(a, v|t) = f(v|a, t) f(a|t)$$

Then one can rewrite (4.37) and (4.38)

$$p(a|t) + \frac{d}{dt} p(a|t) = f(a|t)\, E\{|V(T)| \,|\, X(t) = a\} \tag{4.39}$$

$$p(a|t) - \frac{d}{dt} p(a|t) = f(a|t)\, E\{V(T) \,|\, X(t) = a\} \tag{4.40}$$

Using (4.35) for any time interval T, one can obtain the mean time of $x(t)$ being over the specified level a. To obtain the result, we use the following simple arguments. Let us divide the total period T into n small nonoverlapping intervals δ_j located around the points t_j, $1 \leq j \leq n$. For some t_j, we can write

$$\Pr\{X(t_j) > a\} = \int_a^\infty f(x|t_j)\, dx \qquad (4.41)$$

Assume that the intervals δ_j's are chosen so small that changing signs by the function $x(t) - a$ can be neglected. Next, introduce the indicator function $\{dt_j\}_+$ such that

$$\Delta_j = \begin{cases} \delta_j & \text{if} \quad x(t) - a > 0 \\ 0 & \text{otherwise} \end{cases} \qquad (4.42)$$

Using this notation, one can write that the total time for which the function $x(t)$ exceeds the level a equals

$$T_a = \sum_{1 \leq j \leq n} \Delta_j \qquad (4.43)$$

and the mean time when the function $x(t)$ exceeds the level a equals

$$E\{T_a\} = \sum_{1 \leq j \leq n} E\{\Delta_j\} \qquad (4.44)$$

At the same time,

$$E\{\Delta_j\} = \delta_j \int_a^\infty f(x|t_j)\, dx \qquad (4.45)$$

Using (4.45) and taking the limit in (4.44), we obtain the expression for the total time for which the process $x(t)$ exceeds the level a:

$$E\{T_a\} = \int_0^T \int_a^\infty f(x|t)\, dx\, dt \qquad (4.46)$$

If one is interested in the average number of intersections n_a during a time interval T, the same simple arguments can be used. Now introduce another indicator function

$$N_j = \begin{cases} 1 & \text{if} \quad x(t) - a > 0 \text{ in the interval } \delta_j \text{ at least once} \\ 0 & \text{otherwise} \end{cases}$$

The total number of intersections during period T equals

$$N_a = \sum_{1 \le j \le n} N_j$$

Again the mean value can be expressed as

$$E(N_a) = \sum_{1 \le j \le n} E\{N_j\} \tag{4.47}$$

where

$$E\{N_j\} = p(a|t_j)\delta_j \tag{4.48}$$

Taking the limit of (4.48) with the substitution of (4.35), we obtain

$$E\{N_a\} = \int_0^T \int_0^\infty vf(a, v|t)\, dv\, dt \tag{4.49}$$

In addition to (4.46), the last expression permits us to write the expression for the mean time t_a for which the process $x(t)$ exceeds the level a during a single intersection. Indeed,

$$t_a = \frac{E\{T_a\}}{E\{N_a\}} \tag{4.50}$$

or, using the corresponding complete expressions,

$$t_a = \frac{\int_0^T \int_a^\infty f(x|t)\, dt}{\int_0^T \int_0^\infty vf(a, v|t)\, dv\, dt} \tag{4.51}$$

All of these results are essentially useful for stationary processes because in this case all of the functions do not depend on the current time, that is, $f(x|t) = f(x)$ and $f(x, v|t) = f(x, v)$. Then all of the previous results can be rewritten in the simpler form, namely,

$$E\{T_a\} = T \int_a^\infty f(x)\, dx \tag{4.52}$$

$$E\{N_a\} = T \int_0^\infty v f(a,v)\, dv \tag{4.53}$$

$$t_a = \frac{\int_a^\infty f(x)\, dx}{T \int_0^\infty v f(a,v)\, dv} \tag{4.54}$$

Naturally, for the stationary process the values of $E\{T_a\}$ and $E\{N_a\}$ depend only on the length of the period T. More precisely, they are proportional to T. The mean time $E\{t_a\}$ for which the process exceeds the level a does not depend on T. For a stationary process one can also introduce the mean number of itersections per unit of time λ_a:

$$\lambda_a = \int_0^\infty v f(a,v)\, dv \tag{4.55}$$

that is, the probability of a level crossing in a unit of time.

4.3.2   Gaussian Stochastic Process

To calculate all of the above-mentioned parameters of the specified level intersection, one needs to know the characteristics of the stochastic processes $f(x|t)$ and $f(x,v|t)$. For stationary processes, one needs to know $f(x)$ and $f(x,v)$. Fortunately, for the most important practical case—the *Gaussian stochastic process* (GSP)—sufficiently simple formulas can be obtained.

Note that the Gaussian process is often taken as the mathematical model of the random change of electrical parameters over time. There are many physical reasons to use this model because the influence of the number of internal and external factors leads to the formation of conditions for the validity of such a model. Indeed, these various factors might often be considered as relatively independent, and the influence of each of them on the resulting process is relatively small. Of course, the correctness of these hypotheses should be checked or verified each time.

We consider only a stationary process for which we know the mean $E\{X\}$ and the variance σ_X^2. For a normal process in a stationary regime, the ordinate distribution is

$$f(x) = \frac{1}{\sigma_x \sqrt{2\pi}} \exp\left[-\frac{(x - \bar{x})^2}{2\sigma_x^2} \right]$$

where $\sigma_X^2 = K_x(0)$ and $K_x(\tau)$ is itself the correlation function.

It is known from the theory of stochastic processes that the ordinate of the GSP and its derivative for the same moment of time are noncorrelated. Thus, the joint density function can be presented as the product of the two separated densities

$$f(x,v) = f(x)f(v) \tag{4.56}$$

or

$$f(x,v) = \frac{1}{\sigma_x\sqrt{2\pi}} \exp\left[-\frac{(x-E\{X\})^2}{2\sigma_x^2}\right] \frac{1}{\sigma_v\sqrt{2\pi}} \exp\left(\frac{-v^2}{2\sigma_v^2}\right) \tag{4.57}$$

Note that the variance σ_V^2 can be expressed through the correlation function of the process as

$$\sigma_V^2 = -\left.\frac{d_2}{d\tau^2}K_x(\tau)\right|_{\tau=0} \tag{4.58}$$

and $v(t)$ equals 0 because the stationary process is considered.

An expression for λ_a can be obtained from (4.55) after the substitution of (4.57)

$$\lambda_a = p(a) = \frac{\sigma_V}{2\pi\sigma_x} \exp\left[-\frac{(a-E\{X\})^2}{2\sigma_x^2}\right] \tag{4.59}$$

The expression for $E\{t_a\}$ can be obtained in an analogous way:

$$\tau = \pi\frac{\sigma_x}{\sigma_v} \exp\left[\frac{(a-x)^2}{2\pi\sigma_x^2}\right]\left[1 - \Phi\left(\frac{a-x}{\sigma_x}\right)\right] \tag{4.60}$$

where $\Phi(x)$ is the normal distribution function.

4.3.3 Poisson Approximation

The crossing of a "high level" threshold by a stochastic process is of great interest for reliability analysis. It is clear that the probability of the crossing in this case should be sufficiently small; that is, such intersections are "rare events." As we mentioned above, the sequence of rare events forms a Poisson stochastic process. We omit the proof of the fact that in this particular case this hypothesis is also valid. Here we accept this as a known fact.

In general, we may assume that the mean number of intersections $E\{N_a\}$ of level a for a specified period T approximately equals the mean number of events λT for some Poisson process with parameter λ. Thus, this parameter can be easily expressed as

$$\lambda = \frac{\mathrm{E}\{N_a\}}{T} \tag{4.61}$$

where $\mathrm{E}\{N_a\}$ is determined by (4.51) or (4.55), depending on the type of stochastic process under consideration.

We will not write the expressions for the Poisson probabilities. The reader can do it easily him/herself. We only write the expressions for the probability of a failure-free operation (i.e., no intersection during the time T) for the nonstationary and stationary cases using the corresponding values of $\mathrm{E}\{N_a\}$. For the nonstationary process one has

$$P_0 = \exp\left[-\int_0^T \int_0^\infty vf(a, v|t)\, dt \right] \tag{4.62}$$

and for the stationary process one has

$$P_0 = \exp\left[-T\int_0^\infty vf(a, v)\, dt \right] \tag{4.63}$$

For Gaussian processes, the probability $P_0(T)$ can easily be written with the use of λ_a from (4.59).

It is difficult to estimate the error obtained via the use of such an approximation for the Gaussian process. The only simple physical explanation lies in he fact that there is practically no correlation between neighboring moments of intersections of the specified "high level." (To check this fact, one should take into consideration the mean time between two neighboring intersections: "too much water has passed under the bridge" after the previous intersection!)

Example 4.5 Consider equipment characterized by a two-dimensional parameter with components X and Y. Both components X and Y are fluctuating in time. Their fluctuations are described as the identical independent stationary Gaussian processes with means equal to 0 and correlation functions

$$K_x(\tau) = K_y(\tau) = \sigma^2 e^{-\alpha|\tau|}\left(\cos \beta|\tau| + \frac{\alpha}{\beta} \sin \beta|\tau|\right) \tag{4.64}$$

The tolerance limit area of the equipment parameter is represented by a sphere with radius a. Find the mean time that the system's parameter is spending inside the tolerance limit area if, at the moment $t = 0$, both $X(t)$ and $Y(t)$ are in the center of the tolerance sphere.

Solution.   Let

$$R(t) = \sqrt{X^2(t) + Y^2(t)} \tag{4.65}$$

and

$$v_x = \frac{dR(t)}{dt} \tag{4.66}$$

Using (4.56), we can write

$$\tau = \frac{\int_a^\infty f(r)\, dr}{\int_0^\infty v_r f(a, v_r)\, dv_r} \tag{4.67}$$

where $f(a, v_r)$ is the joint density of the distribution of the two r.v.'s R and v_r for $R = a$.

Consider an arbitrary period of time T. There are, on average, $\lambda_a T$ intersections, and the vector parameter represented by the point (X, Y) is outside the specified tolerance zone during the mean time $E\{t_a\}\lambda_a T$. Consequently, the system parameter will be inside the tolerance zone, on average, during the time $T[1 - E\{t_a\}\lambda_a]$. The mean time that the parameter spends inside the tolerance zone is

$$T' = \frac{T[1 - E\{t_a\}\lambda_a]}{T\lambda_a} = \frac{1}{\lambda_a} - E\{t_a\} \tag{4.68}$$

Using (4.56), we can write

$$T' = \frac{\int_0^a f(r)\, dr}{\int_0^\infty v_r f(a, v_r)\, dv_r}$$

Now we find the corresponding densities and compute the final result in a compact and constructive form. At first, notice that the r.v.'s X and Y are independent and have normal distributions, so

$$f(x, y) = \frac{1}{2\pi\sigma^2} \exp\left[-\frac{1}{2\sigma^2}(x^2 + y^2)\right] \tag{4.69}$$

Further,

$$f(r) \, dr = \Pr\{r < R < r + dr\} = \iint_{r < \sqrt{x^2+y^2} < r+dr} f(x,y) \, dx \, dy$$

$$= \frac{1}{2\pi\sigma^2} \int_0^{2\pi} \int_r^{r+dr} e^{-r^2/2\sigma^2} r \, dr \, d\varphi = \frac{r}{\sigma^2} e^{-r^2/2\sigma^2} \, dr \qquad (4.70)$$

that is, $f(r)$ is a Rayleigh density.

To determine the density $f(r, v_r)$, we need to consider a system of four normally distributed r.v.'s: X, Y, $v_x = dX(t)/dt$, and $v_y = dY(t)/dt$. For a Gaussian process, all of these r.v.'s are independent. The variances of v_x and v_y are identical and equal

$$-\frac{d^2 K_x(\tau)}{d\tau^2}\bigg|_{\tau=0} = \sigma^2(\alpha^2 + \beta^2) \qquad (4.71)$$

Thus, v_x and V_y do not depend on the coordinates of (X, Y) and have a circle normal distribution; that is, the projection of the vector (v_x, v_y) on the direction of R has a normal distribution with variance (4.71). Thus, the two-dimensional density $f(r, v_r)$ can be expressed in the following way:

$$f(r, v_r) = f(r)f(v_r) = \frac{r}{\sigma^2} \exp\left(-\frac{r^2}{2\sigma^2}\right) \frac{1}{\sqrt{2\pi\sigma^2(\alpha^2 + \beta^2)}}$$

$$\times \exp\left[-\frac{v_r^2}{2\sigma^2(\alpha^2 + \beta^2)}\right]$$

After substitution of (4.70) and (4.72) into (4.68), we obtain the final result

$$T' = \frac{\sigma}{a} \sqrt{\frac{2\pi}{\alpha^2 + \beta^2}} \left(e^{a^2/2\sigma^2} - 1\right)$$

This example shows that the use of stochastic process theory to find reliability indexes is not a simple task. But difficult practical problems always need the use of more or less complicated mathematical tools. Note also that besides the technical complexity of the solution there are also some special needs concerning the input data. Such data are not always available.

CONCLUSION

We presented only a very brief description of the problem which could be explained by our intention to consider, primarily, system reliability. The problem related to the degradation of mechanical constructions under the random load and fluctuation of their physical parameters and the strength are special branches of modern reliability theory. Each of these branches must occupy a separate book.

The reader can find some appropriate formulations of the problem of reliability of mechanical systems and many useful related results, for example, in Kapur and Lamberson (1977). We would also like to mention Bolotin (1975). In Becker and Jensen (1977) one finds an analysis of a similar mathematical problem related to the reliability of electrical equipment under a stochastic fluctuation of parameters. Mechanical problems in reliability engineering are considered in Konyonkov and Ushakov (1975). Some results concerning the reliability of mechanical systems are contained in Ushakov (1985, 1994).

Interesting results concerning accumulations of random shocks can be found in Barlow and Proschan (1975). Elegant mathematical results can be obtained with the use of the Kolmogorov equations if the process of the parameter fluctuation can be described as a Markov process.

One can find a lot of interesting results in the extensive literature on noise analysis in radio equipment. This powerful branch of applications was stimulated by the pioneering work of Rice (1944, 1945).

At last, we would like to mention that this problem must be considered on a serious physical level. This chosen mathematical model must correspond to a real object, either electronic equipment or a mechanical construct. Writing a set of abstract models covering this subject area seems to be a hopeless task. Besides, it is not a simple task to find the appropriate statistical data for the models dealing with the random behavior of real parameters.

REFERENCES

Barlow, R. E., and F. Proschan (1975). *Statistical Theory of Reliability and Life Testing*. New York: Holt, Rinehart, and Winston.

Becker, P. W., and F. Jensen (1977). *Design of Systems and Circuits for Maximum Reliability and Maximum Production Yield*. New York: McGraw-Hill.

Bolotin, V. V. (1975). *Application of Methods of Probability Theory and Reliability Theory for Construction Design* (in Russian). Moscow: Stroiizdat.

Gertsbakh, I. B., and Kh. B. Kordonsky (1969). *Models of Failure*. Berlin: Springer.

Kapur, K. C., and L. R. Lamberson (1977). *Reliability in Engineering Design*, New York: Wiley.

Konyonkov, Yu. K., and I. A. Ushakov (1975). *Aspects of Electronic Equipment Reliability Under Mechanical Stress* (in Russian). Moscow: Sovietsko Radio.

Rice, S. O. (1944, 1945). Mathematical analysis of random noise. *Bell Syst. Tech. J.* vol. 23, no. 3, and vol. 24, no. 1.

Sveshnikov, A. A. (1968). *Applied Methods of the Stochastic Function Theory*. Moscow: Nauka.

Ushakov, I. A., ed. (1985). *Reliability of Technical Systems: Handbook* (in Russian). Moscow: Radio i Sviaz.

Ushakov, I. A. (1994). *Handbook of Reliability Engineering*. New York: Wiley.

Vinogradov, O. G. (1991). *Introduction to Mechanical Reliability: A Designer's Approach*. New York: Hemisphere.

EXERCISES

4.1 The distributions of both a strength and a load are normal. The mean of the load is known: $L = 10$ conditional units and the standard deviation $\sigma_g = 2$. Find the parameters of the distribution of the strength S and σ_f which deliver a probability of failure-free operation equal to $R = 0.995$.

4.2 The distributions of both the strength and the load are exponential with parameters $1/S$ and $1/L$, respectively. $L = 1$ conditional unit. Find S which delivers a probability of failure-free operation equal to $R = 0.995$.

4.3 The strength's distribution is normal with unknown parameters S and σ_f and a known coefficient of variation $k = 0.04$. The distribution of the load is exponential with $L = 1$ conditional unit. Find the parameter S which delivers $R = 0.999$.

SOLUTIONS

4.1 First of all, notice that the problem as formulated here is incorrect: one should know in advance the mean of the strength a or its standard deviation σ_g or the coefficient of variation $k = \sigma_g^1/S^2$. Without this correction the problem has no unique answer.

   Let us assume that one knows $\sqrt{k} = 0.04$. The problem can be solved by sequential iterations. For choosing a first value of S, notice that because of the requirement, $R = 0.995$, there must be at least more than $L + 2.5\sigma_g$. Choose $S^{(1)} = L + 3\sigma_g = 16$. Then

$$\sigma_f^1 = \sqrt{k\left(S^{(1)}\right)^2} = \sqrt{(0.04)(256)} \approx 3.2$$

Now it is clear that this level of strength is unacceptable. Choose the next value, for instance,

$$S^{(2)} = L + 3\sigma_g + 3\sigma_f^{(1)} \approx 26$$

This value of $S^{(2)}$ leads to

$$\sigma_f^{(1)} = \sqrt{k\left(S^{(2)}\right)^2} = \sqrt{(0.04)(676)} \approx 5.2$$

Check the above obtained result

$$P = \Phi\left(\frac{26 - 10}{2.0 + 5.2}\right) = \Phi(2.22) \approx 0.987$$

Thus, the value S is still smaller than one needs to deliver $R = 0.995$. The procedure continues. (We leave it to the reader to obtain the final numerical result.)

4.2   From (4.9) one can write

$$S = \frac{LR}{1 - R} = \frac{1 \cdot (0.995)}{0.005} = 199 \text{ conditional units}$$

This coefficient of safety is too large. The assumption that both strength and load distributions are exponential is unrealistic in practice. At the least, this is quite unreasonable as a distribution of strength.

4.3   For a highly reliable construction, one can use (4.12) or (4.13). This gives

$$R \approx 1 - \exp\left[-\frac{1}{2}\left(\frac{2S}{L} - \frac{0.04S^2}{L^2}\right)\right] = 0.999$$

or

$$\exp\left[-\left(\frac{S}{L} - \frac{0.02S^2}{L^2}\right)\right] = 0.001$$

The latter can be rewritten as

$$0.02S^2 - S \approx 6.9$$

We leave to it to the reader to complete the solution.

CHAPTER 5

DISTRIBUTIONS WITH MONOTONE INTENSITY FUNCTIONS

For a quantitative characterization of reliability, we must know the failure distributions. Such detailed and complete information is not always available in engineering practice. Fortunately, in some cases we do not need to know the particular type of distribution, it is enough to know only some parameters of the distribution and the fact that this distribution belongs to some special class of distributions. In this case we can often obtain bounds on the reliability indexes based, for example, on the known mean and variance or other similar parameters of the distribution. Concerning the distributions, we need only know that they belong to the class of distributions with a monotone failure rate. There are several main classes of such distributions and these are described below.

5.1 DESCRIPTION OF THE MONOTONICITY PROPERTY OF THE FAILURE RATE

A very natural phenomenon of reliability as it changes over time is often encountered: the longer an item is functioning, the worse the residual reliability properties become. For many practitioners this phenomenon seems almost to be unique. Indeed, deterioration, fatigue, and other similar physical processes lead to a worsening reliability. Such phenomena (and their associated distributions) are called *aging*.

193

But there also exists another property: if an item works for a long period of time, we become more sure of its reliability. Sometimes this property follows from a physical phenomenon connected with a change in the chemical and mechanical features of an item: a penetration of one material into another through contacting surfaces, a strengthening of the joining materials, a "self-fitting" of frictional parts, and so forth. Sometimes this is connected with a "burning-out" effect. This phenomenon is called *younging*. As an example of the latter property, consider a mixture of two equal parts of items: one with a constant MTTF equal to 100 hours and another with a constant MTTF equal to 900 hours. We observe an item chosen at random from this mixed group of items. At the moment $t = 0$ the MTTF equals

$$T = T_1 p + T_2(1 - p) = 100(0.5) + 900(0.5) = 500 \text{ hours}$$

and the probability of a failure-free operation, say during 200 hours, equals

$$\Pr\{\xi \geq 200| \text{ starting at } t = 0\} = 0.5$$

But if it is known that at $t = 101$ hours the item is still functioning, the values of both reliability indexes under the condition that the new trial starts at $t = 101$ hours change:

$$T^* = T_2 = 799 \text{ hours} \quad \text{and} \quad \Pr\{\xi \geq 200| \text{ starting at } t = 101\} = 1$$

Both values for the used item are larger than for the new item on the average. Of course, there is no change in the item itself. We have only new information which allows us to make a posteriori a new conclusion about the item's reliability. An analogous example was considered in Chapter 1 when the mixture of exponential distributions was analyzed.

Notice that we observe a similar effect in "burning-out." It is normal practice to use some stress tests (temperature shocks, accelerated vibration, etc.) for selecting technologically weak items. The same effect is observed when weak units and manufacturing defects are eliminated during a high-failure "infant mortality" period under normal conditions.

It is the appropriate time to recall the ancient Greek myth about the Spartans who killed their weak and ill infants by throwing them from a high rock into a canyon. They did this to ensure that their remaining children would be healthy and strong. (We must state that this is only a myth: it was not a custom in the ancient democracy. As new sources claim, rich, free citizens of Greece replaced their weak and ill infants with the healthy babies of poor families.)

Of course, there are no "immortal" items. First of all, if a failure rate in the initial phase, an increase in the failure rate at some point is Many items have the failure rate function of a "U-shaped form 2.2). Second, even a probability distribution with a decreasing failure rate has $P(\infty) = 1 - F(\infty) = 0$. (Of course, this puts a special condition on the decreasing failure rate function.) The exponential distribution is the boundary distribution between distributions with increasing and decreasing rates.

One of the basic characteristics in further analysis is the "conditional instantaneous density" of the time-to-failure distribution. For this conditional density, we usually use the terms *failure rate* or *failure intensity*. The strict mathematical definition of this, as we mentioned above, is

$$\lambda(t) = \frac{f(t)}{P(t)} \tag{5.1}$$

Thus, in reliability terms, this is the instantaneous failure distribution density at time t under the condition that the item has not failed until t. A better explanation can be presented in terms of an "element of probability." $\lambda(t)\Delta$ is the probability of an unrepairable unit failure in the interval of time $[t, t + \Delta]$ under the condition that the unit has not failed by moment t. This conditional density changes continuously with time.

Sometimes it is useful to consider the function:

$$\Lambda(t) = \int_0^t \lambda(x) \, dx \tag{5.2}$$

Integration of (5.1) and (5.2) yields

$$P(t) = \exp\left[-\int_0^t \lambda(x) \, dx \right] = e^{-\Lambda(t)} \tag{5.3}$$

In this chapter we consider only the simplest properties of distributions with a monotone failure rate. A more detailed analysis of the subject can be found in Barlow and Proschan (1975).

We do not consider the U-shaped $\lambda(t)$'s or the nonmonotonic ones. Notice that a nonmonotonic $\lambda(t)$ is not very unusual at all. The following example from Barlow and Proschan (1975) can be analyzed in very simple terms.

Example 5.1 Consider an unrepaired system consisting of two different units in parallel. Each unit has an exponentially distributed TTF. For this system

$$P(t) = 1 - \left(1 - e^{-\lambda_1 t}\right)\left(1 - e^{-\lambda_2 t}\right)$$

and

$$\lambda(t) = \frac{\lambda_1 e^{-\lambda_1 t} + \lambda_2 e^{-\lambda_2 t} - (\lambda_1 + \lambda_2)e^{-(\lambda_1 + \lambda_2)t}}{e^{-\lambda_1 t} + e^{\lambda_2 t} - e^{-(\lambda_1 + \lambda_2)t}} \tag{5.4}$$

To find the maximum of (5.4) directly by differentiation is a boring problem. We may analyze it in more simple terms. From a physical viewpoint, if the parallel system is functioning for a very long period of time, the most probable situation is that there is only one unit which has survived. If so, this unit, on average, is the most reliable one. Moreover, the longer the period of observation, the higher the conditional probability that the survivor is the best unit. Suppose that, in our case, $\lambda_1 < \lambda_2$. Thus, for the system $\lambda(t) \to \lambda_1$. At the same time, for any parallel system $\lambda(0) = 0$. Show that at some t the function $\lambda(t)$ is larger than λ_1:

$$\lambda(t) = \frac{\lambda_1 e^{-\lambda_1 t} + \lambda_2 e^{-\lambda_2 t} - (\lambda_1 + \lambda_2) e^{-(\lambda_1 + \lambda_2)t}}{e^{-\lambda_1 t} + e^{\lambda_2 t} - e^{-(\lambda_1 + \lambda_2)t}} > \lambda_1 \qquad (5.5)$$

The inequality (5.5) easily transforms into

$$\lambda_1 e^{-\lambda_1 t} + \lambda_2 e^{-\lambda_2 t} - (\lambda_1 + \lambda_2) e^{-(\lambda_1 + \lambda_2)t} > \lambda_1 e^{-\lambda_1 t} + \lambda_1 e^{\lambda_2 t} - \lambda_1 e^{-(\lambda_1 + \lambda_2)t}$$

and after the simple transformations

$$\frac{\lambda_2 - \lambda_1}{\lambda_2} > e^{-\lambda_1 t}$$

The last inequality is valid starting from

$$t_0 = -\frac{1}{\lambda_1} \ln \frac{\lambda_2 - \lambda_1}{\lambda_2} \qquad (5.6)$$

Thus, the function $\lambda(t)$ for the system starts from 0, then intersects the level λ_1 from below, and after this reaches its maximum and exceeds the limit value of λ_1 from above. From (5.5) one can see that $\lambda(t)$ is monotonically increasing if $\lambda_1 = \lambda_2 = \lambda$. Figure 5.1 presents the $\lambda(t)$ behavior over time for three characterizing proportions between λ_1 and λ_2.

Below we consider the distributions with an increasing failure rate (IFR d.f.'s) though this is only one (and the most narrow class) of the "aging" distributions. The reader can find other subclasses of the "aging" distributions, as well as the "younging" distributions, in the original interpretation, in the excellent book by Barlow and Proschan (1975).

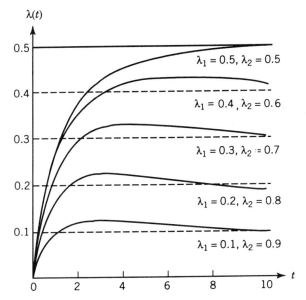

Figure 5.1. Example of a nonmonotone failure rate function for a duplicate system of two different units both with an exponential distribution of time to failure.

5.2   UNIT WITH IFR DISTRIBUTION OF TTF

The evaluation of a unit's indexes is equivalent to finding the parameters of the corresponding distribution of the unit's TTF. If we do not know any additional information about the distribution, the general evaluation is a Chebyshev inequality of the type

$$\Pr\{|\xi - E\{\xi\}| \geq \varepsilon\} \leq \frac{\sigma^2}{\varepsilon^2} \tag{5.7}$$

where ε is an arbitrary positive value.

This inequality is very well known in probability theory. To give the reader a sense of the result, we follow the proof given in Gnedenko (1988). By definition,

$$\Pr\{|\xi - E\{\xi\}| \geq \varepsilon\} = \int_{|x - E\{\xi\}| \geq \varepsilon} dF(x)$$

Because in the domain of integration $(1/\varepsilon)|x - \mathrm{E}\{\xi\}| \geq 1$,

$$\int_{|x-\mathrm{E}\{\xi\}|\geq\varepsilon} dF(x) \leq \frac{1}{\varepsilon^2}\int_{|x-\mathrm{E}\{\xi\}|\geq\varepsilon}(x - \mathrm{E}\{\xi\})^2 dF(x)$$

$$\leq \frac{1}{\varepsilon^2}\int_{-\infty}^{\infty}(x - \mathrm{E}\{\xi\})^2 dF(x) = \frac{\mathrm{Var}\{\xi\}}{\varepsilon^2} = \frac{\sigma^2}{\varepsilon^2}$$

This completes the proof.

Inequality (5.7) is universal and so is not too constructive for practical purposes (as with any universal tool). For instance, one sees that (5.7) only makes sense when $\varepsilon > \sigma$. In other words, this estimate is not true in some area around the mean. But notice that, at the same time, at the distribution's tails, the estimate is very rough. Suppose that additional information is available. Then we can obtain narrower bounds. Consider the class of IFR d.f.'s. We first prove several additional statements.

Theorem 5.1   The graph of an IFR d.f. $P(t)$ crosses the graph of an arbitrary exponential function $e^{-\lambda t}$ at most once from above. If these two functions do not cross, $P(t)$ lies strictly under this exponential d.f. (see Figure 5.2).

Proof. The intensity function $\lambda(t)$ for the IFR distribution might increase infinitely or be bounded above by some number λ^*. In the first case, there exists a moment t_0 when

$$\Lambda(t_0) = \int_0^{t_0}\lambda(x)\, dx = \lambda t_0$$

and, for any $y \geq t_0$,

$$\Lambda(y) \geq \lambda y$$

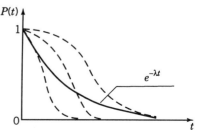

Figure 5.2. Explanation of the contents of Theorem 5.1: possible types of relationships between the exponential function and different IFR distributions of time to failure.

because $\lambda(t)$ increases. Thus,

$$P(t)\begin{cases} \geq e^{-\lambda t} & \text{for} \quad t \leq t_0 \\ \leq e^{-\lambda t} & \text{for} \quad t > t_0 \end{cases} \tag{5.8}$$

In the second case, for any $t > 0$, we have $\Lambda(t) \leq \lambda^*$ and, consequently, $\Lambda(t)$ never crosses λt. Thus, for any t,

$$P(t) < e^{-\lambda t} \tag{5.9}$$

From (5.8) and (5.9) it follows that the "right tail" of the IFR distribution decreases faster than the corresponding tail of the exponential function.

Corollary 5.1 If an IFR d.f. $P(t)$ has a first derivative different from 0 at $t = 0$, say

$$\left.\frac{d}{dt}P(t)\right|_{t=0} = -\alpha$$

then $P(t)$ lies everywhere below $e^{-\alpha t}$.

Proof. The proof follows from the fact that $\Lambda(t) > \alpha t$ for all t. Some hint of a graphical explanation can be found in Figure 5.2.

Corollary 5.2 An IFR d.f. $P(t)$ necessarily crosses $e^{-\lambda t}$ from above once if both distributions have the same MTTF equal to T.

Proof. By Theorem 5.1 both d.f.'s have to intersect once or not intersect at all. The second statement contradicts the corollary, so we need to check this. Suppose that both d.f.'s do not intersect. This means that

$$\int_0^\infty P(t)\,dt < \int_0^\infty e^{-\lambda t}\,dt = \frac{1}{\lambda} = T$$

which contradicts the statement concerning the equality of the MTTFs. For a graphical explanation, refer to Figure 5.2.

Theorem 5.2 For an IFR d.f. $P(t)$, the function

$$\sqrt[t]{P(T)}$$

decreases with increasing t.

Proof. $\Lambda(t)$ is convex, so $\log P(t) = -\Lambda(t)$ is concave. But then

$$\frac{\log P(t) - \log P(0)}{t - 0}$$

decreases with increasing t. Consequently,

$$\frac{1}{t}\log\frac{P(t)}{P(0)}$$

decreases in t. After substituting $P(0) = 1$ and using an exponential transformation, the proof of the theorem follows.

This theorem produces the following interesting corollaries.

Corollary 5.3   For an IFR d.f. $P(t)$,

$$P(x) \le [P(t)]^{x/t}$$

for all $x > t$.

This allows us to predict (i.e., to compute a lower bound) the probability of a failure-free operation of an IFR unit for a specified time, if we know the value of $P(t)$ for a smaller interval of time. This corollary can be of great use for an application in testing IFR units during a short testing period.

Corollary 5.4   For an IFR d.f. the initial moments of all orders are finite.

Proof. Indeed, for any t,

$$\int_t^\infty x^r P(x)\, dx \le \int_t^\infty x^r\left\{[P(t)]^{1/t}\right\}^x dx = \int_t^\infty x^r e^{-\beta x}\, dx < \infty$$

where $[P(t)]^{1/t}$ is replaced by $e^{-\beta}$. The reader knows that the exponential d.f. has the moments of all orders.

The last corollary shows that arguments about the properties of "aging" units, which seem to be just qualitative statements, have led to very strong restrictions on the moments of an IFR d.f. Incidentally, note that the coefficient of variation of the IFR d.f. is always less than 1.

Now, using all of the above results, the following important characteristics of IFR d.f.'s can be obtained.

Theorem 5.3 If ξ_p is the quantile of an IFR d.f. $P(t)$, $\Pr\{x \geq \xi_p\} = p$, then

$$P(t) \begin{cases} \geq e^{-\alpha\xi_p} & \text{for} \quad t \leq \xi_p \\ \leq e^{-\alpha\xi_p} & \text{for} \quad t \geq \xi_p \end{cases}$$

where

$$\alpha = -\frac{\ln(1-p)}{\xi_p}$$

Proof. An exponential function can be found which goes through the point $(\xi_p, 1-p)$. The parameter of the exponent can be found from the equation

$$e^{-\alpha\xi_p} = 1 - p$$

We then use Theorem 5.1 to complete the proof.

Theorem 5.4 A lower bound for an IFR d.f. is determined by

$$P(t) \geq \begin{cases} e^{-t/T} & \text{for} \quad t < T \\ 0 & \text{for} \quad t \geq T \end{cases}$$

where t is the MTTF

$$T = \int_0^\infty P(t)\, dt$$

Proof. We first present a rigorous proof. For an IFR distribution the function $\Lambda(t)$ is an increasing convex function, so by Jensen's inequality

$$E\{\Lambda(\xi)\} \leq \Lambda(E\{\xi\}) = \Lambda(T) \tag{5.10}$$

Denote $P(t) = y$ and rewrite

$$E\{\Lambda(\xi)\} = E\{-\ln P(t)\} = E\{-\ln y\} = \int_0^1 \ln y\, dy = 1 \tag{5.11}$$

From (5.10) and (5.11)

$$E\{\Lambda(t)\} = 1 \leq \Lambda(T) = -\ln P(T)$$

immediately follows where

$$P(T) \geq e^{-1} \tag{5.12}$$

or, equivalently,

$$[P(T)]^{1/T} \geq e^{-1/T}$$

Now from Corollary 5.3 for $t < T$ we can write

$$[P(t)]^{1/t} \geq [P(T)]^{1/T} \geq e^{-1/T}$$

and, finally, for $t < T$ the required result is obtained

$$P(t) \geq e^{-t/T}$$

The same result can be derived from simple explanations based on a graphical presentation (see Figure 5.3).

The first inequality follows immediately from a comparison of the exponential function $e^{-t/T}$ and a degenerate function $G(t)$ with the same MTTF

$$G(t) = \begin{cases} 1 & \text{for} \quad t \leq T \\ 0 & \text{for} \quad t > T \end{cases}$$

The degenerate function (i.e., a distribution of a constant value) is the boundary distribution for the class of IFR d.f.'s. By Theorem 5.1 the degenerate function crosses the exponential function from above at point $t = T$. All strictly IFR d.f.'s may cross the graph of a given exponent only for $t > T$ which follows from the equality of the MTTFs. The second inequality is trivial because $P(t)$ is a nonnegative function. Notice that a lower bound is reached by the exponential d.f. for $t \leq T$ and by the degenerate d.f. for $t > T$.

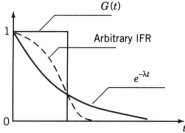

Figure 5.3. Explanation of the proof of Theorem 5.4: relationships among IFR, exponential, and degenerate reliability functions.

Theorem 5.5   An upper bound for an IFR d.f. is determined as

$$P(t) \leq \begin{cases} 1 & \text{for} \quad t^* \leq T \\ e^{-\omega_t t} & \text{for} \quad t^* > T \end{cases} \tag{5.13}$$

where ω_t depends on t^* and is found from the condition

$$\int_0^{t^*} e^{-\omega_t t} \, dt = T$$

or, equivalently,

$$1 - \omega_t T = e^{-\omega_t t^*}$$

Proof. The first inequality in (5.13) is trivial and follows from the definition of a d.f. The second inequality is equivalent to the statement that for $t^* > T$ the IFR function $P(t)$ crosses the graph of the function $E^*(t)$ from above, which is the exponential function truncated from the right at point t^*,

$$E^*(t) = \begin{cases} e^{-\omega_t t^*} & \text{for} \quad t \leq t^* \\ 0 & \text{for} \quad t > t^* \end{cases}$$

at some point $t < t^*$ if both $P(t)$ and $E^*(t)$ have the same MTTF.

This fact can be proved immediately by assuming the contrary. Suppose that there is no such crossing. Then $P(t)$ lies above $E^*(t)$ everywhere, but then

$$\int_0^\infty P(t) \, dt > \int_0^\infty E^*(t) \, dt$$

which contradicts our suggestion about the equality of their MTTFs. A graphical explanation of (5.13) is given in Figure 5.4.

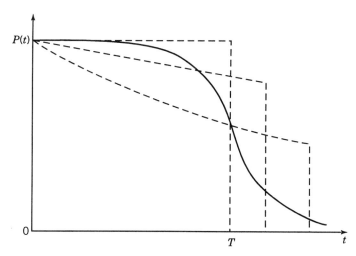

Figure 5.4. Explanation of the proof of Theorem 5.5: finding ω_t by constructing the exponents truncated from the right.

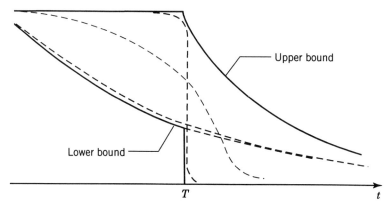

Figure 5.5. Area of possible values of IFR reliability functions with the same MTTF and samples of different IFR reliability functions.

As a result, we have lower and upper bounds for the IFR function $P(t)$ which are represented in Figure 5.5. In this figure $P_1(t)$ is the function with a coefficient of variation close to 0, and $P_1(t)$ is the function with a coefficient of variation close to 1.

Theorem 5.6   An upper bound for the quantile ξ_p of the IFR distribution is expressed by its MTTF, T, and corresponding probability p, $p = 1 - \Pr\{t > \xi_p\}$, as

$$\xi_p \leq \left(-\frac{\ln(1-p)}{p} \right) T$$

Proof. Notice first that, from $\Pr\{t > \xi_p\} = P(\xi_p) = 1 - p$,

$$P(\xi_p) = e^{\ln(1-p)}$$

Now the chain of obvious inequalities based on the previous results can be written as

$$T = \int_0^\infty P(t)\, dt \geq \int_0^{\xi_p} P(t)\, dt \geq \int_0^{\xi_p} \left[P(\xi_p) \right]^{(x/\xi_p)} dx$$

$$= \int_0^{\xi_p} \exp\left[\frac{\ln(1-p)}{\xi_p} x \right] dx$$

and simple integration gives us

$$\int_0^{\xi_p} \exp\left[\frac{\ln(1-p)}{\xi_p} x\right] dx = \frac{\xi_p}{\ln(1-p)} (\exp[\ln(1-p)] - 1)$$

$$= \frac{\xi_p}{\ln(1-p)} [(1-p) - 1]$$

$$= \frac{p\xi_p}{-\ln(1-p)}$$

which produces the desired result.

Theorem 5.7 A lower bound for the quantile ξ_p of the IFR distribution is expressed by T and p as

$$\xi_p \geq \begin{cases} [-\ln(1-p)]T & \text{for} \quad 1-p \geq e^{-1} \\ T & \text{for} \quad 1-p \leq e^{-1} \end{cases}$$

Proof. We prove the first of these inequalities separately for $\xi_p \leq T$ and $\xi_p \geq T$. For the first case

$$1 - p = P(\xi_p) \geq e^{-1} \geq \exp\left(-\frac{\xi_p}{T}\right)$$

For the second case with the use of Theorem 5.2, we immediately write

$$1 - p = P(\xi_p) \geq \exp\left(-\frac{\xi_p}{T}\right)$$

Thus, the desired inequality is valid in both cases.

The second inequality, which is valid for the condition

$$1 - p = P(\xi_p) \leq e^{-1}$$

follows immediately if we recall (5.12). Thus,

$$P(\xi_p) \leq e^{-1} \leq P(T)$$

which corresponds to the desired condition $\xi_p \geq T$.

Corollary 5.5   For the median M of an IFR d.f., the following bounds are valid:

$$\left(-\ln\tfrac{1}{2}\right)T \le M \le \left(-2\ln\tfrac{1}{2}\right)T$$

or

$$\frac{M}{2\ln 2} \le T \le \frac{M}{\ln 2}$$

Proof. The proof follows automatically from Theorems 5.6 and 5.7 after the substitution $p = 1/2$.

If instead of the MTTF, we know the variance of the IFR distribution, the bounds can be improved. We do not consider these more complex cases and advice the reader to refer to Barlow and Proschan (1975). for an excellent discussion of the subject.

5.3   SYSTEM OF IFR UNITS

As we saw above, an IFR type of distribution of a unit TTF leads to interesting and constructive results. An extension of these results appears when we consider systems consisting of units with IFR types of TTF d.f.'s.

We will formulate all of the results in the form of theorems because each of them requires a mathematical proof. First of all, we prove a lemma, which is simple but very important for future considerations.

Lemma 5.1   If (1) the function $f(x)$ is monotonic, restricted, and nonnegative on the positive semiaxis, (2) the function $g(x)$ is absolutely integrable on the positive semiaxis, (3) the latter function is such that $g(x) \ge 0$ for $x < a$ and $g(x) \le 0$ for $x > a$, and (4)

$$\int_0^\infty g(x)\, dx = 0$$

then, if $f(x)$ decreases (increases), the following inequality is true:

$$\int_0^\infty f(x)g(x)\, dx \le (\ge)0$$

Proof. The proof can be presented as a chain of simple transformations:

$$\int_0^\infty f(x)g(x)\,dx = \int_0^a f(x)g(x)\,dx + \int_a^\infty f(x)g(x)\,dx$$

$$\le (\ge)\int_0^a \left[\max_{0 \le x \le a} f(x)\right]g(x)\,dx + \int_a^\infty \left[\min_{x > a} f(x)\right]g(x)\,dx$$

$$= f(a)\int_0^a g(x)\,dx + f(a)\int_a^\infty g(x)\,dx = f(a)\int_0^\infty g(x)\,dx$$

$$= 0$$

The sense of the lemma is clear from Figure 5.6 where an increasing function $f(x)$ is shown. Obviously, the square S_1 is taken in the resulting expression with less "weight" than the square S_2, and thus the sum turns out to be negative.

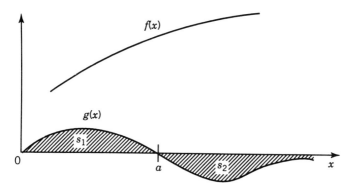

Figure 5.6. Graphical explanation of the proof of Lemma 5.1.

5.3.1 Series System

Theorem 5.8   A lower bound on the probability of a failure-free operation of a series system of IFR units with known MTTF is

$$P(t) = \prod_{1 \le i \le n} P_i(t) \ge \begin{cases} \exp\left(-t \sum_{1 \le i \le n} \frac{1}{T_i}\right) & \text{for} \quad t \le t^* \\ 0 & \text{for} \quad t > t^* \end{cases}$$

where $t^* = \min T_i$.

Proof. The proof immediately follows from Corollary 5.2 by a simple substitution.

This lower bound is very important in practice because it gives a guaranteed estimate of the real but unknown value of the reliability index.

Theorem 5.9   An upper bound on the probability of a failure-free operation of a series system of IFR units with known MTTF is

$$
P(t) \leq \begin{cases}
1 & \text{for } \quad t \leq \min_{1 \leq i \leq n} T_i = T_1 \\[2mm]
\exp\left(-\omega_{t_1^*}^{(1)} t\right) & \text{for } \quad T_1 \leq t < T_2 \\[2mm]
\exp\left[-\left(\omega_{t_1^*}^{(1)} + \omega_{t_2^*}^{(2)}\right) t\right] & \text{for } \quad T_2 \leq t < T_3 \\[2mm]
\cdots\cdots\cdots\cdots\cdots\cdots\cdots\cdots\cdots \\[2mm]
\exp\left(-\sum_{1 \leq i \leq n} \omega_{t_i^*}^{(i)} t\right) & \text{for } \quad t \geq T_n
\end{cases}
$$

where T_i, $1 \leq i \leq n$, are ordered MTTFs and each $\omega_{t_i^*}^{(i)}$ is found from the equation

$$
1 - \omega_{t_i^*}^{(i)} T_i = \exp\left(-\omega_{t_i^*}^{(i)} t\right) \tag{5.15}
$$

Proof.   The proof follows immediately from (5.13) of Theorem 5.5.

Theorem 5.10   An upper bound on the probability of a failure-free operation of a series system of IFR units with known $\alpha = \lambda(0)$ is

$$
P(t) \leq \prod_{1 \leq i \leq n} e^{-\alpha_i t} = \exp\left(-t \sum_{1 \leq i \leq n} \alpha_i t\right) \tag{5.16}
$$

for $t \leq \min T_i$.

Proof.   The proof follows immediately from Corollary 5.1.

Theorem 5.11   The MTTF of a series system of independent IFR units has the following bounds:

$$
\frac{1}{\displaystyle\sum_{1 \leq i \leq n} \frac{1}{T_i}} \leq T_{\text{syst}} \leq \min_{1 \leq i \leq n} T_i \tag{5.17}
$$

Proof.   An upper bound follows trivially from the obvious statement that for any t the system is less reliable than any of its units:

$$
P_i(t) \geq \prod_{1 \leq i \leq n} P_i(t)
$$

Therefore,

$$T_i = \int_0^\infty P_i(t)\, dt \geq \int_0^\infty \prod_{1 \leq i \leq n} P_i(t)\, dt = T_{\text{syst}}$$

We may use Lemma 5.1 to obtain a lower bound. We show that the replacement of an arbitrary unit with an IFR distribution of TTF with a unit with an exponentially distributed TTF, which has the same MTTF, leads to a decrease of the series system's MTTF. Suppose that such a replacement is done for the nth unit of the system. We need to prove that

$$\int_0^\infty \prod_{1 \leq i \leq n} P_i(t)\, dt \geq \int_0^\infty e^{-t/T_n} \prod_{1 \leq i \leq n-1} P_i(t)\, dt$$

or, equivalently,

$$\Delta = \int_0^\infty \left[P_n(t) - e^{-t/T_n} \right] \prod_{1 \leq i \leq n-1} P_i(t)\, dt \geq 0$$

Note that, by Theorem 5.1, $P_n(t)$ crosses $\exp(-t/T_n)$ once and from above and, by assumption, both these functions have the same MTTF. Thus,

$$P_n(t) - e^{-t/T_n}$$

corresponds to the function $g(x)$ of Lemma 5.1. At the same time, the function

$$\prod_{1 \leq i \leq n-1} P_i(t)$$

corresponds to the decreasing function $f(x)$ in Lemma 5.1. Thus, by Lemma 5.1, $\Delta \geq 0$, and the desired intermediate statement is proved.

The systematic replacement of all system units with an IFR distribution with units with a corresponding exponential distribution produces

$$T_{\text{syst}} \geq \int_0^\infty \prod_{1 \leq i \leq n} e^{-t/T_i}\, dt = \int_0^\infty \exp\left(-t \sum_{1 \leq i \leq n} \frac{1}{T_i} \right) dt = \frac{1}{\displaystyle\sum_{1 \leq i \leq n} \frac{1}{T_i}}$$

Thus, the theorem is proved.

The upper bound can be improved if we possess additional information about the distribution $P_i(t)$, for example, if we know the first derivatives in $t = 0$.

Theorem 5.12   The upper bound of the series system MTTF can be written as

$$T_{\text{syst}} \le \frac{1 - \exp\left[- \min_{1 \le i \le n} a_i \sum_{1 \le i \le n} \lambda_i(0)\right]}{\sum_{1 \le i \le n} \lambda_i(0)} \qquad (5.18)$$

where a_i is determined from the condition

$$\int_0^{a_i} e^{-\lambda_i(0)t}\, dt = T_i$$

Proof. Consider an exponential distribution truncated from the right:

$$E_i^*(t) = \begin{cases} e^{-\lambda_i(0)t} & \text{for} \quad t \le a_i \\ 0 & \text{for} \quad t > a_i \end{cases}$$

This distribution $E_n^*(t)$ has the same MTTF as the initial distribution $P_n(t)$. Hence, $E_n^*(t)$ crosses $P_n(t)$ from above (see Figure 5.7).

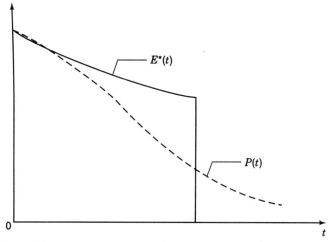

Figure 5.7. Graphical explanation of the proof of Theorem 5.12: intersection of the IFR reliability function and the exponential function truncated from the right where their derivatives in $t = 0$ are equal.

In the expression for the system MTTF, replace the unit with distribution $P_n(t)$ by $E_n^*(t)$. The new system MTTF is

$$\int_0^\infty E_n^*(t) \prod_{1 \le i \le n-1} P_i(t)\, dt = \int_0^{a_n} e^{-\lambda_n(0)t} \prod_{1 \le i \le n-1} P_i(t)\, dt$$

We now find the value of Δ:

$$\Delta = \int_0^\infty E_n^*(t) \prod_{1 \le i \le n-1} P_i(t)\, dt - \int_0^\infty \prod_{1 \le i \le n} P_i(t)\, dt$$

$$= \int_0^\infty \left[E_n^*(t) - P_n(t) \right] \prod_{1 \le i \le n-1} P_i(t)\, dt$$

Again we can use Lemma 5.1, noticing that $E_n^*(t) - P_n(t)$ corresponds to the function $g(x)$ from Lemma 5.1 and $\prod_{1 \le i \le n-1} P_i(t)$ corresponds to the decreasing function from Lemma 5.1. Thus, by Lemma 5.1, $\Delta \ge 0$, that is, the replacement of any IFR unit, say the nth, with a unit with distribution $E_n^*(t)$ might only increase the system's MTTF.

Thus, the systematic replacement of units in the above-described manner leads to the final result

$$T_{\text{syst}} \le \int_0^{\min_{1 \le i \le n} a_i} \exp\left[-t \sum_{1 \le i \le n} \lambda_i(0) \right] dt$$

$$= \frac{1 - \exp\left[-\min_{1 \le i \le n} a_i \sum_{1 \le i \le n} \lambda_i(0) \right]}{\sum_{1 \le i \le n} \lambda_i(0)}$$

and this completes the proof.

5.3.2 Parallel Systems

Theorem 5.13 An upper bound for the probability of failure of a parallel system of IFR units can be expressed as

$$Q(t) = \prod_{1 \le i \le m} Q_i(t) \le \begin{cases} \prod_{1 \le i \le m} \left(1 - e^{-t/T_i}\right) & \text{for} \quad t \le t^* \\ 1 & \text{for} \quad t > t^* \end{cases} \tag{5.19}$$

where

$$t^* = \min_{1 \le i \le m} T_i$$

Proof. The proof follows directly from Theorem 5.4.

For the probability of a failure-free operation, the following lower bound follows from (5.19):

$$P(t) \geq \begin{cases} 1 - \prod_{1 \leq i \leq m} \left(1 - e^{-t/T_i}\right) & \text{for} \quad t \leq t^* \\ 0 & \text{for} \quad t > t^* \end{cases} \tag{5.20}$$

Theorem 5.14 A lower bound for the probability of failure of a parallel system of IFR units has the form

$$Q(t) \geq \begin{cases} \prod_{1 \leq i \leq m} \left[1 - \exp\left(-\omega_{t_i^*}t\right)\right] & \text{for} \quad t \geq \tilde{t}^* \\ 0 & \text{for} \quad t \leq \tilde{t}^* \end{cases} \tag{5.21}$$

where

$$\tilde{t}^* = \max_{1 \leq i \leq m} T_i$$

Proof. The proof follows immediately from Theorem 5.5.

For the probability of a failure-free operation, the following upper bound follows from (5.21):

$$P(t) \leq \begin{cases} 1 & \text{for} \quad t \leq \tilde{t}^* \\ 1 - \prod_{1 \leq i \leq m} \left[1 - \exp\left(-\omega_{t_i^*}t\right)\right] & \text{for} \quad t \geq \tilde{t}^* \end{cases} \tag{5.22}$$

Theorem 5.15 The MTTF of a parallel system of independent IFR units has the bounds

$$\max_{1 \leq i \leq m} T_i \leq T_{\text{syst}} \leq \sum_{1 \leq i \leq m} T_i - \sum_{1 \leq i < j \leq m} \frac{1}{\frac{1}{T_i} + \frac{1}{T_j}}$$

$$+ \cdots + (-1)^{n+1} \frac{1}{\sum_{1 \leq i \leq m} \frac{1}{T_i}} \tag{5.23}$$

Proof. This proof is analogous to the proof of Theorem 5.11 and so we omit it.

Again note that the lower bound is trivial and can be instantly found for a degenerate distribution, that is, for the case when all T_i's are constant.

We now mention that, for a parallel system, the MTTF is larger if the unit failure distributions have larger variances. (In qualitative terms, this result is close to that obtained for dependent units.) It seems paradoxical that an unstable production of some units for a parallel structure is better than a stable production: we used to think that stability is almost always better than instability. But there is no enigma at all if one notices that the random time to failure of a parallel system is the maximum of the unit's random time to failure.

Theorem 5.16 A lower bound for the MTTF of a system consisting of units with an IFR distribution, for which we know the first derivative in $t = 0$, is

$$
\begin{aligned}
T_{\text{syst}} \geq \sum_{1 \leq i \leq m} T_i - \sum_{1 \leq i < j \leq m} \frac{1}{\lambda_i(0) + \lambda_j(0)} \\
\times \left[1 - \exp\left(-\min\{T_i, T_j\}[\lambda_i(0) + \lambda_j(0)] \right) \right] + \cdots \\
+ (-1)^{m+1} \frac{1}{\sum_{1 \leq i \leq m} \lambda_i(0)} \left(1 - \exp\left[\min_{1 \leq i \leq m} T_i \sum_{1 \leq i \leq m} \lambda_i(0) \right] \right)
\end{aligned}
$$

Proof. The proof here is analogous to that of Theorem 5.12. We only need to notice that the probability of a failure-free operation for this case after all substitutions of $P_i(t)$ for $E_i^*(t)$ has the form

$$
P(t) = 1 - \prod_{1 \leq i \leq m} \left[1 - E_i^*(t) \right]
$$

where $E_i^*(t)$ is defined in Theorem 5.12.

5.3.3 Other Monotone Structures

Instead of writing detailed formulas with the simple substitution of IFR d.f.'s $P(t)$ for degenerate or exponential d.f.'s, we mention only that one can obtain a lower bound for the system reliability by substituting lower bounds of the corresponding units' failure-free probabilities. Analogously, after the substitution of the upper bounds of the units' probabilities, a lower bound for the system probability is obtained. The reader can find some related results in Barlow and Proschan (1975).

CONCLUSION

This relatively new branch of reliability theory was initiated by Barlow and Proschan and became widely known after their book [Barlow and Proschan (1975)] was published. First papers on the properties of distributions with a monotone failure rate appeared in the previous decade [Barlow, Marshall, and Proschan (1963); Barlow and Marshall (1964); Solovyev and Ushakov (1967); Gnedenko, Belyaev, and Solovyev (1969); among others].

We would like to present here a simple but important result concerning repairable systems [Ushakov (1966)]. The stationary interval availability coefficient of a system with an "aging" distribution $F(t)$ of TTF has lower and upper bounds of the form

$$K\left(1 - \frac{t_0}{T}\right) \leq R(t_0) \leq Ke^{-t_0/T}$$

where K is the stationary availability coefficient and T is the mean of the distribution $F(t)$. These bounds can be easily obtained with the help of Lemma 5.1. Indeed, $R(t_0)$ for any distribution $F(t)$ can be written as

$$R(t_0) = KP^*(t_0)$$

where $P^*(t_0)$ is the distribution of a stationary residual time:

$$P^*(t) = \frac{1}{T}\int_t^\infty P(x)\, dx$$

Substitution of degenerate and exponential d.f.'s into the latter expression and application of Lemma 5.1 produce the necessary result. This result and some others can be found in Gnedenko, Belyaev, and Solovyev (1969). Some new results can be found in Gnedenko (1983).

A collection of practical results on aging units and systems consisting of aging units is presented in Ushakov (1985, 1994). This problem is especially important in practice when one possesses only very restricted statistical information but has some reasonable physical arguments about the possible behavior of a time-to-failure distribution.

REFERENCES

Barlow, R. E., and A. W. Marshall (1964). Bounds for distributions with monotone hazard rate. I and II. *Ann. Math. Statist.*, vol. 35.

Barlow, R. E., and A. W. Marshall (1965). Tables of bounds for distributions with monotone hazard rate. *J. Amer. Statist. Assoc.*, vol. 60.

Barlow, R. E., and F. Proschan (1975). *Statistical Theory of Reliability and Life Testing.* New York: Holt, Rinehart, and Winston.

Barlow, R. E., A. W. Marshall, and F. Proschan (1963) Properties of probability distributions with monotone hazard rate. *Ann. Math. Statist.*, vol. 34.

Gnedenko, B. V., ed. (1983). *Mathematical Aspects of Reliability Theory* (in Russian). Moscow: Radio i Sviaz.

Gendenko, B. V., Yu. K. Belyaev, and A. D. Solovyev (1969). *Mathematical Methods of Reliability Theory*. New York: Academic.

Solovyev, A. D., and I. A. Ushakov (1967). Some bounds on a system with aging elements (in Russian). *Automat. Comput. Sci.*, no. 6.

Ushakov, I. A. (1966). An estimate of reliability of a system with renewal for a stationary process (in Russian). *Radiotekhnika*, no. 5.

Ushakov, I. A., ed. (1985). *Reliability of Technical Systems: Handbook* (in Russian). Moscow: Radio i Sviaz.

Ushakov, I. A., ed. (1994). *Handbook of Reliability Engineering*. New York: Wiley.

EXERCISES

5.1 Two units have the same mean T. One unit has a uniform d.f. and another has an exponential d.f. Which one will deliver the larger probability of failure-free operation at moment $t = T$? At moment $t = 2T$?

5.2 Consider the Erlang d.f. of a high order (e.g., $n = 10$). Explain (without exact proof) how $\lambda(t)$ behaves.

5.3 One has to choose a system for the continuous performance of an operation during 100 hours. There are two possibilities: to choose for this purpose a system with an MTTF of 200 hours or to choose another system with an MTTF of 300 hours. Which system should be chosen.

5.4 What kind of $\lambda(t)$ has the system depicted in Figure E5.3?

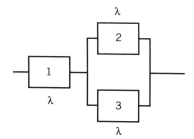

Figure E5.3.

SOLUTIONS

5.1 See Figure E5.1.

5.2 Consider a clear physical example where such a distribution appears: A standby redundancy group of n identical and independent units. One

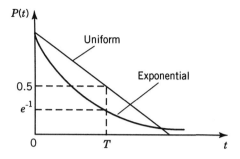

Figure E5.1.

knows that a large number of random variables has an approximately normal distribution (at least far from the "tails"). One knows (see the approximation for a highly reliable redundant group) that $\lambda(0) = 0$ and $\lambda(t)$ is increasing by t and convex near 0. Then consider large t. If the redundant group is still operating, the probability that there is only one up unit is increasing in t. But one unit with an exponentially distributed TTF has a constant failure rate. So, the time diagram for $\lambda(t)$ has the form shown in Figure E5.2.

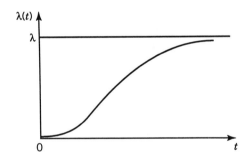

Figure E5.2.

5.3 The problem as formulated here is incorrect: everything depends on the kind of distribution. If both distributions are exponential, then one should choose the second system. If both systems have an almost constant TTF, there is no difference between them although, from a common viewpoint (with no particular sense in this case!), everybody will again choose the second system. This might be, as a matter of fact, unreasonable if the first system is, for instance, cheaper. But if the first system has any "realistic" distribution of TTF (exponential, normal, etc.) and the second one has a "two-mass" distribution, that is,

$$\text{TTF}_2 = \begin{cases} 0 & \text{with probability } p \\ \tau_2 & \text{with probability } 1 - p \end{cases}$$

where $\tau_2 > 300$, the solution is not unique.

Consider an exponential d.f. with $\lambda = 1/200$. For this distribution $PFFO_1(100 \text{ hours}) = e^{-1}$. For the second case let $p = 0.9$ and $\tau_2 = 3000$ hours. This corresponds to $MTTF_2 = 300$ hours. In this case $PFFO_2(100 \text{ hours}) = 0.1$ which is worse than the exponential distribution considered above.

Now assume that $p = 0.5$ and $\tau_2 = 600$ hours. Then $PFFO_2(100 \text{ hours}) = 0.5$, which is better than the previous case. For other distributions one can obtain similar conclusions (with other numerical results).

5.4 One should repeat all of the arguments used in the solution of Exercise 5.2 taking into account that:

- Unit 1 might be the cause of the system failure during all periods of time.
- For a large time period, the parallel connection of units 2 and 3 with probability close to 1 will consist of only one unit; thus the entire system also will almost surely consist of two series units: unit 1 and one of the units 2 or 3.

The solution is represented graphically in Figure E5.4.

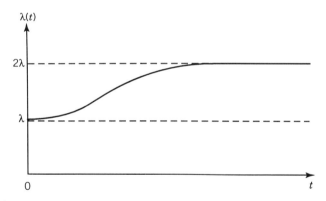

Figure E5.4.

CHAPTER 6

REPAIRABLE SYSTEMS

In engineering practice, one of the most important objects under investigation is a *repairable system*. In general, repairable systems might be analyzed with the help of Monte Carlo simulation. There are no essential analytical results for the most general mathematical models except for some very particular cases. The most important analytical models frequently used in practice are Markov models. For these models all the system units' random TTFs and repair times are assumed to be exponential. (More accurately, each random duration of being in any state has an exponential distribution.) These assumptions might be far from valid, and so each time their appropriateness must be carefully considered. Note that if the suggestion about exponentially distributed TTFs is admissible (especially for electronic equipment), it seems artificial for the repair time. Indeed, the residual repair time should depend on the time already spent. We have discussed this issue earlier. But as we will show below, sometimes the assumptions of a distribution's exponentiality produce acceptable numerical results that can be utilized in engineering design. At any rate, Markov models are very popular for practical engineering problems because of their clarity and mathematical simplicity.

6.1 SINGLE UNIT

6.1.1 Markov Model

We first consider the simplest possible repairable system: a single unit. At any moment in time, the unit is in one of two states: it is either operating or

218

it has failed. The transition graph is presented in Figure 6.1. Here state 0 denotes an operating state, and state 1 corresponds to a failed state. This graph has a simple interpretation. When in state 0, the unit might go to state 1 or stay at the current state. Leaving state 0 occurs with an intensity λ, and leaving state 1 occurs with an intensity μ.

The unit transition process can be described as an alternative renewal process. It is represented by a sequence of mutually independent r.v.'s ξ (a unit up time) and η (a unit repair time). Both ξ and η have exponential distributions with parameters λ and μ, respectively. A sample time diagram is presented in Figure 6.2.

Using the graph of Figure 6.1, we can easily construct the following formula:

$$P_0(t + \Delta t) = (1 - \lambda \Delta t)P_0(t) + \mu \Delta t P_1(t) \qquad (6.1)$$

This expression means that the transition process may appear in state 0 at moment $t + \Delta t$ under the following conditions:

- It was there at moment t and did not leave during the interval Δt.
- At moment t it was in state 1 and moved to state 0 during the interval Δt.

The conditional probability of leaving state 0 equals $\lambda \Delta t$, and the conditional probability of leaving state 1 equals $\mu \Delta t$.

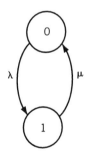

Figure 6.1. Transition graph for a renewable unit.

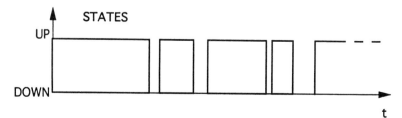

Figure 6.2. Time diagram for a renewable unit.

From (6.1) we obtain

$$\frac{P_0(t + \Delta t) - P_0(t)}{\Delta t} = -\lambda P_0(t) + \mu P_1(t) \tag{6.2}$$

In the limit as $\Delta t \to 0$, we obtain

$$\frac{d}{dt} P_0(t) = -\lambda P_0(t) + \mu P_1(t) \tag{6.3}$$

This represents the simplest example of Kolmogorov's equation. This equation expresses a condition of dynamic equilibrium. To solve it with respect to any $P_k(t)$, we need to have one more equation. It is clear that another equation cannot be obtained in the same manner: it would be linearly dependent on the first one and, consequently, would not be useful in obtaining the solution. The second equation which should be chosen is the so-called *normalization equation*:

$$P_0(t) + P_1(t) = 1 \tag{6.4}$$

which means that at any moment the unit must be in one of two possible states.

We also need to determine the initial condition for the solution of the system of differential equations. In this simple case the problem can easily be solved in general when $P_0(t_0) = p$ and, consequently, $P_1(t_0) = q$, $p + q = 1$.

This problem can be solved with the help of different methods. We will use the Laplace–Stieltjes transform (LST) to make the presentations in the book uniform.

Recall that the LST $\varphi(s)$ of the function $f(t)$ is defined as

$$\varphi(s) = \int_0^\infty f(t) \, e^{-st} \, dt \tag{6.5}$$

[In this context we consider functions $f(t)$ defined over the positive axis.]

Nonstationary Availability Coefficient   The system (6.2) and (6.4) for this case has the LST:

$$-p + s\varphi_0(s) = -\lambda \varphi_0(s) + \mu \varphi_1(s)$$
$$\varphi_0(s) + \varphi_1(s) = \frac{1}{s} \tag{6.6}$$

or, in canonical form,

$$(\lambda + s)\varphi_0(s) - \mu \varphi_1(s) = p$$
$$s\varphi_0(s) + s\varphi_1(s) = 1 \tag{6.7}$$

Thanks to the LST, the system of linear differential equations turns into a system of algebraic equations. To solve (6.7), we can use Cramér's rule:

$$\varphi_0(s) = \frac{\begin{vmatrix} p & -\mu \\ 1 & s \end{vmatrix}}{\begin{vmatrix} \lambda + s & -\mu \\ s & s \end{vmatrix}} = \frac{ps + \mu}{s^2 + (\lambda + \mu)s} \tag{6.8}$$

To invert this LST, we have to present it in the form of a sum of terms of type a/s or $b/(s + \alpha)$. The inverse functions for these terms are a constant and an exponential function, respectively.

To present the solution (6.8) in the desired form, we should find the roots of the denominators of (6.8). They are: $s_1 = 0$ and $s_2 = -(\lambda + \mu)$. Now we can write

$$\varphi_0(s) = \frac{A}{s - s_1} + \frac{B}{s - s_2} = \frac{A}{s} + \frac{B}{s + \lambda + \mu} \tag{6.9}$$

where A and B are the unknown constants to be determined. To find them, we should note that two polynomials with similar denominators are equal if and only if the coefficients of their numerators are equal. Thus, we set the two representations equal:

$$\frac{A}{s} + \frac{B}{\lambda + \mu + s} = \frac{ps + \mu}{s(\lambda + \mu + s)} \tag{6.10}$$

And so we obtain a new system for A and B by equalizing the coefficients of the polynomials:

$$\begin{aligned} A + B &= p \\ A(\lambda + \mu) &= \mu \end{aligned} \tag{6.11}$$

It is easy to find

$$A = \frac{\mu}{\lambda + \mu}$$

$$B = p - \frac{\mu}{\lambda + \mu} = \frac{\lambda p - \mu(1 - p)}{\lambda + \mu} \tag{6.12}$$

Thus, the LST of interest can be written as

$$\varphi_0(s) = \frac{\mu}{\lambda + \mu} \frac{1}{s} + \frac{\lambda p - \mu(1 - p)}{\lambda + \mu} \frac{1}{\lambda + \mu + s} \tag{6.13}$$

Finally, the nonstationary availability coefficient, that is, the inverse LST of (6.10), is

$$K(t) = P_0(t) = \frac{\mu}{\mu + \lambda} + \frac{\lambda p - \mu(1 - p)}{\lambda + \mu} e^{-(\lambda + \mu)t} \qquad (6.14)$$

If the original system state is operational, that is, if $P_0(t) = 1$, the solution is

$$K(t) = \frac{\mu}{\lambda + \mu} + \frac{\lambda}{\lambda + \mu} e^{-(\lambda + \mu)t} \qquad (6.15)$$

The function $K(t)$ showing the time dependence of the system availability is presented in Figure 6.3.

Stationary Availability Coefficient It is clear that if $t \to \infty$, $K(t)$ approaches the stationary availability coefficient K:

$$K = \frac{\mu}{\lambda + \mu} = \frac{T}{T + \tau} \qquad (6.16)$$

where $T = 1/\lambda$ is the unit's MTTF and $\tau = 1/\mu$ is the unit's mean time to repair (MTR).

We should notice that, in general, such a method of obtaining a stationary availability coefficient is not excusable in a computational sense. For this purpose, one should write a system of linear algebraic equations in a direct way without the use of the system of differential equations. It is important to realize that the stationary regime represents static equilibrium. This means that all derivatives $dP_k(t)/dt$ are equal to 0 because no states are changing in

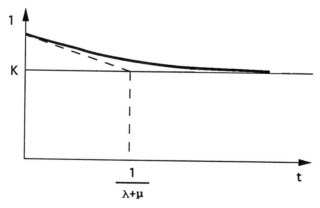

Figure 6.3. Time dependence of nonstationary availability coefficient $K(t)$ for exponential distributions of TTF and repair time.

time, "on the average." Consequently, all P_k's must be constant. It is also clear that the initial conditions (the original state of the unit at moment $t = 0$) also will not make any sense. This assumption leads directly to the following system of algebraic equations:

$$-\lambda P_0 + \mu P_1 = 0$$
$$P_0 + P_1 = 1$$

(6.17)

where

$$P_k = \lim_{t \to \infty} P_k(t)$$

(6.18)

are the stationary probabilities of interest.

Again, the solution can be obtained with Cramér's rule

$$P_0 = K = \frac{\begin{vmatrix} 0 & \mu \\ 1 & 1 \end{vmatrix}}{\begin{vmatrix} -\lambda & \mu \\ 1 & 1 \end{vmatrix}} = \frac{\mu}{\lambda + \mu} = \frac{T}{T + \tau}$$

(6.19)

Of course, we mention Cramér's rule not as a computational tool, but rather as a methodological reference. Everyone might choose his or her own method for this particular computational task.

Probability of a Failure-Free Operation   Considering previous reliability indexes, we assumed that both unit states are transient. But if one needs indexes such as the probability of a failure-free operation during a specified time interval, or the MTTF, the transition graph should be reconstructed. In these cases the unit failure state has to be absorbing.

The transition graph for this case is presented in Figure 6.4. There is no transition from state 1 back to state 0, that is, $\mu = 0$. In this case we have the

Figure 6.4. Nontransitive graph for computation of the MTTF of a renewable unit with an exponentially distributed TTF.

equation

$$\frac{d}{dt}P_0(t) = -\lambda P_0(t) \qquad (6.20)$$

This differential equation can again be solved with the help of the LST. First, we write the following algebraic equation with the natural initial condition $P_0(t) = 1$:

$$-1 + s\varphi_0(s) = -\lambda\varphi_0(s) \qquad (6.21)$$

and then solve it to obtain

$$P_0(t) = e^{-\lambda t} \qquad (6.22)$$

Mean Time to Failure To find the unit's MTTF, we should recall that the mean of nonnegative r.v.'s can be found as

$$E\{X\} = \int_0^\infty P(x)\, dx \qquad (6.23)$$

where $P(t) = 1 - F(t)$. Using the previous notation, we take $P(t) = P_0(t)$. At the same time, we can write

$$E\{X\} = \int_0^\infty \bar{F}(x)e^{-sx}\, dx\big|_{s=0} \qquad (6.24)$$

It follows that, to find the MTTF, we can use the solution for $P_0(t)$ in terms of the LST and substitute $s = 0$. In fact, it is even sometimes simpler to solve a corresponding system of equations directly with the substitution $s = 0$. Considering a single unit, there is no technical difference:

$$T = \frac{1}{\lambda + s}\bigg|_{s=0} = \frac{1}{\lambda} \qquad (6.25)$$

Notice that if we need to find the MTR, it is necessary to start from state 1 and choose state 0 as absorbing.

We present an in-depth analysis of this simple case in order to make future explanations of more complex models more understandable. We do this to avoid explanations below with unnecessary additional details. The same purpose drives us to use a homogeneous mathematical technique for all routine approaches (though, in general, we try to use various methods because our main purpose is to present ideas and not results).

6.1.2 General Distributions

Many results can be obtained for a renewal unit. We remark that it might be very useful for the reader to review Section 1.6.5.

Consider an alternative renewal process $\{\xi, \eta\}$ starting with a subinterval of type ξ; that is, at $t = 0$ an r.v. ξ_1 starts. This process can be considered as a model of the operation of a socket with installed units which is replaced after failure. In this case ξ is the random TTF and η is the random repair time. Let $F(t)$ and $G(t)$ be the distributions of the r.v.'s ξ and η, respectively. Let us call $\theta_k = \xi_k + \eta_k$ the kth cycle of operation of the socket. The distribution of θ can be written as

$$B(t) = \int_0^t F(t - x)\, dG(x) = \int_0^t G(t - x)\, dF(x) \qquad (6.26)$$

Nonstationary Availability Coefficient This reliability index means that at moment t, a unit is in an up state; that is, one of the r.v.'s ξ covers the point t on the time axis (see Figure 6.5): Consider a renewal process formed with $\{\theta\}$ and denote a renewal function of this process by $H(t)$. Then, using the results of Section 1.5.2, we immediately obtain the following integral equation:

$$K(t) = 1 - F(t) + \int_0^t \left[1 - F(t - x)\right] dH(x) \qquad (6.27)$$

In other words, (6.27) means that either no failures have occurred or—if failures have occurred—the last cycle θ is completed by moment $x, 0 < x \le t$, and a new r.v. ξ is larger than the remaining time, $\xi > t - x$. The function $H(t)$ in this case is

$$H(t) = \sum_{\forall k} B^{*k}(t)$$

where $B^{*k}(t)$ is the k-order convolution of $B(t)$. Thus, in general, $K(t)$ could be found with the help of (6.27).

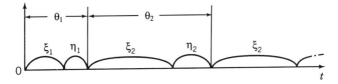

Figure 6.5. Time diagram for an alternative renewal process describing a unit operation.

Stationary Availability Coefficient   Intuitively, it becomes clear that (6.27) has a limit when time is increasing: $K(t) \to K$. (Strictly speaking, the involved distributions must be continuous.) Indeed, applying the Smith theorem (Section 1.5.2), we obtain

$$K = \lim_{t \to \infty} K(t) = \frac{1}{E\{\theta\}} \int_0^\infty [1 - F(x)] \, dx = \frac{E\{\xi\}}{E\{\xi\} + E\{\eta\}} \quad (6.28)$$

On a heuristic level, this result can be explained by the following arguments. Consider some interval of time L such that the number of cycles on it n is sufficiently large. Then

$$L = \xi_1 + \eta_1 + \xi_2 + \eta_2 + \cdots + \xi_n + \eta_n$$

The index K is the probability that an arbitrary moment will be covered by an interval of type ξ. It is clear that this probability is proportional to the total portion of time occupied by all intervals of type ξ:

$$K = \frac{\displaystyle\sum_{1 \le i \le n} \xi_i}{\displaystyle\sum_{1 \le i \le n} \xi_i + \sum_{1 \le i \le n} \eta_i} = \frac{\dfrac{1}{n} \displaystyle\sum_{1 \le i \le n} \xi_i}{\dfrac{1}{n} \displaystyle\sum_{1 \le i \le n} \xi_i + \dfrac{1}{n} \sum_{1 \le i \le n} \eta_i}$$

and, if n is large, one may replace each sum with the coefficient $1/n$ for the mean of the respective r.v.

$$K = \frac{E\{\xi\}}{E\{\xi\} + E\{\eta\}} \quad (6.29)$$

Nonstationary Interval Availability Coefficient   Again, we can write the integral equation

$$R(t, t_0) = 1 - F(t + t_0) + \int_0^t [1 - F(t + t_0 - x)] \, dH(x) \quad (6.30)$$

The explanation of (6.30) is similar to the explanation of (6.27).

Stationary Interval Availability Coefficient   Again, we use the Smith theorem and write

$$R(t_0) = \lim_{t \to \infty} R(t, t_0) = \frac{1}{E\{\theta\}} \int_{t_0}^\infty P(x) \, dx \quad (6.31)$$

It is convenient to rewrite (6.31) in the form

$$R(t_0) = K \Pr\{\zeta \geq t_0\} = K \frac{1}{E\{\xi\}} \int_{t_0}^{\infty} P(x) \, dx \qquad (6.32)$$

where ζ is the residual time of the renewal process formed with the r.v.'s $\{\xi\}$.

From (6.32) it becomes clear that $R(t_0)$ differs from $R_{wrong}(t_0) = KP(t_0)$. In engineering practice, nevertheless, $R_{wrong}(t_0)$ is often erroneously used. We should emphasize that ξ and its residual time ζ are statistically equivalent only for an exponentially distributed r.v. Consequently, in this case (and only in this case!),

$$R(t_0) = KP(t_0) = R(t_0) = Ke^{-\lambda t}$$

For a highly reliable unit, (6.32) can be written in the convenient form of two-sided bounds if $F(t)$ is "aging." For this purpose we use a result from Chapter 5. Recall that

$$\int_t^{\infty} D_T(x) \, dx \leq \int_t^{\infty} F(x) \, dx \leq \int_t^{\infty} e^{-x/T} \, dx$$

where $F(t)$ is an "aging" distribution with mean T and $D_T(t)$ is a degenerate distribution, that is, a constant T. Then it follows that

$$1 - \frac{t_0}{T} \leq \tilde{P}(t_0) = \Pr\{\zeta \geq t_0\} \leq e^{-t_0/T}$$

where ζ is the residual value of the renewal process formed with $\{\xi\}$.

For a highly reliable unit, we can write a very simple and very convenient approximation

$$\tilde{P}(t_0) \approx 1 - \frac{t_0}{T}$$

Thus, for the index of interest, we write

$$\frac{1}{1 + \dfrac{\tau}{T}} \left(1 - \frac{t_0}{T} \right) \leq R(t_0) \leq \frac{1}{1 + \dfrac{\tau}{T}} e^{-t_0/T} \qquad (6.33)$$

and for a highly reliable unit

$$R(t_0) \approx 1 - \frac{\tau + t_0}{T} \qquad (6.34)$$

6.2 REPAIRABLE SERIES SYSTEM

6.2.1 Markov Systems

Consider a series system of n independent units. Assume that distributions of the TTF, $F_i(t)$, and distributions of the repair time, $G_i(t)$, are exponential:

$$F(t) = 1 - e^{-\lambda t} \qquad G(t) = 1 - e^{-\mu t}$$

Here λ_i and μ_i are the parameters of the distributions, or the intensities of failure and repair, respectively.

Reliability indexes depend on the usage of the system's units during the system's idle time. We consider two main regimes of system units in this situation:

1. After a system failure, a failed unit is shipped to a repair shop and all of the remaining system units are switched off. In other words, the system failure rate equals 0 during repair. In this case only one repair facility is required and there is no queue for repair.

2. After a system failure, a failed unit is shipped to a repair shop but all the remaining system units are kept in an operational state. Each of them can fail during the current repair of the previously failed unit (or units). In this case several repair facilities might be required. If the number of repair facilities is smaller than the number of system units, a queue might form at the repair shop.

System with a Switch-Off During Repair   The transition diagram for this system is presented in Figure 6.6. We will not write the equations for this case. As much as possible, we will try to use simple verbal explanations.

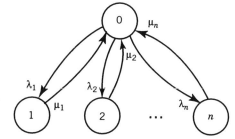

Figure 6.6. Transition graph for a series system which is switching off during idle time.

1. Probability of Failure-Free Operation

Any exit from state 0 leads to failure. Hence,

$$P(t) = \exp\left(-\sum_{1 \le i \le n} \lambda_i t\right) = e^{-\Lambda t} \qquad (6.35)$$

where

$$\Lambda = \sum_{1 \le i \le n} \lambda_i$$

Thus, by this reliability characteristic, the system is equivalent to a single unit with a failure rate Λ.

2. MTTF

If $P(t) = e^{-\Lambda t}$, the MTTF of the system equals $T_{\text{syst}} = 1/\Lambda$. No comments are needed.

3. Mean Repair Time

Let us consider a general case where all units differ by their repair time $1/\mu_i$. The current repair time of the system depends on which unit has failed. The distribution of the system's repair time can be represented in the form

$$\Pr\{\eta \le t\} = \sum_{1 \le i \le n} p_k e^{-\mu_k t} \qquad (6.36)$$

where p_k is the probability that the kth unit is under repair. The probability p_k can be easily found as

$$p_k = \frac{\lambda_k}{\displaystyle\sum_{1 \le i \le n} \lambda_i} \qquad (6.37)$$

Notice that the distribution of the system's repair time has a decreasing intensity function; that is, with the growth of the current repair time, the residual repair time becomes larger and larger.

We consider this phenomenon in more detail. Write (6.36) in the form

$$r(t) = \sum_{1 \le k \le n} p_k \, e^{-\mu_k t} = \exp\left(-\int_0^t \mu(x)\, dx\right)$$

From here we find

$$\mu(t) = -\frac{dr(t)}{r(t)\, dt} = \frac{\displaystyle\sum_{1 \le k \le n} \mu_k p_k \, e^{-\mu_k t}}{\displaystyle\sum_{1 \le k \le n} p_k \, e^{-\mu_k t}}$$

Now we note that $\mu(t)$ is a monotone function. For $t = 0$, a simple qualitative analysis gives us

$$\mu(t) = \frac{\displaystyle\sum_{1 \le k \le n} \mu_k p_k}{\displaystyle\sum_{1 \le k \le n} p_k} = \sum_{1 \le k \le n} \mu_k p_k = \mathrm{E}\{\mu\}$$

Now, as $t \to \infty$,

$$\lim_{t \to \infty} \mu(t) = \frac{\mu_{k^*} p_{k^*} \, e^{-\mu_{k^*} t}}{p_{k^*} \, e^{-\mu_{k^*} t}} = \mu_{k^*}$$

where k^* corresponds to the subscript of a minimal μ_k. Obviously, the average value is larger than the minimum. This function is never below the minimal μ_k. Hence, $\mu(t)$ decreases from the average value of μ at $t = 0$ to the minimal value among all μ's. It can be shown that this decrease is monotone.

Of course, from (6.36) and (6.37), it follows immediately that

$$T_{\text{syst}} = \mathrm{E}\{\eta_{\text{syst}}\} = \sum_{1 \le i \le n} \tau_i \frac{\lambda_i}{\displaystyle\sum_{1 \le i \le n} \lambda_i} = \frac{1}{\Lambda} \sum_{1 \le i \le n} \lambda_i \tau_i$$

where $\tau_i = 1/\mu_i$ is the MTTF of the ith units.

4. Nonstationary Availability and Interval Availability Coefficients

We are able to find these reliability indexes only with the help of general methods of renewal process theory, in spite of the exponentiality of a TTF distribution. One can also use standard Markov methods applied to the transition graph presented in Figure 6.6. The corresponding system of equations for the availability coefficient is

$$P'_k(t) = -\mu_k P_k(t) + \lambda_k P_0(t) \qquad 1 \le k \le n$$

$$\sum P_k(t) = 1 \qquad \text{taken for } 0 \le k \le n$$

and the initial condition $P_0(0) = 1$. We will not solve these equations here. But if $K(t)$ is found, then—because of the exponentiality of the TTF distribution—$R(t, t_0) = K(t) e^{-\Lambda t}$.

5. Stationary Availability Coefficient

With known T_{syst} and τ_{syst}, this index can be found in a standard way as $K = (T_{\text{syst}})/(T_{\text{syst}} + \tau_{\text{syst}})$. Note that in this particular case it is convenient to write

$$K = \frac{1}{1 + \sum_{1 \le i \le n} \lambda_i \tau_i} \tag{6.38}$$

6. Stationary Interval Availability Coefficient

Because of the exponential distribution of the system TTF, we can use the expression $R(t_0) = KP(t_0)$ where $P(t_0)$ is defined in (6.35).

Notice that if all μ_i are constant (equal to μ), the above-described system is transformed into a single repairable unit with an intensity of failure equal to

$$\Lambda = \sum_{1 \le i \le n} \lambda_i$$

and an intensity of repair μ. In most practical cases it is enough for the first stages of design to put $\mu = E\{\mu\}$ in this model and to use this approximation instead of using the exact model. We remark that in most practical cases, when the equipment has a modular construction, the mean time of repair might be considered almost equal for different system units. But an even more important argument for such a suggestion is that a mathematical model must not be too accurate at the design stage when one does not have accurate input data.

System Without Switch-Off During Repair   First, consider a series system of n different repairable units when there are n repair facilities in a workshop; that is, each unit might be repaired independently. The units' failures are assumed independent. In this case the system can be considered

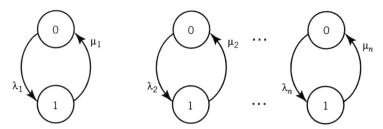

Figure 6.7. Set of transition graphs for independently operating units of a series system.

as a set of independent repairable units. A set of corresponding transition diagrams is presented in Figure 6.7. In this case

$$P_{\text{syst}}(t_0) = \prod_{1 \leq i \leq n} P_i(t_0) = e^{-\Lambda t}$$

$$T_{\text{syst}} = \frac{1}{\Lambda}$$

$$K_{\text{syst}}(t) = \prod_{1 \leq i \leq n} K_i(t) \tag{6.39}$$

$$K_{\text{syst}} = \prod_{1 \leq i \leq n} K_i = \prod_{1 \leq i \leq n} \frac{T_i}{T_i + \tau_i} = \prod_{1 \leq i \leq n} \frac{1}{1 + \lambda_i \tau_i}$$

In this case it is not a simple task to find τ_{syst} in a direct way. But if we use the direct definition of K_{syst},

$$\frac{T_{\text{syst}}}{T_{\text{syst}} + \tau_{\text{syst}}} = K_{\text{syst}}$$

then

$$\tau_{\text{syst}} = \frac{1 - K}{K} T_{\text{syst}} \tag{6.40}$$

where all variables on the left side are known.

In more complex cases when, for instance, the number of repair facilities k is less than n, the results concerning reliability indexes cannot be obtained so simply, especially if we consider a system with different units (this is the most realistic practical case, by the way). In this case there is no way other than to construct a transition graph, to write a system of linear differential equations, and then to solve them.

6.2.2 General Distribution of Repair Time

If the TTFs of all the units remain exponentially distributed, the main simple results can be obtained practically in the same form as for the Markovian model.

System With Switch-Off During Repair   First of all, $P_{syst}(t_0)$ and T_{syst} remain the same as in the previous case. The mean repair time is defined with the help (6.36). Consequently, the stationary availability and interval availability coefficients can be expressed in standard form. At the same time, nonstationary indexes can be found with the help of the general methods of renewal process theory. The model of the investigated operation process forms an alternative process $\{\xi^*, \eta^*\}$. Each ξ^* is an exponential r.v. with parameter Λ and η^* is an r.v. with a complex "weighted" d.f.

$$G^*(t) = \Pr\{\eta^* \le t\} = \frac{1}{\Lambda} \sum_{1 \le i \le n} \lambda_i G_i(t)$$

For analytical purposes it is more reasonable to use Monte Carlo simulation. We would like to emphasize again that a detailed exploration of a nonstationary regime is usually a task far removed from practical needs because of the insufficiency of the input data.

System Without Switch-Off During Repair   In this case $P_{syst}(t_0)$ and T_{syst} remain the same as in the previous cases. Even such a stationary index as K can be found only for the case when the number of repair facilities equals the number of system units, that is, when all system units are totally independent. In this case K is defined as

$$K = \frac{1}{1 + \Lambda_{syst}\tau_{syst}}$$

If the system units are dependent through the lack of repair facilities, we recommend the use of Monte Carlo simulation for the computation of nonstationary indexes.

But if K or $K(t)$ is known, to find $R(t_0)$ or $R(t, t_0)$ is a simple task because of the exponentiality of the system TTF: $R(t_0) = KP(t_0)$ and $R(t, t_0) = K(t)P(t_0)$.

6.2.3 General Distributions of TTF and Repair Time

This case is especially difficult if one considers nonstationary indexes. They can only be found with the help of Monte Carlo simulation. Let each unit of the system be described with the help of an alternative renewal process. The superposition of these processes is not an alternative renewal stochastic process. The new process has more sophisticated structure: it does not have the regeneration moments that appeared when one considered a system of units with exponential TTF.

But to find the stationary coefficient K, we can use the idea that stationary probabilities do not depend on the distribution of the repair time. Therefore, one can use a Markov model with the following parameters for each unit:

$$\lambda_i = \frac{1}{T_i} = \frac{1}{\int_0^\infty [1 - F_i(t)]\, dt} \quad \text{and} \quad \mu_i = \frac{1}{\tau_i} = \frac{1}{\int_0^\infty [1 - G_i(t)]\, dt}$$

If the number of repair facilities equals the number of system units (all units are totally independent), the system stationary availability coefficient can be found as

$$K_{\text{syst}} = \prod_{1 \le i \le n} \frac{T_i}{T_i + \tau_i}$$

The stationary interval availability coefficient can also be found with the help of the following arguments. For each unit we can easily find the conditional stationary probability of a failure-free operation under the condition that a unit is in an operational state at the starting moment:

$$\tilde{P}_i(t) = \Pr\{\zeta \ge t_0\} = \frac{1}{T_i} \int_{t_0}^\infty P(x)\, dx$$

Then for the system

$$\tilde{P}_{\text{syst}}(t_0) = \prod_{1 \le i \le n} \frac{T_i}{T_i + \tau_i} \tilde{P}_i(t_0)$$

Now it is possible to write $R(t_0)$ as

$$R_{\text{syst}}(t_0) = K_{\text{syst}} \tilde{P}_{\text{syst}}(t_0) = \prod_{1 \le i \le n} \frac{1}{T_i + \tau_i} \int_{t_0}^\infty P_i(x)\, dx$$

Also, we can again use the two-sided bounds (6.33) if the unit TTF distributions are "aging":

$$\prod_{1 \le i \le n} \frac{1}{1 + \dfrac{\tau_i}{T_i}} \left(1 - \frac{t_0}{T_i}\right) \le R_{\text{syst}}(t_0) \le \prod_{1 \le i \le n} \frac{1}{1 + \dfrac{\tau_i}{T_i}} \exp\left(-t_0 \sum_{1 \le i \le n} \frac{1}{T_i}\right)$$

$$(6.41)$$

Naturally, using (6.34) for a highly reliable system, we can write

$$R_{\text{syst}}(t_0) \approx 1 - \sum_{1 \le i \le n} \frac{\tau_i + t_0}{T_i} \qquad (6.42)$$

Of course, analogous approximations could be written for all of the above-considered cases in this section.

6.3 REPAIRABLE REDUNDANT SYSTEMS OF IDENTICAL UNITS

6.3.1 General Markov Model

Let us consider a redundant system consisting of k main operating units and $n = n_1 + n_2 + n_3$ redundant units. Here we use the following notation:

- n_1 is the number of active redundant units in the same regime as the main operating units; each unit has a failure rate λ;
- n_2 is the number of units in an underloaded on-duty regime; each unit has a failure rate λ', $\lambda' = \nu\lambda$, where ν is the so-called loading coefficient, $0 < \nu < 1$;
- n_3 is the number of standby units; each unit has $\lambda'' = 0$.

A failed unit is shipped to the repair shop. A failed operating unit is replaced with an active redundant unit. Instantly, this unit is replaced by a unit which is in an underloaded regime, and, in turn, the latter is replaced by a standby unit. An analogous chainlike procedure is performed with a failed unit of other levels of redundancy. There are l repair facilities, $1 \le l \le n + k$. All units have an exponential repair time distribution with the same parameter μ.

Let H_j denote a system state with j failed units. Obviously, the system can change its state H_j only for one of two neighboring states: H_{j-1} after the repair of a failed unit or H_{j+1} after a new failure. Hence, this process is described with the help of a linear transition graph (see Figure 6.8) and belongs to the birth and death process (see Section 1.6).

The transition from state H_j to state H_{j+1} within a time interval $[t, t + \Delta]$ occurs with probability $\Lambda_j \Delta t + o(\Delta t)$. The transition to state H_{j-1} in the same time interval occurs with probability $M_j \Delta t + o(\Delta t)$. With probability $1 - \Lambda_j \Delta t - M_j \Delta t - o(\Delta t)$, no change occurs. For underloaded units, the coefficient of loading is ν, $0 < \nu < 1$.

A system with n redundant units has $n + k + 1$ states $H_0, H_1, H_2, \ldots, H_{n+k}$. States H_{n+j} with $n + j$ failed units, $1 \le j \le k$, are states corresponding to a system failure. After an exit into the first system failure state, H_{n+1}, the process develops further: it may move to the next

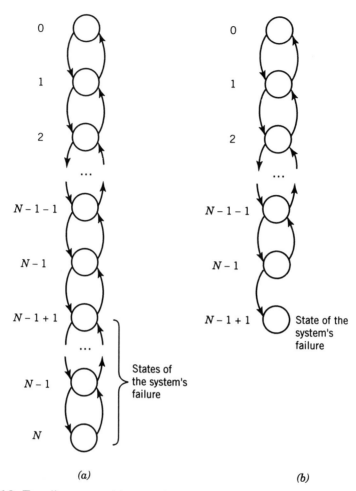

0

1

2

...

$N-1-1$

$N-1$

$N-1+1$

...

$N-1$

N

States of the system's failure

0

1

2

...

$N-1-1$

$N-1$

$N-1+1$ State of the system's failure

(a)   *(b)*

Figure 6.8. Two linear transition graphs for a redundant system: (*a*) without an absorbing state; (*b*) with an absorbing state.

system failure state, H_{n+2}, and so on, or it may return to the up state, H_n. If the state H_{n+1} is absorbing, the system of equations must be changed: all absorbing states must have no transition to the set of operational states. This new system of equations can be used for calculating the probabilities of successful operation, the interval availability coefficient, the MTTF, and the MTBF.

If we consider the nonstationary and stationary availability coefficients, the state H_{n+k} is reflecting. The corresponding system of equations can be used for calculating the nonstationary availability and/or interval availability coefficients.

Considering the transition graph without an absorbing state, for a state H_j, $0 \leq j \leq n = n_1 + n_2 + n_3 + k$, one may write Λ_j and M_j:

$$\Lambda_0 = k\lambda + n_1\lambda + n_2\nu\lambda$$
$$\Lambda_1 = k\lambda + n_1\lambda + n_2\nu\lambda = \Lambda_0$$
$$\vdots$$
$$\Lambda_{n_3} = \Lambda_0$$
$$\Lambda_{n_3+1} = k\lambda + n_1\lambda + (n_2 - 1)\nu\lambda$$
$$\Lambda_{n_3+2} = k\lambda + n_1\lambda + (n_2 - 2)\nu\lambda$$
$$\vdots$$
$$\Lambda_{n_3+n_2} = k\lambda + n_1\lambda$$
$$\Lambda_{n_3+n_2+1} = k\lambda + (n_1 - 1)\lambda$$
$$\Lambda_{n_3+n_2+2} = k\lambda + (n_1 - 2)\lambda$$
$$\vdots$$
$$\Lambda_{n_3+n_2+n_1} = k\lambda$$
$$\Lambda_{n_1+n_2+n_3+1} = (k - 1)\lambda$$
$$\Lambda_{n_1+n_2+n_3+2} = (k - 2)\lambda$$
$$\vdots$$
$$\Lambda_{n_1+n_2+n_3+k} = 0$$

and for all M_j, $0 \leq j \leq n + k = n_1 + n_2 + n_3 + k$,

$$M_1 = \mu, \; M_2 = 2\mu, \ldots, M_l = l\mu, \qquad M_{l+1} = l\mu, \ldots, M_{n+k} = l\mu$$

The system with the absorbing H_{n+1} state is the system which operates until a first failure. This system can be analyzed with the following system of differential equations:

$$\frac{dp_j(t)}{dt} = \Lambda_{j-1}p_{j-1}(t) - (\Lambda_j + M_j)p_j(t) + M_{j+1}p_{j+1}(t) \qquad 0 \leq j \leq n + 1$$
(6.43)

$$\Lambda_{-1} = \Lambda_{n+1} = M_0 = M_n = \cdots = M_{n+k} = 0 \qquad (6.44)$$

where $p_j(t)$ is the probability that the system is in state H_j at moment t. The normalization condition is

$$\sum_{0 \leq j \leq n+1} p_j(t) = 1$$

The system with the reflecting H_{n+k} state can be described by the following system of differential equations:

$$\frac{dp_j(t)}{dt} = \Lambda_{j-1} p_{j-1}(t) - (\Lambda_j + M_j) p_j(t) + M_{j+1} p_{j+1}(t) \qquad 0 \le j \le n + k$$

$$\Lambda_{-1} = \Lambda_{n+k} = M_0 = M_{n+k+1} = 0$$

with normalization equation

$$\sum_{0 \le j \le n+k} p_j(t) = 1$$

Because our goal is not to write down formulas for very general models but to show the methodology and methods, we hope that the reader can use the corresponding equations from Section 1.6 dedicated to the death and birth process.

Precise formulas for such a general case are almost always long and complicated. If one deals with highly reliable systems, we recommend the reader refer to Chapter 12. (If one deals with an unreliable system, we recommend a redesign of the system, not a useless calculation!)

The next section is devoted to general methods of analysis of repairable systems.

6.4 GENERAL MARKOV MODEL OF REPAIRABLE SYSTEMS

6.4.1 Description of the Transition Graph

From the very beginning, we would like to emphasize that a Markov model is an idealization of a real process. Our main problem is not to solve the system of mathematical equations but rather to identify the real problem, to determine if the real problem and the model are an appropriate fit to each other. If, in fact, they are a good fit, then a Markov model is very convenient.

Now let us assume that we can construct the transition graph which describes a system's operation. This graph must represent a set of mutually exclusive and totally exhaustive system states with all of their possible one-step transitions. Using some criterion of system failure, all of these states can be divided into two complementary disjoint subsets, *up states* and *down states*. A transition from the subset of up states to the subset of down states may occur only when an operating unit fails. An inverse transition may occur only if a failed unit is renewed by either a direct repair or by a replacement. Let us consider a system with n units. Any system state may be denoted by a binary vector

$$\mathbf{s} = (s_1, \ldots, s_n)$$

where s_i is the state of the ith unit. We set $s_i = 1$ if the unit is operational and $s_i = 0$ otherwise. The transition from $(s_1, \ldots, s_i = 1, \ldots, s_n)$ to $(s_1, \ldots, s_i = 0, \ldots, s_n)$ means that the ith unit changes its state from up to down. The *transition rate* (or the *transition intensity*) for this case equals the ith unit's failure rate.

A transition from system state $(s_1, \ldots, s_i = 0, \ldots, s_n)$ to state $(s_1, \ldots, s_i = 1, \ldots, s_n)$ means that the ith unit was in a failed state and was renewed . The transition rate for this case equals the ith unit's repair rate. These kinds of transitions are most common. For Markovian models we assume that only one unit may fail (or be renewed) at a time. (If several units may change states simultaneously, for example, under a group repair, we will consider this separately.) Of course, there are other possible interpretations of states and transitions. For instance, $s_i = 1$ may be a state before monitoring or switching, and $s_i = 0$ is the same state after the procedure. We denote these transitions from state to state on transition graphs with arrows. The rates (intensities) are denoted as weights on the arrows. The graph structure is determined by the operational and maintenance regime of the system's units and the system itself. After the transition graph has been constructed, it can be used as a visual aid to determine different reliability indexes. An example of such a transition graph for a system consisting of three different units is presented in Figure 6.9.

6.4.2 Nonstationary Coefficient of Availability

Let $E(k)$ denote the subset of the entire set of system states which includes states from which a direct transition to state k is possible, and let $e(k)$ denote the subset to which a direct transition from state k is possible. The union $E(k) \cup e(k)$ is the subset of system states that have a direct connection to or from state k (see Figure 6.10). For each state k of the transition graph, we can write the following differential equation:

$$\frac{d}{dt} p_k(t) = -p_k(t) \sum_{i \in e(k)} \Lambda_{ik} + \sum_{i \in E(k)} \Lambda_{ik} p_i(t) \tag{6.45}$$

where Λ_{ik} is the intensity of the transition from i to k, and $p_i(t)$ is the probability that the system is in state H_i at moment t.

The transition graph and the system of differential equations can be interpreted as those which describe a dynamic equilibrium. Indeed, imagine that each node i is a "basin" with "liquid" which flows to each other node j (if there is an arrow in the corresponding direction). The intensity of the flow is proportional to λ_{ij} (specified in each direction) and to a current amount of "liquid" in the source $p_i(t)$.

If there are n states, we can construct n differential equations. To find the nonstationary coefficient of availability, we take any $n - 1$ equations, add the

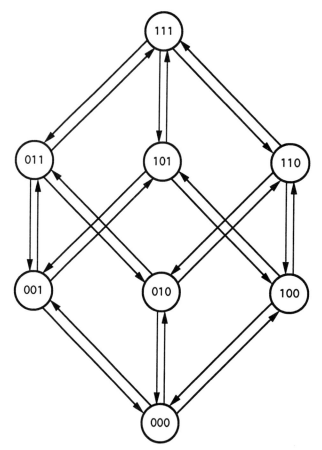

Figure 6.9. Transition graph for a system consisting of three different renewable units.

normalization condition

$$\sum_{1 \le i \le n} p_i(t) = 1 \tag{6.46}$$

and add initial conditions of the type $p_i(0) = p_i$ where $p_i(0)$ is the probability that the system is in state i at $t = 0$. In turn, the p_i's are probabilities that conform to a normalization condition similar to (6.46). If $p_i = 1$ for some i, then $p_j = 0$ for all j, $j \ne i$. In most problems the initial system state is the state when all units are up.

To find the nonstationary availability coefficient, we can use the Laplace–Stieltjes transform (LST). Then the system of n linear differential

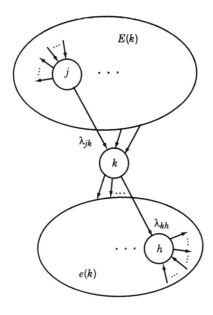

Figure 6.10. Fragment of a transition graph for compiling a system of differential equations.

equations transforms into the system of linear algebraic equations:

$$s\varphi_k(s) - p_k = -\varphi_k(s) \sum_{i \in e(k)} \Lambda_{ki} + \sum_{i \in E(k)} \Lambda_{ik}\varphi_i(s) \qquad (6.47)$$

$$\sum_{1 \le i \le n} s\varphi_i(s) = 1 \qquad (6.48)$$

where $\varphi_i(s)$ is the LST for $p_i(t)$:

$$\varphi_i(s) = \int_0^\infty p_i(t) e^{-st} dt \qquad (6.49)$$

For writing a system of algebraic equations directly in terms of the LST, one can construct a special graph which is close to the one depicted in Figure 6.11. This new graph includes a state (distinguished by shadowing) which "sends" to each state i of the graph a "flow" equal to the value of $p_i(0)$. Recall that this is the probability that the system is in state i at time $t = 0$. At the same time, each state "sends" to this special state a "flow" equal to s (argument of the LST). The construction of this graph can be clarified by a comparison with the previous one depicted in Figure 6.10.

In (6.47) we use any $n - 1$ equations of the total number n, because the entire group of equations is linearly dependent. This is always true when we consider a transition graph without absorbing states. In this case, in order to find all n unknown $\varphi_i(s)$'s, we must use the normalization equation (6.46).

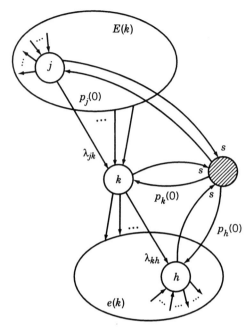

Figure 6.11. Fragment of a transition graph for compiling a system of algebraic equations in LST terms.

This system of equations may be written in canonical form:

$$b_{11}\varphi_1(s) + b_{11}\varphi_2(s) + \cdots + b_{1n}\varphi_n(s) = c_1$$
$$b_{21}\varphi_1(s) + b_{22}\varphi_2(s) + \cdots + b_{2n}\varphi_n(s) = c_2$$
$$\vdots \qquad\qquad (6.50)$$
$$b_{n1}\varphi_1(s) + b_{n2}\varphi_2(s) + \cdots + b_{nn}\varphi_n(s) = c_n$$

where b_{ij} is the coefficient of the jth term of the ith row and c_i is the corresponding constant.

To solve this system of linear equations, we can apply Cramér's rule:

$$\varphi_i(s) = \frac{D_i(s)}{D(s)} \qquad (6.51)$$

where $D(s)$ is the determinant of the system and $D_i(s)$ is the determinant of the matrix formed by replacing the ith column of D by the vector $(c_1, c_2, \ldots, c_n)$. Once more, we repeat that the reference to Cramér's rule is made for explicit explanations, not as a recommendation for computation.

We then find the LST of the availability coefficient:

$$\varphi(s) = \sum_{i \in E} \varphi_i(s) = \frac{1}{D(s)} \sum_{i \in E} D_i(s) \tag{6.52}$$

where E is a subset of up states. We can use the following procedure to invert this LST.

1. Write $\varphi(s)$ in the form

$$\varphi(s) = \frac{A_0 + A_1 s + A_2 s^2 + \cdots + A_n s^n}{B_0 + B_1 s + B_2 s^2 + \cdots + B_{n+1} s^{n+1}} \tag{6.53}$$

where A_j and B_j are known coefficients.

2. Find the polynomial roots:

$$B_0 + B_1 s + B_2 s^2 + \cdots + B_{n+1} s^{n+1} = 0$$

Let these roots be $b_1, b_2, \ldots, b_{n+1}$. Thus,

$$B_0 + B_1 s + \cdots + B_{n+1} s^{n+1} = \prod_{1 \le j \le n+1} (s - b_j)$$

3. Write $\varphi(s)$ in the form of a sum of simple fractions:

$$\varphi(s) = \frac{\beta_1}{s - b_1} + \frac{\beta_2}{s - b_2} + \cdots + \frac{\beta_{n+1}}{s - b_{n+1}} \tag{6.54}$$

where the β_i's are coefficients to be found.

4. Rewrite $\varphi(s)$ in the form

$$\varphi(s) = \frac{\displaystyle\sum_{1 \le j \le n} \beta_j \prod_{i \ne j} (s - b_i)}{(s - b_1)(s - b_2) \cdots (s - b_{m+1})}$$

After elementary transformations, we obtain

$$\varphi(s) = \frac{\alpha_0 + \alpha_1 s + \alpha_2 s^2 + \cdots + \alpha_m s^m}{(s - b_1)(s - b_2) \cdots (s - b_{m+1})}$$

where the α_i's are expressed through different β_j's and b_j's.

5. Polynomials of the form $\varphi(s)$ and of the form of (6.53) are equal if and only if

$$A_0 = \alpha_0, \ A_1 = \alpha_1, \ A_2 = \alpha_2, \ldots, A_n = \alpha_n$$

The $\alpha_i s$'s are defined from these equations.

6. After we have found α_i's, the inverse LST is applied to $\varphi(s)$ in the form of (6.54)

$$\varphi(s) = \sum_{1 \le j \le n+1} \frac{\beta_j}{s - b_j} \Leftrightarrow K(t) = \sum_{1 \le j \le n+1} \beta_j e^{b_j t}$$

REMARK. If $\alpha(s)$ has multiple roots for the denominator, that is, if several b_j's are equal, then (6.54) may be rewritten as

$$\varphi(s) = \sum_{1 \le i \le n'} \frac{\beta_j}{(s - b_j)^k}$$

where k is the number of roots equal to b_j and n' is the number of different roots. To all terms of the form

$$\frac{\beta_j}{(s - b_j)^k}$$

the corresponding inverse LST is applied:

$$\frac{\beta_j}{(s - b_j)^k} \Leftrightarrow \beta_j \frac{t^{k-1}}{(k - 1)!} e^{b_j t}$$

6.4.3 Probability of Failure-Free Operation

To determine the probability of a failure-free operation, absorbing states are introduced into the transition graph. They are the system's failure states. Transitions from any absorbing state are impossible, which means that all transition intensities out of an absorbing state are 0. We can change the domain of summation in the previous equations in a way which is equivalent to eliminating the zero transition rates. Using the previous notation, we can write for an operational state k:

$$\frac{d}{dt} P_k(t) = -P_k(t) \sum_{i \in e(k)} \Lambda_{ki} + \sum_{i \in E(k)} \Lambda_{ik} P_i(t)$$

If the transition graph has m operational states, we can construct m differential equations. (In this case the equations are not linearly dependent. Of course, we may use the normalization condition as one of the equations in this new system, excluding any one of the differential equations.) These equations and initial conditions are used to find the probability of a failure-free operation of the system.

We again use the LST to find the following system of linear differential equations:

$$s\varphi_k(s) - p_k = -\varphi_k(s) \sum_{i \in e(k)} \Lambda_{ki} + \sum_{i \in E(k)} \Lambda_{ik}\varphi_i(s) \qquad (6.55)$$

for all $k \in E$. The solution of this system of equations can be found with the help of the same methodology as before.

6.4.4 Determination of the MTTF and MTBF

Recall that

$$T = \int_0^\infty P(t)\, dt$$

If $\varphi(s)$ is the LST for the probability of failure-free operation of the system, then

$$T = \left[\int_0^\infty e^{-st}P(t)\, dt\right]_{s=0} = \varphi(s)|_{s=0}$$

Thus, we can find the MTTF (or MTBF) by solving the following system:

$$-p_k = \varphi_k(0) \sum_{i \in e(k)} \Lambda_{ki} + \sum_{i \in E(k)} \Lambda_{ik}\varphi_i(0)$$

for all $k \in E$. Note once more that this system was derived from (6.55) by the substitution of $s = 0$. To find the MTTF, one sets the initial conditions as $p_i(0) = 1$, where i is the subscript of a state in which the system is totally operable. Obviously, $p_j(0) = 0$ for all the other states. To find the MTBF, we set the initial conditions in the form $p_i^*(0) = p_i^*$ where the p_i^*'s in this case are the conditional stationary probabilities of the states i that belong to E^*. The latter is a subset of the up states which the process visits first after the system renewal.

The conditional stationary probabilities p_i^*'s can be obtained from the unconditional ones as

$$p_i^* = \frac{p_i(0)}{\displaystyle\sum_{j \in E^*} p_j(0)}$$

Example 6.1 Consider a repairable system of two different units in parallel (Figure 6.12). The parameters of the units are λ_1, λ_2, μ_1, and μ_2. Both units can be repaired independently. The transition graph is presented in Figure 6.13. Here H_0 is the state with both units operational; H_1 (H_2) is the state where the first (second) unit failed; H_{12} is the state where both units failed.

Let $p_k(t)$ equal the probability of the kth state at moment t. There are two systems of equations to calculate the reliability indexes. If the system's failure state H_{12} is reflecting, the system of equations is

$$\frac{d}{dt}p_0(t) = -(\lambda_1 + \lambda_2)p_0(t) + \mu_1 p_1(t) + \mu_2 p_2(t)$$

$$\frac{d}{dt}p_1(t) = \lambda_1 p_0(t) - (\lambda_2 + \mu_1)p_1(t) + \mu_2 p_{12}(t)$$

$$\frac{d}{dt}p_2(t) = \lambda_2 p_0(t) - (\lambda_1 + \mu_2)p_2(t) + \mu_1 p_{12}(t)$$

$$p_0(t) + p_1(t) + p_2(t) + p_{12}(t) = 1$$

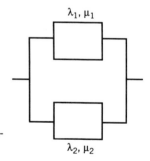

Figure 6.12. Repairable system of two different independent units connected in parallel.

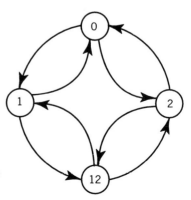

Figure 6.13. Transition graph for the system depicted in Figure 6.12.

If state H_{12} is absorbing,

$$\frac{d}{dt}p_0(t) = -(\lambda_1 + \lambda_2)p_0(t) + \mu_1 p_1(t) + \mu_2 p_2(t)$$

$$\frac{d}{dt}p_1(t) = \lambda_1 p_0(t) - (\lambda_2 + \mu_1)p_1(t)$$

$$\frac{d}{dt}p_2(t) = \lambda_2 p_0(t) - (\lambda_1 + \mu_2)p_2(t)$$

$$\frac{d}{dt}p_{12}(t) = \lambda_2 p_1(t) + \lambda_1 p_2(t)$$

The corresponding solutions in the form of the Cramér determinants are

$$T = -\frac{\begin{vmatrix} 1 & 1 & 1 & 0 \\ -(\lambda_1 + \lambda_2) & \mu_1 & \mu_2 & p_0(0) \\ \lambda_1 & -(\lambda_0 + \mu_1) & 0 & p_1(0) \\ \lambda_2 & 0 & -(\lambda_1 + \mu_2) & p_2(0) \end{vmatrix}}{\begin{vmatrix} -(\lambda_1 + \lambda_2) & \mu_1 & \mu_2 \\ \lambda_1 & -(\lambda_2 + \mu_1) & 0 \\ \lambda_2 & 0 & -(\lambda_1 + \mu_2) \end{vmatrix}}$$

$$T^{(0)} = -\frac{\begin{vmatrix} 1 & 1 & 1 \\ \lambda_1 & -(\lambda_0 + \mu_1) & 0 \\ \lambda_2 & 0 & -(\lambda_1 + \mu_2) \end{vmatrix}}{\begin{vmatrix} -(\lambda_1 + \lambda_2) & \mu_1 & \mu_2 \\ \lambda_1 & -(\lambda_2 + \mu_1) & 0 \\ \lambda_2 & 0 & -(\lambda_1 + \mu_2) \end{vmatrix}}$$

$$K = -\frac{\begin{vmatrix} 1 & 1 & 1 & 1 \\ -(\lambda_1 + \lambda_2) & \mu_1 & \mu_2 & 0 \\ \lambda_1 & -(\lambda_0 + \mu_1) & 0 & \mu_2 \\ \lambda_2 & 0 & -(\lambda_1 + \mu_2) & \mu_1 \end{vmatrix}}{\begin{vmatrix} 1 & 1 & 1 & 1 \\ -(\lambda_1 + \lambda_2) & \mu_1 & \mu_2 & 0 \\ \lambda_1 & -(\lambda_2 + \mu_1) & 0 & \mu_2 \\ \lambda_2 & 0 & -(\lambda_1 + \mu_2) & \mu_1 \end{vmatrix}}$$

The solutions are not presented in closed form because of their length and complexity.

6.5   TIME REDUNDANCY

Considering reliability indexes, we emphasize that so-called *time redundancy* might be a very effective measure of a system's reliability improvement. This type of redundancy can be used in two main cases:

- The required time of operation completion by an absolutely reliable system is less than the time admissible for operation performance.
- System failures leading to short idle periods might be ignored in the sense of successful operation performance.

These problems are solved with special mathematical methods differing from the usual ones used in other reliability problems. Let us consider several main types of systems with time redundancy.

6.5.1   System with Instant Failures

Consider a system performing an operation of duration t_0. System failures are very short, practically instantaneous. The flow of these failures can be successfully described by a point renewal process. Each failure interrupts a system's successful operation, and the system is forced to restart its operation from the beginning. In other words, we assume that a failure destroys the result of an operation. For restarting the operation in an attempt to complete the required performance, the system must have a time resource.

Such situations are encountered in practice if one considers a computer operating with short errors which destroy a current result. A computer performs a task which requires t_0 units of failure-free time for its successful performance. Thus, if there is a time resource, a computer can perform its operation even after the appearance of some error.

We assume that the total time for the system performance is $T > \theta$. Let us also assume that the system begins to operate at the moment $t = 0$ when it is "new." The distribution of the TTF is $F(t)$. Let $R_0(\theta|T)$ denote the probability that during interval $[0, T]$ there will be at least one period between failures exceeding the required value θ, and let $P(t) = 1 - F(t)$.

The system performs its operation successfully during time T if two events occur:

- There are no failures during time interval $[0, \theta]$.
- A failure has occurred at $x < \theta$, but, after this moment, the system successfully performs its operation during the remaining time $T - x$.

The latter event is complex. First, a failure might occur at any moment of time between 0 and θ, and, second, at the moment of a failure the process starts from the beginning but for a smaller time interval. This verbal explanation leads us to the recurrent expression

$$R_0(\theta|T) = P(\theta) + \int_0^\theta R_0(\theta|T - x)\, dF(x) \qquad (6.56)$$

If the remaining interval is smaller than θ, the operation cannot be performed successfully. This leads to the condition

$$R_0(\theta|x < \theta) = 0$$

Equations of such a recurrent type are usually solved numerically. We will not provide a mathematical technique for this solution.

Above we considered a situation where a system begins to operate at moment $t = 0$. Now let us assume that a system is in an on-duty regime and a request for starting the operation arrives in a random time. More exactly, we assume that we consider a stationary process, and a random time from the request arrival to a system failure is a residual time. Such a situation is typical of many military systems which must be ready at all times to perform their duties: no enemy in modern times informs you about the beginning of hostile actions.

Let the tilde denote a distribution of the residual time. In this case the expression of interest is not changed significantly. We give it without explanation because of its obviousness:

$$\tilde{R}_0(\theta|T) = \tilde{P}(\theta) + \int_0^\theta R_0(\theta|T - x)\, dF(x) \qquad (6.57)$$

where the function R_0 under the integral must be taken from (6.56) with the corresponding condition.

Of course, in this case we must again write the condition

$$\tilde{R}(\theta|x < t_0) = 0$$

which means that a system cannot successfully perform its operation if the time resource is smaller than the required time of operation.

6.5.2 System with Noninstant Failures

If failures are noninstant, one must take into account the lengths of idle periods between up periods. Let $G(t)$ denote a distribution of idle time. If a failure has occurred within the first interval $[0, t_0]$, a random period of idle time is needed to restore the system. In general, there are no restrictions on the length of the idle time y. Thus, we must consider the possibility that this value changes within the entire interval $[x, T - x]$. At the same time, if the system spent x units of time for unsuccessful operation and then y units of time for restoration, only $T - x - y$ units of time remain to perform the operation.

This verbal description permits us to write a recurrent expression

$$R(\theta|T) = P(\theta) + \int_0^\theta \left[\int_0^{T-x} R(t_0|T - x - y)\, dG(y) \right] dF(x) \quad (6.58)$$

where $R(\theta|x < t_0) = 0$.

Now we consider the above-analyzed system which is operating in an on-duty regime. In principle, the explanation of the equation remains similar to the previous case. We must additionally take into account the fact that the system at an arbitrary stationary moment of time can be found in one of two possible states: up or down. We only explain the situation where a system at the beginning of operation is in a down state. In this case one first observes a residual restoration time and after this a system is considered as "new." Again, we use a tilde to denote the distribution of a residual value. The expression for this case can be written in the form

$$\tilde{R}(t_0|T) = K\left[\tilde{P}(\theta) + \int_0^\theta \left[\int_0^{T-x} R(t_0|T - x - y)\, dG(y) \right] dF(x) \right]$$

$$+ k \int_0^T R(\theta|T - x)\, d\tilde{G}(x)$$

where K is the availability coefficient and $k = 1 - K$. Recall that $K = t/(T + \tau)$ where T is the MTTF and τ is the MTR.

6.5.3 System with a Time Accumulation

Some systems must accumulate time of successful operation during a total period of performance. Of course, in this case we consider an alternating process of up and down periods. Denote the probability that a system will accumulate more than θ units of successful operation during period T as $S(\theta|T)$. For this probability one can consider two events that lead to success:

- A system works without failures during time t_0 from the beginning.
- A system failed at moment $x < \theta$, was repaired during time y, and during the remaining interval of $T - x - y$ tries to accumulate $\theta - x$ units of time of successful operation. This description leads us to the recurrent expression

$$S(\theta|T) = P(\theta) + \int_0^\theta \left[\int_0^{T-x} S(t_0 - x|T - x - y)\, dG(y) \right] dF(x)$$

This expression is correct for the case where a system starts to perform at $t = 0$.

If a system is in an on-duty regime and begins to accumulate time of successful operation at a stationary arbitrary moment, one must take into account that a system may occur at an up or down state. Each of the corresponding periods is represented by a residual time. The expression for the probability that a system will accumulate more than θ units of successful operation during period T as $S(\theta|T)$ starts to perform at an arbitrary moment is

$$\tilde{S}(\theta|T) = K\left[\tilde{P}(\theta) + \int_0^\theta\left[\int_0^{T-x}S(\theta - x|T - x - y)\,dG(y)\right]d\tilde{F}(x)\right]$$

$$+ k\int_0^T R(\theta|T - y)\,d\tilde{G}(y) \tag{6.59}$$

where $\tilde{F}(x) = 1 - \tilde{P}(x)$ is the distribution of a time of failure-free operation, $\tilde{G}(x)$ is the distribution of a repair time, and $S(\theta|T - x)$ is taken from (6.58). Expression (6.59) is correct with the additional condition $S(x|y < x) = 0$.

6.5.4 System with Admissible Down Time

A system is considered to be successfully operating if during period T there will be no down time larger than η. This case in some sense is a "mirror" for that considered on page 249. We will omit the details and write the recurrent expression immediately:

$$Q(\eta|T) = P(T) + \int_0^T\left[\int_0^\eta Q(\eta|T - x - y)\,dG(y)\right]dF(x)$$

This expression is correct under an additional condition:

$$Q(\eta|x \le \eta) = 1$$

The same system may be considered in an on-duty regime. We again will omit the details and write the recurrent expression

$$\tilde{Q}(\eta|T) = K\left[\tilde{P}(T) + \int_0^T\left[\int_0^\eta Q(\eta|T - x - y)\,dG(y)\right]d\tilde{F}(x)\right]$$

$$+ k\int_0^\eta Q(\eta|T - y)\,dG(y)$$

This expression is correct under an additional condition:

$$Q(\eta|x \leq \eta) = 1$$

This subject as a whole requires a much more detailed discussion. There are many interesting detailed models concerning, for instance, computer systems. The reader who is interested in the subject can refer to Kredentser (1978), Cherkesov (1974), and Ushakov (1985, 1994). Some applications of these methods to oil and gas transportation systems can be found in Rudenko and Ushakov (1989).

CONCLUSION

The models of repairable systems discussed in this chapter concern some ideal schemes: switches are supposed to be absolutely reliable; monitoring of the operation of the system's units is continuous; after repair, units are considered to be as good as new; and so forth. Besides, when using Markov models, one must assume that all distributions of failure-free intervals and repair times are exponentially distributed.

All of these assumptions seem to make such kinds of models practically useless. But the same can be said about *any* mathematical model: a mathematical model is only a reflection of a real object or real process. Each mathematical model may only be used if the researcher understands all of the model's limitations.

First of all, Markov models are very simple though simplicity is not a good excuse for their use. But using Markov models for highly reliable systems very often gives the desired practical results in reliability prediction.

Next, the lack of some realistic assumptions concerning switching and monitoring may be taken into account. (We try to show this in the next chapter.) This point is really very serious and must be taken into consideration. To demonstrate the importance of continuous monitoring of redundant units, let us consider a simple example.

A repairable system consists of n units in parallel (i.e., this is a group of one main and $n - 1$ loaded redundant units). A system unit has an exponentially distributed TTF. Redundant units are checked only at the moment of failure of the main operating unit. At this moment all failed units are repaired instantaneously! If there is at least one nonfailed redundant unit, this unit replaces a failed main unit and the system continues to operate under its initial conditions. It seems that such a system with instant repair should be very reliable. But this system has no control over the system's unit states.

Find the MTTF of this system on the basis of simple explanations. A main unit has failed, on average, in T units of time, and with probability $1/n$ up to this moment all of the remaining $n - 1$ units have failed. It is clear that such a system will work, on average, nT units of time until a failure. But as the reader will recall, a standby redundant group of n units without repair has the same MTTF!

It is difficult to find out who wrote the pioneering works in this area. The reader can find a review in the next chapter dedicated to renewal duplicated systems—a particular case of redundancy with repair. The reader can find general information on this question in a number of books on reliability, some of which are listed at the end of this book. For a brief review, we refer the reader to the *Handbook of Reliability Engineering* by Ushakov (1994).

Time redundancy represents a separate branch of renewal systems, closely related to the theory of inventory systems with continuous time. The reader can find many interesting models for reliability analysis of such systems in Cherkesov (1974) and Kredentser (1978). The reader can find applications of these methods to gas and oil pipelines with intermediate storage in Rudenko and Ushakov (1989). General methods of time redundancy are briefly presented in Ushakov (1985, 1994). An interesting discussion on repairable systems can be found in Ascher and Feingold (1984).

REFERENCES

Ascher, H., and H. Feingold (1984). *Repairable Systems Reliability: Modelling, Inference, Misconceptions and Their Causes*. New York: Marcel Dekker.

Cherkesov, G. N. (1974). *Reliability of Technical Systems with Time Redundancy*. Sovietskoe Radio: Moscow.

Gertsbakh, I. B. (1984). Asymptotic methods in reliability theory: a review. *Adv. in Appl. Probab.*, vol 16.

Gnedenko, B. V., Yu. K. Belyaev, and A. D. Solovyev (1969). *Mathematical Methods of Reliability Theory*. New York: Academic.

Gnedenko, D. B., and A. D. Solovyev (1974). A general model for standby with renewal. *Engrg. Cybernet.* (USA), vol. 12, no. 6.

Gnedenko, D. B., and A. D. Solovyev (1975). Estimation of the reliability of complex renewable systems. *Engrg. Cybernet.* (USA), vol. 13, no. 3.

Kredentser, B. P. (1978). *Prediction of Reliability of Systems with Time Redundancy* (in Russian). Kiev: Naukova Dumka.

Rudenko, Yu. N., and I. A. Ushakov (1989). *Reliability of Energy Systems* (in Russian). Novosibirsk: Nauka.

Solovyev, A. D. (1972). Asymptotic distribution of the moment of first crossing of a high level by birth and death process. *Proc. Sixth Berkeley Symp. Math. Statist. Probab.*, Issue 3.

Ushakov, I. A., ed. (1985). *Reliability of Technical Systems: Handbook* (in Russian). Moscow: Radio i Sviaz.

Ushakov, I. A., ed. (1994). *Handbook of Reliability Engineering*. New York: Wiley.

EXERCISES

6.1 A system has an exponentially distributed TTF with a mean $t = 100$ hours and a repair time having a general distribution $G(t)$ with a mean $\tau = 0.5$ hour. Find the system's stationary interval availability coefficient for the operation during 0.5 hour.

6.2 Construct a transition graph for a repairable system consisting of two main units, one loaded redundant unit which can replace instantaneously each of them, and three spare units. After a main unit has failed, a loaded redundant unit replaces it. In turn, one of the spare units replaces the redundant unit. Failed units are subjected to repair after which they become as good as new. All units are identical, each with a failure rate equal to λ. There are two repair facilities, each of which can repair only one failed unit at a time. The intensity of repair by a repair facility is equal to μ. After a total exhaustion of all redundant units, repair is performed over the entire system with intensity M.

6.3 Construct a transition graph for the system depicted in Figure E6.2.

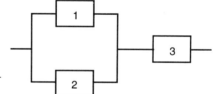

Figure E6.2. Structure diagram for the system described in Exercise 6.3.

SOLUTIONS

6.1 The stationary availability coefficient depends only on the mean and not on the type of distribution of the TTF and repair time. Thus, $K = (100)/(100 + 0.5) = 0.995$. If the system is found within a failure-free interval, which is exponentially distributed, then the probability of successful operation of length t_0 beginning at an arbitrary moment of time can be written as

$$\lim_{t \to \infty} P(t, t + t_0) = P(t_0) = e^{-t_0/T}$$

Finally, after substituting the corresponding numerical data, one has

$$\exp(-(0.5/100)) = 0.995$$

and

$$R(t_0 = 0.5) = (0.995)(0.995) = 0.99$$

6.2   The solution is depicted in Figure E6.1

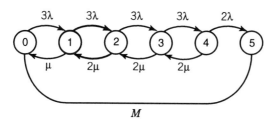

Figure E6.1. Transition graph for the system described in Exercise 6.2.

6.3   See Figure E6.3.

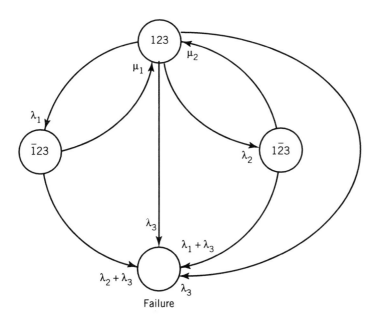

Figure E6.3. Transition graph for the system described in Exercise 6.3.

CHAPTER 7

REPAIRABLE DUPLICATED SYSTEM

Duplication refers to the particular case of redundancy where there is a single redundant unit to support a single working (main) unit. We distinguish this particular case for both practical and methodological reasons. First of all, when a designer feels that the reliability of some unit is low (sometimes this understanding may occur on a purely intuitive level), duplication is a simple way to improve it. Indeed, if a failure may occur with a relatively small probability, it is generally not necessary to have more than one redundant unit. In general, the number of redundant units depends on the desired value of the system's reliability index and/or on permissible economical expenditures.

From a methodological viewpoint, duplication presents the clearest way to explain certain special mathematical tools, their idiosyncrasies, and their ability to treat a real technical problem. It allows for the possibility of following mathematical transformations in detail. (Unfortunately, nobody has either the capacity or the desire to present similar detailed explanations for more complicated cases.)

7.1 MARKOV MODEL

As we have pointed out, a duplicated renewal system is one of the most frequently encountered structures in engineering practice. In the reliability analysis of electronic equipment (at least, in the first stages of design), the distributions of the time to failure and of the repair time are usually assumed exponential. In this case Markov models are adequate mathematical models to describe such systems. We note that the final results obtained with Markov models are usually acceptable in a wide variety of practical cases (especially when applied to highly reliable systems).

7.1.1 Description of the Model

Consider a duplicated system consisting of two identical units. Usually, the following assumptions are made:

- The system units are mutually independent.
- After a failure of the operating unit, its functions are immediately assumed to be performed by the redundant unit.
- Repair (renewal) of a failed unit begins immediately.
- A repaired unit is considered to be a new unit.
- The switching device is considered absolutely reliable.

Two important aspects of a renewal system should also be taken into account: the regime of the redundant unit and the attributes of the repair workshop.

The following regimes of a redundant unit characterized by failure rate λ' might be considered:

1. The redundant unit operates under the same conditions as an operational unit; that is, their failure rates are equal, $\lambda = \lambda'$.
2. The redundant unit is in a completely idle state, that is $\lambda' = 0$.
3. The redundant unit is in an intermediate state between completely idle and operational, that is, $0 < \lambda' < \lambda$.

The first case is often referred to as *internal redundancy*, the second as *standby redundancy*, and the third as *waiting redundancy*.

The renewal regime might be distinguished by the number of repair facilities (places for repair, the number of technicians special equipment), that is, by the number of failed units which can be repaired simultaneously. We consider two cases:

1. An unrestricted renewal when the number of repair facilities equals the number of possible failed units (in this particular case, two facilities are enough).
2. An extremely restricted renewal with a single repair facility.

The transition graphs describing these models are presented in Figure 7.1 (there are two of them: with and without an absorbing state). Corresponding particular cases for different regimes of redundant units and different attributes of the repair shop are reflected in Figure 7.2.

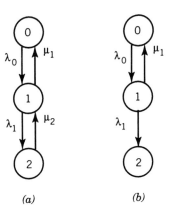

Figure 7.1. Transition graph for a renewable duplicated system: (*a*) state 2 is reflecting; (*b*) state 2 is
absorbing.

(a)   *(b)*

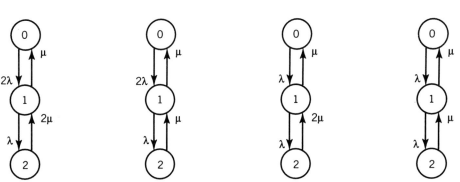

Figure 7.2. Transition graphs for four main models of a renewable duplicated system:
(*a*) a loaded redundant unit, two repair facilities; (*b*) a loaded redundant unit, one
repair facility; (*c*) an unloaded redundant unit (spare unit), two repair facilities;
(*d*) an unloaded redundant unit, one repair facility.

Using the above-described technique, corresponding systems of equations
for obtaining the various reliability indexes can be easily written. In this
particular simple case, the solutions can be obtained in a general form. The
final results for particular cases can be derived easily.

7.1.2 Nonstationary Availability Coefficient

The system of differential equations (in canonical form) with the initial
conditions $P_0(0) = 0$ is

$$\frac{d}{dt}P_0(t) = -\lambda_0 P_0(t) + \mu_1 P_1(t)$$

$$\frac{d}{dt}P_1(t) = \lambda_0 P_0(t) - (\lambda_1 + \mu_1)P_1(t) + \mu_2 P_2(t) \tag{7.1}$$

$$1 = P_0(t) + P_1(t) + P_2(t)$$

$$P_0(0) = 1$$

The LST of (7.1) is

$$(\lambda_0 + s)\varphi_0(s) - \mu_1\varphi_1(s) = 1$$
$$- \lambda_0\varphi_0(s) + (s + \lambda_1 + \mu_1)\varphi_1(s) + \mu_2\varphi_2(s) = 0 \qquad (7.2)$$
$$s\varphi_0(s) + s\varphi_1(s) + s\varphi_2(s) = 1$$

Notice that the availability coefficient equals

$$K(t) = P_0(t) + P_1(t) = 1 - P_2(t)$$

Thus, to find the LST of $K(t)$, we can find [see the last line in (7.2)]

$$\varphi_0(s) + \varphi_1(s) = \frac{1}{s} - \varphi_2(s)$$

From (7.2) it is easy to write

$$\varphi_2(s) = \frac{\begin{vmatrix} \lambda_0 + s & -\mu_1 & 1 \\ -\lambda_0 & \lambda_1 + \mu_1 + s & 0 \\ s & s & 1 \end{vmatrix}}{\begin{vmatrix} \lambda_0 + s & -\mu_1 & 0 \\ -\lambda_0 & \lambda_1 + \mu_1 + s & -\mu_2 \\ s & s & s \end{vmatrix}}$$

$$= \frac{\lambda_0\lambda_1}{s\left[s^2 + s(\lambda_0 + \lambda_1 + \mu_1 + \mu_2) + \lambda_0\lambda_1 + \lambda_0\mu_2 + \mu_1\mu_2\right]}$$

Thus,

$$\varphi_0(s) + \varphi_1(s) = \frac{1}{s} - \varphi_2(s)$$

$$= \frac{s^2 + s(\lambda_0 + \lambda_1 + \mu_1 + \mu_2) + \lambda_0\mu_2 + \mu_1\mu_2}{s\left[s^2 + s(\lambda_0 + \lambda_1 + \mu_1 + \mu_2) + \lambda_0\lambda_1 + \lambda_0\mu_2 + \mu_1\mu_2\right]}$$

Now we should refer to the technique described in Chapter 6 in the section on Markov processes:

1. Represent the LST as the sum of simple fractions

$$\frac{s^2 + s(\lambda_0 + \lambda_1 + \mu_1 + \mu_2) + \lambda_0\mu_2 + \mu_1\mu_2}{s\left[s^2 + s(\lambda_0 + \lambda_1 + \mu_1 + \mu_2) + \lambda_0\lambda_1 + \lambda_0\mu_2 + \mu_1\mu_2\right]}$$

$$= \frac{A}{s - s_1} + \frac{B}{s - s_2} + \frac{C}{s - s_3}$$

where A, B, and C are unknown.

2. Find the roots of the denominator. The first two roots of the denominator are conjugate, that is

$$S_{1,2} = -\frac{\alpha}{2} \pm \sqrt{\frac{\alpha^2}{4} - \beta}$$

where, in turn,

$$\alpha = \lambda_0 + \lambda_1 + \mu_1 + \mu_2$$
$$\beta = \lambda_0\lambda_1 + \lambda_0\mu_2 + \mu_1\mu_2$$

and $s_3 = 0$.

3. Find the unknown values A, B, and C by equalizing the polynomial coefficients of the numerators.

4. Apply the inverse LST to obtain simple fractions with the numerators A, B, and C found above to obtain the final result.

After these transformations the result is obtained

$$K(t) = 1 - \frac{\lambda_0\lambda_1}{s_1 - s_2}\left[1 - \frac{1}{s_1 s_2}\left(s_1 e^{s_2 t} - s_2 e^{s_1 t}\right)\right] \qquad (7.3)$$

Obviously, if $s_1 = s_2$, l'Hospital's rule must be used.

Now any result of interest can be obtained by substituting the appropriate values of λ and μ. In general, the solution for a duplicate renewal system can be obtained in a closed form, but this solution is not very compact, even for the simplest case.

Of course, we should notice that, for active redundancy and unrestricted repair, the final result can be written immediately with the use of the appropriate result for a single unit:

$$K(t) = 1 - (1 - K^*(t))^2 = 1 - \left[\frac{\lambda}{\lambda + \mu}\left(1 - e^{-(\lambda+\mu)t}\right)\right]^2$$

where $K^*(t)$ is the nonstationary availability coefficient of a single unit. This result is obvious because both units are supposed to be mutually independent.

7.1.3 Stationary Availability Coefficient

The solution can be derived from (7.1) by putting the derivatives equal to 0. The same result can be directly obtained from the corresponding transition graph by writing the equilibrium equations:

$$-\lambda_0 P_0 + \mu_1 P_1 = 0$$
$$\lambda_0 P_0 - (\lambda_1 - \mu_1) P_1 + \mu_2 P_2 = 0 \qquad (7.4)$$
$$P_0 + P_1 + P_2 = 1$$

The solution is

$$K = 1 - P_2 = 1 - \frac{\begin{vmatrix} -\lambda_0 & \mu_1 & 0 \\ \lambda_0 & -(\lambda_1 + \mu_1) & 0 \\ 1 & 1 & 1 \end{vmatrix}}{\begin{vmatrix} -\lambda_0 & \mu_1 & 0 \\ \lambda_0 & -(\lambda_1 + \mu_1) & \mu_2 \\ 1 & 1 & 1 \end{vmatrix}} = 1 - \frac{\lambda_0 \lambda}{\lambda_0 \lambda_1 + \lambda_0 \mu_2 + \mu_1 \mu_2}$$

$$= \frac{\lambda_0 \mu_2 + \mu_1 \mu_2}{\lambda_0 \mu_2 + \mu_1 \mu_2 + \lambda_0 \lambda_1} = \frac{1}{1 + \dfrac{\lambda_0 \lambda_1}{\lambda_0 \mu_2 + \mu_1 \mu_2}}. \qquad (7.5)$$

To obtain values of this reliability index for different cases, the specific λ's and μ's should be substituted. The results for the four most important cases depicted in Figure 7.2 are presented in Table 7.1. In this table we used the notation $\gamma = \lambda/\mu$. For highly reliable systems with $\lambda \ll 1$, all of the expressions in Table 7.1 can be easily transformed to obtain the approximations given in Table 7.2. The expressions in Table 7.2 allow one to give an understandable explanation of all of the effects. Naturally, the worst value of K gives the case of active redundancy and restricted repair (the failure rate

TABLE 7.1 Availability Coefficient for Four Main Models of a Renewable Duplicated System

	(a)	(b)
(A)	$\dfrac{1}{1 + \dfrac{\gamma^2}{1 + 2\gamma}}$	$\dfrac{1}{1 + \dfrac{2\gamma^2}{1 + 4\gamma}}$
(B)	$\dfrac{1}{1 + \dfrac{\gamma^2}{2(1 + \gamma)}}$	$\dfrac{1}{1 + \dfrac{\gamma^2}{1 + \gamma}}$

(A) Loaded redundant unit; (B) unloaded redundant unit; (a) two repair facilities; (b) one repair facility.

TABLE 7.2 Approximation for Availability Coefficient for Highly Reliable Duplicated System

	(a)	(b)
(A)	$1 - \dfrac{\gamma^2}{1 + 2\gamma}$	$1 - \dfrac{2\gamma^2}{1 + 4\gamma}$
(B)	$1 - \dfrac{\gamma^2}{2(1 + \gamma)}$	$1 - \dfrac{\gamma^2}{1 + \gamma}$

(A) Loaded redundant unit; (B) unloaded redundant unit; (a) two repair facilities; (b) one repair facility.

λ_0 is the largest and the repair rate μ_2 is the smallest). The case of unrestricted repair yields a mean repair time of less than one-half of the restricted repair time ($1/2\mu$ and $1/\mu$, respectively). Below in this section we will show that the MTTF of highly reliable systems of active redundancy is one-half of the MTBF for standby redundancy.

Of course, for two independent units, that is, when the redundancy is active and the repair is unrestricted, we can write

$$K = 1 - (1 - K^*)^2 = 1 - \left[\frac{\lambda}{\lambda + \mu}\right]^2$$

using the availability coefficient for the single unit K^*. Then

$$K = \frac{2\lambda + \mu^2}{\lambda^2 + 2\lambda + \mu^2} = \frac{1}{1 + \dfrac{\lambda^2}{2\lambda + \mu_2}} = \frac{1}{1 + \dfrac{\gamma^2}{1 + 2\gamma}}$$

The intermediate case with either the "underloaded" redundant unit (when $\lambda < \lambda_0 < 2\lambda$) or with the "dependent" repair when ($\mu < \mu_2 < 2\mu$) can be easily obtained numerically from the general expression (7.5). Of course, this index can be realized as

$$K = \lim_{t \to \infty} K(t)$$

but this is not effective when $K(t)$ is not available.

7.1.4 Probability of Failure-Free Operation

To find this probability, it is necessary to construct a system of differential equations using the graph of Figure 7.1b with absorbing state 2. In this case the equations are not linearly dependent. For the initial conditions $P_0(0) = 1$, the system of linear differential equations is

$$\frac{d}{dt}P_0(t) = -\lambda_0 P_0(t) + \mu_1 P_1(t)$$

$$\frac{d}{dt}P_1(t) = \lambda_0 P_0(t) - (\lambda_1 + \mu_1)P_1(t) \qquad (7.6)$$

$$P_0(0) = 1$$

The LST of (7.6) is

$$(\lambda_0 + s)\varphi_0(s) - \mu_1\varphi_1(s) = 1$$
$$-\lambda_0\varphi_0(s) + (\lambda_1 + \mu_1 + s)\varphi_1(s) = 0 \qquad (7.7)$$

and the solution has the form

$$\varphi_0(s) + \varphi_1(s) = \frac{\begin{vmatrix} 1 & -\mu_1 \\ 0 & \lambda_1 + \mu_1 + s \end{vmatrix} + \begin{vmatrix} \lambda_0 & 1 \\ -\lambda_0 & 0 \end{vmatrix}}{\begin{vmatrix} \lambda_0 + s & -\mu_1 \\ -\lambda_0 & \lambda_1 + \mu_1 + s \end{vmatrix}}$$

$$= \frac{s + \lambda_0 + \lambda_1 + \mu_1}{s^2 + s(\lambda_0 + \lambda_1 + \mu_1) + \lambda_0\lambda_1} \qquad (7.8)$$

Applying the procedure that we used to obtain (7.3), we find

$$P^{(0)}(t) = \frac{1}{s_1^* - s_2^*}\left(s_1^* e^{s_2^* t} - s_2 e^{s_1^* t}\right) \qquad (7.9)$$

where the superscript (0) stands for the initial conditions $P_0(0) = 1$ and also

$$s_{1,2}^* = -\frac{\alpha^*}{2} \pm \sqrt{\frac{(\alpha^*)^2}{4} - \beta}$$

$$\alpha^* = \lambda_0 + \lambda_1 + \mu_1$$

$$\beta^* = \lambda_0\lambda_1$$

If we are interested in the system's PFFO immediately after its repair, it is necessary to set $P_1(0) = 1$. The corresponding system of linear algebraic equations in the LST is

$$(\lambda_0 + s)\varphi_0(s) - \mu_1\varphi_1(s) = 0$$
$$-\lambda_0\varphi_0(s) + (\lambda_1 + \mu_1 + s)\varphi_1(s) = 1$$

and the solution in the LST is

$$\varphi_0(s) + \varphi_1(s) = \frac{\begin{vmatrix} 0 & -\mu_1 \\ 1 & \lambda_1 + \mu_1 + s \end{vmatrix} + \begin{vmatrix} \lambda_0 & 0 \\ -\lambda_0 & 1 \end{vmatrix}}{\begin{vmatrix} \lambda_0 + s & -\mu_1 \\ -\lambda_0 & \lambda_1 + \mu_1 + s \end{vmatrix}}$$

$$= \frac{\lambda_0 + \mu_1}{s^2 + s(\lambda_0 + \lambda_1 + \mu_1) + \lambda_0\lambda_1} \tag{7.10}$$

Notice that the denominator in (7.10) is the same as in (7.8), so we can use the roots (eigenvalues) obtained above. Omitting routine transformations, we may write the final result for this case

$$P^{(1)}(t) = \frac{1}{s_1^* - s_2^*}\left[(s_1^* - \lambda_0 - \lambda_1)e^{s_2^* t} - (s_2^* - \lambda_0 - \lambda_1)e^{s_1^* t}\right] \tag{7.11}$$

where the superscript (1) indicates the corresponding initial conditions.

7.1.5 Stationary Coefficient of Interval Availability

The task can be solved by setting the initial conditions: $P_0(0) = P_0$ and $P_1(0) = P_1$ in (7.6) where P_0 and P_1 are the stationary probabilities obtained from (7.4). The P_i's can be found from (7.4) separately as

$$P_0 = \frac{\begin{vmatrix} 0 & \mu_1 & 0 \\ 0 & -(\lambda_1 + \mu_1) & \mu_2 \\ 1 & 1 & 1 \end{vmatrix}}{\begin{vmatrix} -\lambda_0 & \mu_1 & 0 \\ \lambda_0 & -(\lambda_1 + \mu_1) & \mu_2 \\ 1 & 1 & 1 \end{vmatrix}} = \frac{\mu_1\mu_2}{\lambda_0\lambda_1 + \lambda_0\mu_2 + \mu_1\mu_2}$$

and

$$P_1 = \frac{\begin{vmatrix} -\lambda_0 & 0 & 0 \\ \lambda_0 & 0 & \mu_2 \\ 1 & 1 & 1 \end{vmatrix}}{\begin{vmatrix} -\lambda_0 & \mu_1 & 0 \\ \lambda_0 & -(\lambda_1 + \mu_1) & \mu_2 \\ 1 & 1 & 1 \end{vmatrix}} = \frac{\lambda_0\mu_2}{\lambda_0\lambda_1 - \lambda_0\mu_2 + \mu_1\mu_2}$$

In other words, the following system needs to be solved:

$$(\lambda_0 + s)\varphi_0(s) - \mu_1\varphi_1(s) = P_0$$
$$-\lambda_0\varphi_0(s) + (\lambda_0 + \lambda_1 + s)\varphi_1 = P_1$$

This index can also be found in a different way, using the Markov property of the process. We can write

$$R(t_0) = P_0 P^{(0)}(t_0) + P_1 P^{(1)}(t_0)$$

where the P_i's are the above-mentioned stationary probabilities of the corresponding states. In this case they are the initial states of the PFFOs $P^{(i)}(t_0)$'s until they reach the absorbing state 2 (the state of the system failure). $P^{(0)}(t_0)$ and $P^{(1)}(t_0)$ are found in (7.10) and (7.11).

We will not obtain the large expression for the nonstationary coefficient, or interval availability, because it is tedious to obtain it. Technically, this task is no different from the previously addressed task.

7.1.6 MTTF and MTBF

From the LSTs (7.8) and (7.10), the desired expressions follow immediately:

$$\text{MTTF} = T^{(0)} = \frac{s + \lambda_0 + \lambda_1 + \mu_1}{s^2(\lambda_0 + \lambda_1 + \mu_1)s + \lambda_0\lambda_1}\bigg|_{s=0} = \frac{1}{\lambda_1} + \frac{1}{\lambda_0} + \frac{\mu_1}{\lambda_0\lambda_1}$$

$$\text{MTBF} = T^{(1)} = \frac{s + \lambda_0 + \mu_1}{s^2(\lambda_0 + \lambda_1 + \mu_1)s + \lambda_0\lambda_1}\bigg|_{s=0} = \frac{1}{\lambda_1} + \frac{\mu_1}{\lambda_0\lambda_1} \qquad (7.12)$$

It is often more reasonable to use (7.7) directly with the substitution of $s = 0$:

$$\lambda_0\theta_0 - \mu_1\theta_1 = 1$$
$$-\lambda_0\theta_0 + (\lambda_1 + \mu_1)\theta_1 = 0$$

where θ_1 and θ_2 are values such that the MTTF $= \theta_1 + \theta_2$. The solution of this equation system yields

$$\theta_0 = \frac{1}{\lambda_1} \quad \text{and} \quad \theta_1 = \frac{\lambda_1 + \mu_1}{\lambda_0\lambda_1} = \frac{1}{\lambda_0} + \frac{\mu_1}{\lambda_0\lambda_1}$$

Of course, this result coincides with (7.12).

The MTBF may be computed in the same manner, and we leave this as an exercise.

To find the system's MTTF, it is sometimes more convenient to use the following arguments. Consider the transition graph in Figure 7.1. Let us find the system's MTTF (this means that at $t = 0$ the system is in state 0). Denote the mean time needed to reach the absorbing state 2 from the initial state 0 as T_{02} and from state 1 as T_{12}. (Apropos, $T_{02} = $ MTTF and $T_{12} = $ MTBF.)

Obviously,

$$T_{02} = \frac{1}{\lambda_0} + T_{12} \tag{7.13}$$

because the process inevitably moves from state 0 to state 1. After this, based on the Markov property, the process can be considered to be starting from state 1.

The process stays in state 1 for an average time $1/(\lambda_2 + \mu_1)$ and then moves either to state 2 or to state 0. It moves to state 0 with probability $\mu_1(\lambda_2 + \mu_1)$ and then starts traveling again from state 0. Hence, we can write

$$T_{12} = 1/(\lambda_2 + \mu_1) + (\mu_1 T_{02})/(\lambda_2 + \mu_1) \tag{7.14}$$

Substituting (7.14) into (7.13) yields

$$\text{MTTF} = T_{02} = \frac{1}{1 - \dfrac{\mu_1}{\lambda_1 + \mu_1}} \left[\frac{1}{\lambda_0} + \frac{1}{\lambda_1 + \mu_1} \right] = \frac{\lambda_0 + \lambda_1 + \mu_1}{\lambda_0 \lambda_1}$$

$$= \frac{1}{\lambda_0} + \frac{1}{\lambda_1} + \frac{\mu_1}{\lambda_0 \lambda_1}$$

From (7.13) it also immediately follows that

$$\text{MTBF} = T_{12} = \frac{1}{\lambda_1} + \frac{\mu_1}{\lambda_0 \lambda_1}$$

Now, on a very understandable and almost verbal level, we can explain the difference between the MTBF (or the MTTF) for repaired duplicated systems of identical units which have a different regime for the redundant unit. For active redundancy $\lambda_0 = 2\lambda$ and for standby redundancy $\lambda_0 = \lambda$. In other words, in the first case, the system stays in state 0, on average, one-half the time that it stays in the second case. This fact can be seen more clearly from the approximate expressions for a highly reliable system when $\gamma = \lambda/\mu \ll 1$:

$$\text{MTTF}_{\text{active}} = \frac{1}{\lambda} + \frac{1}{2\lambda} + \frac{\mu}{2\lambda^2} \approx \frac{1}{2\gamma\lambda} \qquad (7.15)$$

and

$$\text{MTTF}_{\text{standby}} = \frac{1}{\lambda} + \frac{1}{\lambda} + \frac{\mu}{\lambda^2} \approx \frac{1}{\gamma\lambda} \qquad (7.16)$$

Incidentally, (7.15) and (7.16) could be explained on the basis of the Rényi theorem. Consider an alternative renewal process describing the operation of a repaired duplicated system.

For a highly reliable system, this process can be approximately represented as a simple renewal process if one neglects small intervals of being at state 1. The system's successful operation period consists of the sum of a random number of intervals of the length $1/\lambda_0$ until the process has jumped to state 2. This random number has a geometrical distribution with parameter $p = \mu_1/(\lambda_1 + \mu_1) \approx 1 - \gamma$. Thus, the sum also has an exponential distribution with parameter $\lambda_0\gamma$. This means that approximately

$$P^0(t_0) \approx P^0(t_0) \approx \exp(-\lambda_0\gamma t_0) \qquad (7.17)$$

We should now remember that for active redundancy $\lambda_0 = 2\lambda$ and for standby redundancy $\lambda_0 = \lambda$.

We wrote all of these solutions in such a detailed form because the LST technique is very important in engineering applications. A certain amount of practice is needed to apply this to practical problem solutions. We believe that the best way to master these approaches is to work out simple exercises.

7.2 DUPLICATION WITH AN ARBITRARY REPAIR TIME

For repairable duplicated systems, models more complicated than the Markovian type can be analyzed. We first consider a model described in the following way:

- Both units are independent and identical.
- The operating unit has an exponential distribution of time to failure $F(t)$ with parameter λ, and the redundant unit has a similar distribution $F_1(t)$, also exponential with parameter λ_1, $0 \leq \lambda_1 \leq \lambda$. (This condition means that the redundant unit might be, in general, in an underloaded regime.)

- The repair time of a failed unit has an arbitrary distribution $G(t)$.
- The repair of a failed unit begins immediately after a failure has occurred.
- After repair, the unit becomes completely new.
- The repaired unit is immediately installed into the system.

It is clear that after an operational unit has failed, the redundant unit replaces it and becomes operational. A system failure occurs if and only if the operating unit fails during the repair of the other unit, that is, when both of the system's units have failed.

Let us find the distribution of the system's time to failure $R_s(t)$. A failure-free operation of the duplicated system during a time period t can be represented as the union of the following mutually disjoint events:

1. The first failure in the system occurs after moment t; what happens with the redundant unit does not play any role. The probability of this event is $\exp[(-\lambda + \lambda_1)t]$.

2. The first failure of either of the two units occurs at some moment $z < t$, the failed unit is not repaired during the interval $(t - z)$, but the unit in the operating position has not failed up to t. The probability of this event is

$$\int_0^\infty (\lambda + \lambda_1) e^{-(\lambda + \lambda_1)z} [1 - G(t - z)] e^{-\lambda(t - z)} \, dz$$

3. The last event is the most complicated. In this case, as some moment $x < t$, the duplicated system comes to the initial state, that is, state 0, where both system's units are operational. This occurs if one of the units has failed during the interval $[z, z + dz]$, the repair has taken time $x - z$, and the operating unit has not failed during repair. After the completion of the repair, the system operates successfully during the remaining period of time $(t - x)$ with probability $R(t - x) = 1 - F_s(t - x)$. The probability of this event is

$$\int_0^\infty R(t - x) \, dx \int_0^x (\lambda + \lambda_1) e^{-(\lambda + \lambda_1)z - \lambda(x - z)} g(x - z) \, dz$$

where $g(t)$ is the density function of the distribution $G(t)$.

Now it is easy to write the final equation for the probability of a system's failure-free operation:

$$R(t) = e^{-(\lambda+\lambda_1)t} + e^{-\lambda t}(\lambda + \lambda_1)\int_0^t e^{-\lambda_1 x}[1 - G(t - x)]\, dx$$

$$+ \int_0^t R(t - x)e^{-\lambda x}(\lambda + \lambda_1)\, dx \int_0^x e^{-\lambda_1 z}g(x - z)\, dz \quad (7.18)$$

Thus, we have an integral equation with a kernel of the type

$$R(t) = A(t) + \int_0^t R(t - x)B(x)\, dx \qquad (7.19)$$

where, in the above case,

$$A(t) = e^{-(\lambda+\lambda_1)t} + e^{-\lambda t}(\lambda + \lambda_1)\int_0^t e^{-\lambda_1 x}[1 - G(t - x)]\, dx$$

$$B(t) = e^{-\lambda t}(\lambda + \lambda_1)\int_0^t e^{-\lambda z}g(t - z)\, dz \qquad (7.20)$$

The recurrent equation (7.18) can be solved by the method of sequential iterations. But we prefer to obtain the solution in the form of the LST as it allows us to investigate the asymptotical behavior of $R(t)$.

If we denote

$$a(s) = \int_0^\infty e^{-st}A(t)\, dt$$

$$b(s) = \int_0^\infty e^{-st}B(t)\, dt$$

$$\Psi(s) = \int_0^\infty e^{-st}\, dG(t)$$

$$\tilde{\varphi}(s) = \int_0^\infty e^{-st}R(t)\, dt$$

then the solution can be represented in the form

$$\tilde{\varphi}(s) = a(s) + \tilde{\varphi}(s)b(s) \qquad (7.21)$$

and, finally, the LST of interest is

$$\tilde{\varphi}(s) = \frac{a(s)}{1 - b(s)} \qquad (7.22)$$

The functions $a(s)$ and $b(s)$ can easily be found from (7.20)

$$a(s) = \frac{s + \lambda + (\lambda + \lambda_1)[1 - \psi(s + \lambda)]}{(s + \lambda + \lambda_1)(s + \lambda)}$$

$$b(s) = \frac{(\lambda + \lambda_1)\Psi(s + \lambda)}{\lambda + \lambda_1 + s}$$

(7.23)

Thus, after substituting (7.23) into (7.22), we obtain

$$\tilde{\varphi}(s) = \frac{s + \lambda + (\lambda + \lambda_1)[1 - \Psi(s + \lambda)]}{(s + \lambda)[s + (\lambda + \lambda_1)(1 - \Psi(\lambda + s))]}$$

(7.24)

Therefore, the general case has been investigated. It is clear that for active redundancy, when $\lambda_1 = \lambda$,

$$\tilde{\varphi}(s) = \frac{s + \lambda + 2\lambda[1 - \Psi(\lambda + s)]}{(\lambda + s)[s + 2\lambda(1 - \Psi(\lambda + s))]}$$

For standby redundancy, when $\lambda_1 = 0$,

$$\tilde{\varphi}(s) = \frac{s + \lambda + \lambda[1 - \Psi(\lambda + s)]}{(\lambda + s)[s + \lambda(1 - \Psi(\lambda + s))]}$$

Since (7.24) is the LST of $R(t) = 1 - F_s(t)$, the MTTF can be derived from this expression directly with the substitution $s = 0$:

$$T_s = \tilde{\varphi}(s)|_{s=0} = \frac{\lambda + (\lambda + \lambda_1)[1 - \Psi(\lambda)]}{\lambda(\lambda + \lambda_1)[1 - \Psi(\lambda)]} = \frac{1}{\lambda} + \frac{1}{(\lambda + \lambda_1)[1 - \Psi(\lambda)]}$$

(7.25)

In deriving the latter expression, we use the memoryless property of the exponential distribution: if an object with an exponentially distributed random TTF has not failed until some moment t, then the conditional probability of the random residual TTF of the object is the same exponential distribution as the original one. Thus, the process of an operation of a duplicated system has the so-called *renewal moments*, that is, such Markov moments at which all of the prehistory of the process has no influence on the future development of the process starting from this moment.

The MTTF of the duplicated system without repair, as it was obtained above, equals

$$T = \frac{1}{\lambda + \lambda_1} + \frac{1}{\lambda}$$

From (7.25) it follows that the effectiveness of redundancy with renewal increases very quickly as $\psi(\lambda) \to 1$. Notice that $\psi(\lambda)$ is not more than the probability that the random TTF of a unit exceeds the duration of its repair. It means that

$$\alpha = 1 - \psi(\lambda) = \Pr\{\xi < \eta\} = \int_0^\infty \left[1 - e^{-\lambda t}\right] dG(t) \qquad (7.26)$$

is the probability of an unsuccessful repair, denoted by α. Here ξ is a random TTF and η is a random repair time. Notice that for the exponential distribution this probability equals $\lambda/(\lambda + \mu) = \gamma/(1 + \gamma)$ where $\gamma = \lambda/\mu$.

Let us make a final remark concerning the system MTTF. It is possible to write a clear and understandable recurrent equation to express T. The period of a system's successful operation can be represented by a sequence of cycles of the type "time to failure of any of the system's units + time of successful repair" which terminates with a system's failure (the cycle with an unsuccessful repair). Arguing similarly as in Section 7.1, the recurrent relationship can be written as

$$T_s = \frac{1}{\lambda + \lambda_1} + \alpha \frac{1}{\lambda} + (1 - \alpha) T_s$$

Finally, for T_s, one obtains

$$T_s = \frac{1}{(\lambda + \lambda_1)\alpha} + \frac{1}{\lambda}$$

It is clear that the MTTF of the duplicated repairable system depends on the distribution $G(t)$. Let us investigate this relationship in more detail. From (7.26) we can derive

$$\alpha = \int_0^\infty \left[1 - \sum_{k \geq 0} \frac{(-\lambda t)^k}{k!}\right] dG(t)$$

$$= \lambda \int_0^\infty t \, dG(t) - \frac{\lambda^2}{2} \int_0^\infty t^2 \, dG(t) + \frac{\lambda^3}{6} \int_0^\infty t^3 \, dG(t)$$

$$= \lambda \, E\{\eta\} - \frac{\lambda^2}{2} E\{\eta^2\} + \frac{\lambda^3}{6} E\{\eta^3\} + \cdots$$

Such a representation is very useful if the system is highly reliable, that is, when $\lambda \tau \ll 1$. Then the following approximation is true:

$$\alpha \approx \lambda\, E\{\eta\} - \frac{\lambda^2}{2} E\{\eta^2\} = \lambda\tau - \frac{\lambda^2}{2}\left[\tau^2 + \text{Var}\{\eta\}\right]$$

where $\tau = E\{\eta\}$. Thus, between several distributions $G(t)$ with the same mean, the probability α is smaller if the variance of the repair time is larger. From this statement it follows that the best repair is characterized by a (practically) constant duration.

Let us give several simple examples.

Example 7.1   Find α when the repair time is constant, $\eta = \tau$. By direct calculations

$$\alpha = \Pr\{\xi \le \tau\} = 1 - e^{-\lambda\tau}$$

and, for the highly reliable system when $\lambda\tau = \gamma \ll 1$,

$$\alpha \approx \lambda\tau - \tfrac{1}{2}(\lambda\tau)^2$$

Example 7.2   Find α when the repair time distribution is exponential. By direct calculations

$$\alpha = \Pr\{\xi \le \tau\} = \frac{\lambda}{\lambda + \mu} = 1 - \frac{\mu}{\lambda + \mu} = 1 - \frac{1}{1 + \lambda\tau}$$

approximately,

$$\alpha \approx \lambda\tau - (\lambda\tau)^2$$

Example 7.3   Find α when the repair time distribution is normal with mean equal to τ and variance equal to σ^2. Find the LST for this distribution, using essentially the same technique that we applied in Section 1.3.3 for obtaining the m.g.f.

$$\tilde{\varphi}(s) = e^{\tau s + s^2 \sigma^2 / 2}$$

We remind the reader that the LST and the m.g.f. differ only by the sign of the argument s. Thus,

$$\alpha = 1 - \varphi(\lambda) = 1 - e^{-(\tau\lambda + \lambda^2 \sigma^2 / 2)}$$

An approximation has the form

$$\alpha \approx \lambda\tau + \tfrac{1}{2}\lambda^2 \sigma^2$$

From these examples we see that with $\lambda\tau \to 0$ the repair (renewal) effectiveness becomes higher and tends to be invariable with respect to the type of $G(t)$. This is true for most reasonable practical applications. For example, to replace a failed bulb may take some 10 seconds, but its lifetime may equal hundreds of hours; to change or even to repair a car's tire takes a dozen minutes which is incomparably less than its average lifetime. This fact leads to new methods of investigation, namely, to asymptotic methods.

Assume that the parameters λ and λ_1 of the model are fixed and then consider a sequence of repair time distributions $G_1, G_2, \ldots, G_n, \ldots$ which changes in such a way that

$$\alpha_n = \int_0^\infty \left[1 - G_n(t)\right] dF(t) \to 0 \tag{7.27}$$

This means that the probability that the operational unit fails during repair goes to 0.

Under this condition the appearance of some limit distribution of a system's failure-free operation is expected. Of course, if $\alpha_n \to 0$ the system's MTTF goes to ∞. To avoid this, we must consider a normalized random TTF, namely, $\alpha\xi$. It is clear that this new r.v. has a constant mean equal to 1 independent of the value of α. The distribution of this r.v. is

$$\Pr\{\alpha\xi > t\} = R\left(\frac{t}{\alpha}\right)$$

The LST of this function is from (7.24)

$$\int_0^\infty e^{-st} R\left(\frac{t}{\alpha}\right) dt = \alpha\tilde{\varphi}(\alpha s)$$

$$= \alpha \frac{\alpha s + \lambda + (\lambda + \lambda_1)\left[1 - \Psi(\alpha s + \lambda)\right]}{(\alpha s + \lambda)\left[\alpha s + (\lambda + \lambda_1)(1 - \Psi(\alpha s + \lambda))\right]} \tag{7.28}$$

Now under the assumption that $\alpha_n \to 0$, we can write

$$\Psi(\lambda) - \Psi(\alpha s + \lambda) = \int_0^\infty e^{-\lambda t}\left(1 - e^{-\alpha st}\right) dG(t)$$

$$\leq \alpha s \int_0^\infty te^{-\lambda} dG(t) \leq \frac{\alpha s}{\lambda} \int_0^\infty \left(1 - e^{-\lambda t}\right) dG(t) = \frac{\alpha^2 s}{\lambda}$$

that is,

$$\Psi(\lambda) - \Psi(\alpha s + \lambda) = \frac{\alpha^2 s}{\lambda}\theta$$

where $0 < \theta < 1$. Therefore, if $\alpha_n \to 0$ uniformly on any finite interval of the domain of s,

$$\alpha\tilde{\varphi}(\alpha s) = \alpha\frac{\alpha s + \lambda + (\lambda + \lambda_1)\left(\alpha + \dfrac{\alpha^2 s}{\lambda}\theta\right)}{(\alpha s + \lambda)\left[\alpha s + (\lambda + \lambda_1)\left(\alpha + \dfrac{\alpha^2 s}{\lambda}\theta\right)\right]} \to \frac{1}{s + \lambda + \lambda_1} \quad (7.29)$$

Because the LST (7.29) corresponds to an exponential d.f., we get the following asymptotic result:

$$\lim_{\alpha \to 0} \Pr\{\alpha\xi > t\} = e^{-(\lambda + \lambda_1)t}$$

For practical problems, this means that for a small value of α the following approximate expression can be used:

$$R(t) \approx e^{-(\lambda + \lambda_1)\alpha t} \quad (7.30)$$

or

$$R(t) \approx e^{-t/T} \quad (7.31)$$

where T has been defined in (7.25). Incidentally, (7.31) is more accurate than (7.30): it has an error of order α in comparison with α^2 associated with the latter.

This can be explained in the following "half-verbal" terms. The random time to failure of a duplicated system consists of a random number of cycles of the type "totally operational system − successful repair" and a final cycle of the type "totally operational system − unsuccessful repair." Stochastically, all cycles of the first type are identical and, by the assumption of the exponentiality of the distribution, are mutually independent (the latter assumption is based on the memoryless property). The only cycle differing from these is the last one. But if to suggest that the number of cycles of the first type is large, on average, the distribution of the system time to failure can be approximated by the exponential distribution.

The use of the approximations (7.30) and (7.31) requires the value of α. This value can be obtained easily in this case. Moreover, if we know that

$G(t)$ has a restricted variance, then in limit $\alpha \approx \lambda\tau$. (Of course, the condition $\alpha \to 0$ is necessary.) In turn, this means that the following can be obtained:

$$R(t) \approx e^{-(\lambda+\lambda_1)\lambda\tau t} \tag{7.32}$$

If the conditions for the variance of $G(t)$ do not hold, the latter expression, of course, is wrong. The reader can verify this with an example of a sequence of two-mass discrete distributions $G_1(t), G_2(t), \ldots, G_n(t)$, each with two nonzero values of probability: at 0 and at some positive point. Let all of the distributions have the same mean but, with increasing n, the probability at the positive point becomes smaller with moving of the point to the right along the axis. The variance in this case is infinitely increasing with increasing n.

7.3 STANDBY REDUNDANCY WITH ARBITRARY DISTRIBUTIONS

For standby redundancy the results can be obtained for the most general case, namely, when both distributions—of a random TTF and of a random repair time—are arbitrary. Let us use the same notation for the distributions: $F(t)$ and $G(t)$. The duplicated system's operation can be graphically represented in Figure 7.3.

The system's operation consists of the following random intervals. The first one endures until the main unit fails; its random length is ξ_0. The second interval and all of the remaining intervals, ξ_k, $k = 2, 3, \ldots$, are successful if and only if each time a random failure-free time of the operating unit ξ_k is longer than the corresponding random repair time of the failed unit η_k. The last interval when a system failure has occurred has a random duration different from all of the previous ones: this is the distribution of the random TTF under the condition that $\xi < \eta$. All of these explanations become transparent if one considers a constant repair time: the first failure-free interval has unconditional distribution of the r.v. ξ; all of the remaining intervals (except the last one) have a conditional distribution under the condition that $\xi > \eta$; and the last one has a conditional distribution under the condition that $\xi < \eta$. In other words, the first of these distributions is positively biased and another is truncated from the right.

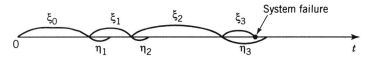

Figure 7.3. Time diagram for duplicated system operation with a standby redundant unit.

Let ξ^* denote a random value representing the system's time to failure starting from the moment ξ_0: $\xi^* = \xi_1 + \xi_2 + \cdots + \xi_k$. Here k is the number of the last cycle when a system failure has occurred. For the distribution of the r.v. ξ^*, the following recurrent equation can be easily written:

$$1 - F^*(t) = 1 - F(t) - \int_0^t [1 - F^*(t - x)] G(x)\, dF(x) \quad (7.33)$$

The first term of the sum reflects the fact that during time t no failure occurs. The expression under the integral means that the first failure occurs in the interval $[x, x + dx]$, but the repair of the failed unit has been completed up to this moment, and from this moment on the system is in the same state as in the previous moment at moment ξ_0. Thus, this is the regeneration moment for the renewal process under consideration.

The final goal is to find the distribution of the random value $\xi_0 + \xi^*$ and to express the probability of the system's successful operation $P(t)$:

$$P(t) = 1 - Q(t) = 1 - \int_0^t F^*(t - x)\, dF(x) \quad (7.34)$$

The numerical solution of (7.33) and (7.34) can be obtained by sequential iteration. But again we will use the LST which is useful for future asymptotic analysis.

Introduce the following notation:

$$\Phi(s) = \int_0^\infty e^{-st}\, dF(t)$$

$$\psi(s) = \int_0^\infty e^{-st} G(t)\, dF(t)$$

$$\tilde{\varphi}(s) = \int_0^\infty e^{-st}\, dF^*(t)$$

$$\varphi(s) = \int_0^\infty e^{-st}\, dQ(t)$$

Then the LST of (7.33) can be written as

$$\tilde{\varphi}(s) = \Phi(s) + \psi(s)[1 - \tilde{\varphi}(s)]$$

and, finally,

$$\tilde{\varphi}(s) = \frac{\Phi(s) - \psi(s)}{1 - \psi(s)} \quad (7.35)$$

Combining (7.35) with (7.34), we get

$$\varphi(s) = \Phi(s)\bar{\varphi}(s) = \Phi(s)\frac{\Phi(s) - \psi(s)}{1 - \psi(s)} \tag{7.36}$$

From this LST, the system's MTTF can be found by setting $s = 0$. But in this case we prefer a more direct way:

$$T_{\text{syst}} = E\{\xi_0\} + E\left\{ \sum_{1 \le k \le \nu} \xi_k \right\} = T + \frac{T}{\alpha}$$

where ν is the random number of cycles which has a geometric distribution with parameter α:

$$\alpha = \int_0^\infty [1 - G(t)]\, dF(t)$$

which is small in practical cases.

Let us investigate the asymptotic behavior of $p(t)$. Suppose that $F(t)$ is fixed and the distribution of the repair time changes by some sequence in such a way that

$$\alpha_n = \int_0^\infty [1 - G_n(t)]\, dF(t) \to 0$$

Let us introduce the corresponding distributions and LSTs: $Q_n(t)$, $\varphi_n(s)$, $\psi_n(s)$, and, additionally, $\chi_n(s)$:

$$\chi_n(s) = \Phi_n(s) - \psi_n(s) = \int_0^\infty e^{-st}[1 - G_n(t)]\, dF(t)$$

Now we evaluate the difference

$$\alpha_n - \chi_n(\alpha_n s) = \int_0^\infty (1 - e^{-\alpha_n st})[1 - G_n(t)]\, dF(t)$$

$$\le \alpha_n s \int_0^\infty t[1 - G_n(t)]\, dF(t)$$

$$\le \alpha_n s \left[C_n \int_0^\infty [1 - G_n(t)]\, dF(t) + \int_{C_n}^\infty t\, dF(t) \right]$$

If in this inequality we let

$$C_n = \frac{1}{\sqrt{\alpha_n}}$$

then both terms in the last set of square brackets go to 0. This leads to the statement:

$$\lim_{n \to \infty} \frac{\chi_n(\alpha_n s)}{\alpha_n} = 1$$

and the limit is uniformly exceeded on any finite area of domain of s.

Now the normalized random variable $\alpha_n \tau$ is considered. The d.f. of this r.v. is

$$\Pr\{\alpha_n \tau < t\} = Q_n\left(\frac{t}{\alpha_n}\right)$$

The LST of this d.f. is

$$\varphi_n(\alpha_n s) = \int_0^\infty e^{-st} \, dQ_n\left(\frac{t}{\alpha_n}\right)$$

For $\alpha_n \to 0$, from (7.36) it follows that

$$\varphi_n(\alpha_n s) = \Phi(\alpha_n s) \frac{\chi_n(\alpha_n s)}{1 - \Phi(\alpha_n s) + \chi_n(\alpha_n s)}$$

$$= \Phi(\alpha_n s) \frac{\dfrac{\chi_n(\alpha_n s)}{\alpha_n}}{\dfrac{1 - \Phi(\alpha_n s)}{\alpha_n} + \dfrac{\chi_n(\alpha_n s)}{\alpha_n}} \to \frac{1}{1 + sT}$$

and the limit is uniformly exceeded on any finite area of domain of s. Consequently,

$$\lim_{n \to \infty} \Pr\{\alpha_n \tau < t\} = 1 - e^{-t/T} \tag{7.37}$$

From (7.37) it follows that for a small value of α the approximation

$$P(t) = \Pr\{\xi_{\text{syst}} > t\} \approx e^{-\alpha t/T}$$

is true.

7.4 METHOD OF INTRODUCING FICTITIOUS STATES

As we considered in Chapter 1, some combinations of exponential distributions can produce distributions with both increasing and decreasing intensity functions, or failure rates. This fact leads to the idea of an approximation of some arbitrary distributions. We will show that such an approximation can allow us to reduce semi-Markov processes to Markov processes.

A mixture of exponential distributions with different parameters leads to a distribution which has the decreasing intensity function

$$F(t) = \sum_{1 \leq i \leq n} p_i\left[1 - e^{-\lambda_i t}\right] = 1 - \sum_{1 \leq i \leq n} p_i e^{-\lambda_i t}$$

$$\lambda_i \neq \lambda_j \quad \forall (i, j) \tag{7.38}$$

$$p_i > 0 \quad \forall i, \ \sum p_i = 1$$

A convolution of n identical exponential distributions $e(t) = \exp(-\lambda t)$ leads to an Erlang distribution of the nth order which can be expressed in the following recurrent way:

$$F_{\text{Erl}}^{(n)}(t) = F_{\text{Erl}}^{(n-1)} * e(t) = \lambda \frac{(\lambda t)^{n-1}}{(n-1)!} e^{-\lambda t} \tag{7.39}$$

If there are exponential functions with different parameters $e_k(t) = \exp(-\lambda_k t)$, then the generalized Erlang d.f. holds

$$F_{\text{Erl}}^{(n)}(t) = e_1 * e_2 * \cdots * e_n(t) = \int_0^t e_1 * e_2 * \cdots * e_{n-1}(t - x)\, de_n(x) \tag{7.40}$$

Both the Erlang and the generalized Erlang distributions belong to the IFR class. Notice that the generalized Erlang d.f. can naturally approximate a wider subclass of distributions belonging to the IFR class.

It is reasonable to remember that the Erlang distribution represents an appropriate mathematical model for standby redundancy. Indeed, the process of a standby redundant group's operation can be described as a sequence of a constant number of periods of a unit's successful operation.

Thus, (7.38) can be used as a possible approximation of the IFR distributions, and (7.39) with (7.40) can be used for the DFR distributions. Of course, such an approximation leads to an increase in the number of states in the stochastic process under consideration. (Nothing can be obtained free, even in mathematics!) But we should mention that the process itself becomes much simpler: it becomes purely Markov. At the same time, such an approximation is good only for systems of a very restricted size.

For simplicity of further illustrations, we will consider only cases where the initial distributions are approximated with the help of combinations of two exponential distributions. We should mention that the problem of determining an appropriate approximation of a distribution with monotone failure rates by the means of (7.38) to (7.40) is a special problem lying outside of the scope of this book.

Now we illustrate the main idea by means of simple examples.

IFR Repair Time and Exponential Time to Failure   For some applied problems it is natural to use the exponential distribution for a random TTF. At the same time, to assume an exponential distribution for the repair time might seem strange: why should the residual time of repair not depend on the time already spent? If a repair involves a routine procedure, a more realistic assumption involves the IFR distribution of this r.v. To make this statement clearer, we consider a repair process as a sequence of several steps: if one step is over, the residual time of repair is smaller because now it consists of a smaller number of remaining steps.

In this case two failure states might be introduced for a unit: state 1 and state 1*, both with an exponentially distributed time remaining in each of them. These sequential states represent the series sequence of two stages of repair (see Figure 7.4a). The total random time of staying in a failed state subset is the sum of two exponentially distributed random variables and, consequently, will have an IFR distribution. Incidentally, in this case (7.40) has the following expression:

$$\frac{\lambda_1\lambda_2}{\lambda_1 - \lambda_2}\left(e^{-\lambda_2 t} - e^{\lambda_1 t}\right) \tag{7.41}$$

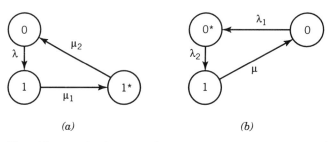

(a)   *(b)*

Figure 7.4. Transition graphs for a multistate model of renewable units: (*a*) with an IFR distributed repair time and an exponentially distributed failure-free time; (*b*) with an IFR distributed failure-free time and an exponentially distributed repair time.

Suppose that for some reason a unit should be considered as having an IFR distribution for its TTF and an exponential distribution of repair time. Then a "dual" transition graph is considered with two operational states: state 0 and state 0* (see Figure 7.4b).

DFR Repair Time and Exponential Time to Failure Sometimes a DFR repair time might be reasonably assumed. For example, a system may consist of two units: one of them takes more time for repair than another although both of them have exponentially distributed random repair times with different parameters. Thus, the system's repair time depends on which of the two units fails. In this case a "weighed" distribution could be a good mathematical model, and one more realistically assumes the DFR distribution of random time.

In this case two failure states are introduced: state 1 and state 1*, both with an exponential distribution but with different parameters. Both states are separate and located on the same layer of the transition graph (see Figure 7.5a). Therefore, the process goes from operational state 0 to state 1 with probability

$$\frac{p_1\lambda}{p_1\lambda + (1 - p_1)\lambda}$$

and to state 1* with probability

$$\frac{(1 - p_1)\lambda}{p_1\lambda + (1 - p_1)\lambda}$$

The total time of staying in a failed state subset has a DFR distribution.

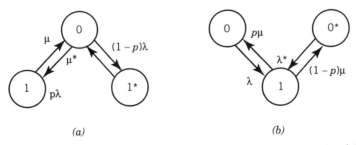

(a)   *(b)*

Figure 7.5. Transition graphs for a multistate model of renewable units: (*a*) with a DFR distributed repair time and an exponentially distributed failure-free time; (*b*) with a DFR distributed failure-free time and an exponentially distributed repair time.

Apparently, for a unit with a DFR distribution of TTF and an exponential distribution of repair time, the transition graph with two operational states and one failure state should be considered (see Figure 7.5b).

Non-Markov Cases Of course, a much more complicated case arises if one considers a unit with two nonexponential distributions. In this case a general, non-Markov process might be analyzed. The Markov approximation seems more reasonable, but, at the same time, even a simple model becomes clumsy. We present four cases without special explanation that can be easily analyzed by the reader. These cases are:

- Both distributions are IFR (Figure 7.6a).
- An IFR distribution of TTF and a DFR distribution of repair time (Figure 7.6b).

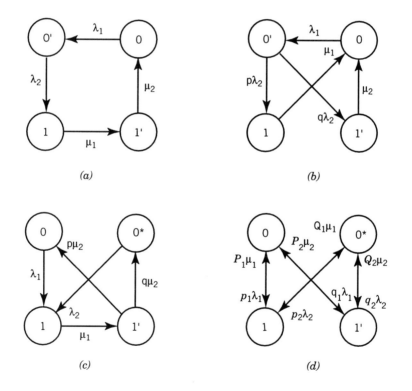

(a)

(b)

(c)

(d)

Figure 7.6. Transition graphs for a multistate model of renewable units: (a) with an IFR distributed repair time and an IFR distributed failure-free time; (b) with a DFR distributed repair time and an IFR distributed failure-free time; (c) with an IFR distributed repair time and a DFR distributed failure-free time; (d) with a DFR distributed repair time and an DFR distributed failure-free time.

- A DFR distribution of TTF and an IFR distribution of repair time (Figure 7.6c).
- Both distributions are DFR (Figure 7.6d).

It should be mentioned that this mathematical scheme allows one to create models for even more complex situations. For example, let us consider the following case. Both distributions are DFR, and the transition graph is close to that presented in Figure 7.6d, but with some differences. Let state 0 correspond to a long average operational time (a small value of intensity λ_0) and let state 0* correspond to a short average operational time (a large value of intensity λ_{0*}). Thus, $\lambda_{0*} > \lambda_0$. States 1 and 1* have the same meanings: the first state is characterized by a repair intensity μ_1 and the second by μ_{1*} where $\mu_{1*} > \mu_1$. Let us assume that a "short" system repair time follows after a "long" time of a successful operation, and, on the contrary, after a "short" up time a repair time is usually "long." This can be explained on a physical level in the following way. A failure after a "normally long" failure-free operation is expected to be "normal" itself; that is, it requires, on average, a smaller time of repair. In the transition graph, it means that $p_0 > q_0 = 1 - p_0$. On the other hand, "short" periods of failure-free operation are supposed to be connected with some kind of "serious" failure, which leads to a "long" repair time. In the transition graph, this means that $p_{0*} > q_{0*} = 1 - p_{0*}$.[1]

Of course, the inverse situation might be considered. Explanations also seem very reasonable: a "long" repair might follow a "long" period of successful operation. Indeed, we expect more failures of redundant units may appear during the longer period of time. As usual, a narrative of the system, which is taken to be a basis for the mathematical model, depends on the concrete actual nature of the system under investigation. For systems consisting of several units such an approximation may lead to difficulties in the construction of the corresponding transition graph.

Let us consider the simple case represented in Figure 7.4a in more detail. Incidentally, this case shows the special behavior of the availability coefficient. The system of differential equations is constructed in the usual way:

$$K_0'(t) = -\lambda_0 K_0(t) + \mu_2 K_2(t)$$

$$K_1'(t) = \lambda_0 K_0(t) - \mu_1 K_1(t)$$

$$1 = K_0(t) + K_1(t) + K_2(t) \tag{7.42}$$

$$K_0(0) = 1$$

---

[1] We put "long" and "short" in quotation marks because we consider r.v.'s with corresponding large and small means, but this does not mean that an r.v. with a larger mean cannot be less than an r.v. with a small mean, and vice versa. For simplicity, we use these terms for r.v.'s.

The LST of (7.42) gives the following system of algebraic equations:

$$(\lambda_0 + s)\varphi_0(s) - \mu_2\varphi_2 = 1$$
$$-\lambda_0\varphi_0(s) + (\mu_1 + s)\varphi_1(s) = 0 \qquad (7.43)$$
$$s\varphi_0(s) + s\varphi_1(s) + s\varphi_2(s) = 1$$

and the solution for $\varphi(s)$ is

$$\varphi_0(s) = \frac{s^2 + (\mu_1 + \mu_2)s + \mu_1\mu_2}{s[s^2 + (\lambda_0 + \mu_1 + \mu_2)s + \lambda_0\mu_1 + \lambda_0\mu_2 + \mu_1\mu_2]} \qquad (7.44)$$

Denote the eigenvalues (roots) of the denominator by s_k:

$$s_{1,2} = -\frac{a}{2} \pm \sqrt{\frac{a^2}{4} - b}$$

$$s_3 = 0$$

where

$$a = \lambda_0 + \mu_1 + \mu_2$$
$$b = \lambda_0(\mu_1 + \mu_2) + \mu_1\mu_2$$

Note that the discriminant of the denominator is negative for any λ_0, μ_1, and μ_2, which leads to the complex roots s_1 and s_2.

A representation of $\varphi_0(s)$ is found in the form

$$\varphi_0(s) = \frac{A}{s} + \frac{B}{s - s_1} + \frac{C}{s - s_2}$$

$$= \frac{(A + B + C)s^2 - [A(s_1 + s_2) + Bs_2 + Cs_1]s + As_1s_2}{s(s - s_1)(s - s_2)} \qquad (7.45)$$

Equations (7.44) and (7.45) lead to the following system of equations:

$$A + B + C = 1$$
$$\mu_1 + \mu_2 = -A(s_1 + s_2) - Bs_2 - Cs_1$$
$$As_1s_2 = \mu_1\mu_2$$

or, taking into account that $s_1 = s_2 = b$, A can be immediately expressed as

$$A = \frac{\mu_1\mu_2}{s_1s_2} = \frac{m_1\mu_2}{b} = \frac{\mu_1\mu_2}{\lambda_0(\mu_1 + \mu_2) + \mu_1\mu_2}$$

The following system is obtained as a result:

$$B + C = 1 - \frac{\mu_1 \mu_2}{b}$$

$$Bs_2 + Cs_1 = a \frac{\mu_1 \mu_2}{b} - \mu_1 - \mu_2$$

Because of the complex conjugate roots s_1 and s_2, one can write

$$Bi\sqrt{b - \frac{a^2}{4}} - Ci\sqrt{b - \frac{a^2}{4}} = 0$$

Thus, for the real parts of these roots,

$$B = C = \frac{1}{2}\left(1 - \frac{\mu_1 \mu_2}{b}\right)$$

Now we can find $K_0(t)$ in the form

$$K_0(t) = \frac{\mu_1 \mu_2}{b} + \frac{1}{2}\left(1 - \frac{\mu_1 \mu_2}{b}\right)\left(e^{-s_1 t} + e^{-s_2 t}\right)$$

For the complex root $s = \alpha + i\beta$,

$$e^{st} = e^{\alpha t}(\cos \beta t + i \sin \beta t)$$

Taking into account that the complex roots s_1 and s_2 are conjugate, we may

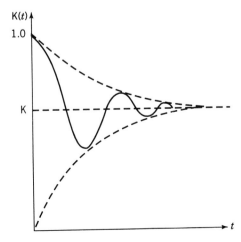

Figure 7.7. Time dependence of the availability coefficient for a unit with an IFR distributed repair time. Here $K = \mu_1 \eta_2 / \beta$ and

$$x = \frac{1}{\pi}\sqrt{\beta - \frac{\alpha^2}{4}}$$

write

$$e^{-s_1 t} + e^{-s_2 t} = 2 e^{-(a/2)t} \cos\sqrt{b - \frac{a^2}{4}}\, t$$

The final result is

$$K_0(t) = \frac{\mu_1 \mu_2}{b} + \left(1 - \frac{\mu_1 \mu_2}{b}\right) e^{-(a/2)t} \cos\sqrt{b - \frac{a^2}{4}}\, t$$

In this particular case the nonstationary availability coefficient is periodically oscillating with a decreasing amplitude (see Figure 7.7).

7.5 DUPLICATION WITH SWITCH AND MONITORING

Because of their relative simplicity, the mathematical models of a duplicated system with renewal allow one to consider some sophisticated cases close to real situations. Indeed, most "classical" mathematical models of redundant systems with repair are based on the assumption that the redundant group of units has an ideal switch which performs its functions reliably, without errors and delays. Moreover, the units are supposed to be totally and continuously monitored; that is, the occurrence of an operating or redundant unit failure becomes known immediately. It is clear that such assumptions are far from real. Sometimes, of course, these factors may be neglected. (But only sometimes!)

When a duplicated system is described by a Markov model, it is possible to provide an analysis of reliability by accounting for some additional factors. Obviously, it does not lead to especially clear and understandable models. At any rate, the solution can be derived.

Below we consider several examples which illustrate how one may construct appropriate mathematical models. We will not present the final results because they are inevitably bulky. A computer must be used for the numerical calculations. But, as is well known, no computer can substitute for the human mind, at least during the first stage of any research: one needs to be able to construct an appropriate mathematical model and only after this may one resort to computer calculation.

7.5.1 Periodic Partial Control of the Main Unit

We start with a simple example. A duplicated system consists of two independent identical units. One of them is in an operating position (the main unit) and the other is in a redundant position. The unit's failure rate depends on the current occupied position: operating or waiting. Let us assume that only part of the main unit can be monitored continuously. The state of the remaining nonmonitored part of the main unit can be checked only periodically. In other words, if a failure has occurred in the nonmonitored part of the main unit, no switching to replace this failed unit is performed. A periodic test discovers that the main unit has failed and only then the switching might be performed. Thus, before this test, the duplicated system remains in a state of "hidden failure." The switching is assumed to be instantaneous. The redundant unit is continuously monitored, so a repair of the failed redundant unit begins instantly after a failure has occurred. Of course, the same thing happens if a failure occurs in the monitored part of the main unit.

During a repair of the main unit, all of its failures—both in the monitored and nonmonitored parts—are deleted. In other words, the repaired unit becomes as good as new. As soon as the failure of the main unit is detected (by any means—continuous or periodical monitoring), the redundant unit is switched into the main position. After repair, the unit becomes the redundant one. If one finds both units have failed, the total system repair is performed. After repair, the system, as a whole, becomes as good as new.

For the use of a Markov model, let us assume that monitoring is provided at randomly chosen moments of time. Moreover, assume that the distribution of the length of the periods between the tests is exponential. We mention that such an assumption is sometimes close to reality: in a computer, tests can be applied between runs of the main programs, and not by a previously set strict schedule.

The transition graph for this case is presented in Figure 7.8. The following notation is used in this example:

- M is the operational state of the main unit.
- M^* is the "hidden failure" of the main unit.
- $\underline{M}$ is the failure state of the main unit.
- R is the operational state of the redundant unit.
- $\underline{R}$ is the failure state of the redundant unit.
- λ_1 is the failure rate of the nonmonitored part of the main unit.
- Λ_1 is the failure rate of the monitored part of the main unit.
- Λ is the failure rate of the redundant unit.
- μ is the intensity of repair of a single unit.
- μ^* is the intensity of repair of the duplicated system as a whole.
- ν is the intensity of periodical tests.

The transition graph presented in Figure 7.8a is almost self-explanatory. Notice that for this case there are two states of system failure: $[M^*\ R]$ and $[\underline{M}\ \underline{R}]$. These failure states are denoted by bold frames in the figure.

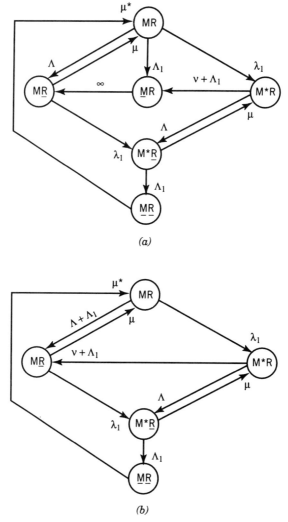

Figure 7.8. Transition graphs for a duplicated system with a partially monitored main unit: (*a*) graph including instantaneous "jumps" (intensity equal to ∞); (*b*) equivalent graph excluding the state in which the system spends no time.

We will not describe the routine procedure of finding the reliability indexes. Our main goal is to build a mathematical model from the verbal description and to clarify all of the needed assumptions.

7.5.2 Periodic Partial Monitoring of Both Units

The duplicated system consists of two identical independent units. One of them is in an operating position (the main unit) and the other is in a redundant position. The unit's failure rate depends on the occupied position: operating or waiting. Let us assume that only a part of each unit can be monitored continuously. (The monitored parts are identical in both units.) The state of the remaining nonmonitored parts of each of these units can be checked only periodically. Tests of the main and redundant units have different periods (intensity).

The switching system is analogous to that described in the previous example. If one knows that both units have failed, the repair is performed until complete renewal of the system. Let us consider two possible means of repair: (a) there are independent repair facilities for each unit, and (b) there is only one repair facility. After repair, the unit becomes as good as new.

The transition graph for this case is presented in Figure 7.9. The following notation is used in this example:

- M is the operational state of the main unit.
- M^* is the "hidden failure" of the main unit.
- $\underline{M}$ is the failure state of the main unit.
- R is the operational state of the redundant unit.
- R^* is the "hidden failure" of the redundant unit.
- $\underline{R}$ is the failure state of the redundant unit.
- λ_1 is the failure rate of the nonmonitored part of the main unit.
- Λ_1 is the failure rate of the monitored part of the main unit.
- λ is the failure rate of the nonmonitored part of the redundant unit.
- Λ is the failure rate of the monitored part of the redundant unit.
- μ is the intensity of repair of a single unit.
- μ^* is the intensity of repair of the duplicated system when there are two failed units.
- ν_1 is the intensity of periodic tests of the main unit.
- ν is the intensity of periodic tests of the redundant unit.

The transition graph presented in Figure 7.9 is almost self-explanatory. We only discuss the following two transitions:

1. $[M\ R^*]$ to $[M\ \underline{R}]$: This transition occurs if (a) an extra failure appears in the continuously monitoring part of the redundant unit or (b) the periodic test has found a "hidden failure."
2. $[M\ R^*]$ to $[M^*\ \underline{R}]$: This transition occurs if a failure appears in the continuously monitoring part of the main unit. Then the main unit is directed to repair and is substituted by the redundant one with a "hidden failure."

These failure states are again denoted by bold frames in the figure.

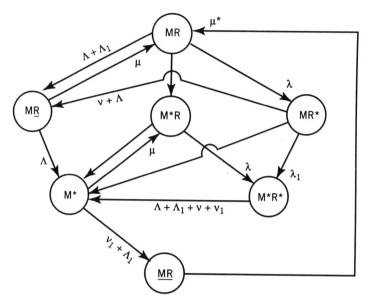

Figure 7.9. Transition graph for duplicated system with partially monitored main and redundant units.

We will again write no equations. This is obviously a routine procedure. As an exercise, consider the system when, after the failure of both units,

We will again write no equations. This is obviously a routine procedure.

As an exercise, consider the system when, after the failure of both units, the system is subjected to a total renewal: the system is repaired as a whole until both units are as good as new (the transition from the system failure state to the failure-free state).

7.5.3 Unreliable Switch

Consider a duplicated system with an unreliable switching device. A switching failure becomes known immediately and its repair begins at once. There is only one repair facility. Repair is performed in accordance with a FIFO (first-in, first-out) rule. If both units have failed, the total repair is performed. The failure of the main unit, occurring during the repair of the switch, leads to a system failure even if the redundant unit is operational. But a switching failure itself does not interrupt the main unit's successful operation. The monitoring of both units is supposed to be continuous and ideal. Repairs of both units and the switch are supposed to be independent.

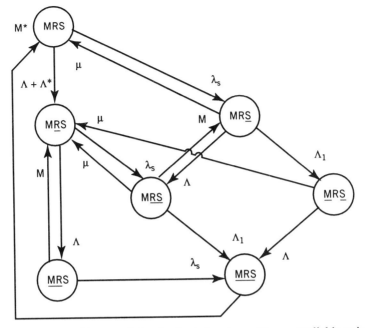

Figure 7.10. Transition graph for duplicated system with an unreliable switch.

The transition graph for this case is presented in Figure 7.10. The following notation is used in this example:

- M is the operational state of the main unit.
- $\underline{M}$ is the failure state of the main unit.
- R is the operational state of the redundant unit.
- $\underline{R}$ is the failure state of the redundant unit.
- S is the operational state of the switch.
- $\underline{S}$ is the failure state of the switch.
- λ_1 is the failure rate of the main unit.
- λ_s is the failure rate of the switch.
- Λ is the failure rate of the redundant unit.
- μ is the intensity of repair of a single unit.
- μ_s is the intensity of repair of the switch.
- μ^* is the intensity of repair of the system as a whole.

As in the previous examples, the transition graph presented in Figure 7.10 is almost self-explanatory. We only discuss the following transitions:

1. $[\underline{M}\ R\ \underline{S}]$ to $[M\ \underline{R}\ S]$: This transition occurs if the switch has been repaired: at the moment of the termination of repair, the redundant unit is instantly directed to the position of the failed main unit, and the latter begins to be considered as a redundant unit directed to repair.

2. $[\underline{M}\ \underline{R}\ S]$ to $[M\ \underline{R}\ \underline{S}]$ and $[\underline{M}\ \underline{R}\ S]$ to $[\underline{M}\ R\ \underline{S}]$: These two transitions depend on which unit is repaired first: if the switching device is in repair, it is impossible to put the repaired redundant unit into the position of the main unit.

These failure states are again denoted by bold frames in the figure.

7.5.4 Unreliable Switch and Monitoring of Main Unit

A duplicated system consists of two identical independent units: main and redundant. The unit failure rate depends on the occupied position. Let us assume that only a part of the main unit can be monitored continuously. The state of the remaining nonmonitored part of the unit can be checked only periodically. The switching device works as described above. Repairs of both the units and the switch are independent. After repair, a unit (or a switch) becomes as good as new.

The transition graph for this case is presented in Figure 7.10. The following notation is used in this example:

- M is the operational state of the main unit.
- M^* is the "hidden failure" state of the main unit.
- $\underline{M}$ is the failure state of the main unit.
- R is the operational state of the redundant unit.
- $\underline{R}$ is the failure state of the redundant unit.
- λ_1 is the failure rate of the nonmonitored part of the main unit.
- Λ_1 is the failure rate of the monitored part of the main unit.
- Λ is the failure rate of the redundant unit.
- Λ_s is the failure rate of the switch.
- μ is the intensity of repair of a single unit.
- μ_s is the intensity of repair of the switch.
- μ^* is the intensity of repair of the system as a whole.
- ν is the intensity of periodical tests of the main unit.

Note that the switching device may be one of the two following main types: (a) as we considered before, a switching failure does not interrupt the system's operation, or (b) a switching failure interrupts the system operation. In the latter case, the switch is a necessary part of the system. This may occur, for example, if the switch plays the role of an interface between the duplicated system's output and the input of another system or subsystem.

Of course, there are many other concrete examples of this type. We can only repeat that our main goal is to explain the methodology of modeling and not to give a list of the results or to make the reader exhausted with boring solutions of bulky equations.

The mathematical techniques used in this section are simple enough. But the results obtained are not always very clear or "transparent" for further analysis: What will happen if one changes some parameters? What will happen if the switching or monitoring methods are changed? Of course, in practical situations an engineer would like to have correct *and* simple formulas to perform a quick and understandable analysis of the designed system. Fortunately, for highly reliable systems (we emphasize again that this is the most important practical case!), it is possible to develop such simple and sufficiently accurate methods. The reader can find such methods in Chapter 13 dedicated to heuristic methods in reliability.

CONCLUSION

It seems that the first paper on the analysis of a duplicated system with repair (renewal) was published by Epstein and Hosford (1960). They solved the problem for a purely Markov model when the distributions—both TTF and repair time—were exponential. They solved the problem with the help of birth and death processes. Their model is also described in Gnedenko and Kovalenko (1987). Here the solution of the same problem for the duplicated system was found for both active and underloaded redundancy.

A systematic investigation of renewal systems, in particular, the duplicated system, may be found in Ushakov (1985, 1994). Belyaev (1962) developed an elegant method of so-called "striping Markov processes" which has allowed one to solve the problem with no assumption of exponentiality on the repair time. Independently, Gaver (1963) obtained practically the same results with the help of traditional methods.

Gnedenko (1964a, 1964b) obtained solutions for the general case when both distributions are arbitrary. Theorems concerning the asymptotic behavior of renewal duplicated systems have been obtained by D. Gnedenko and Solovyev (1974, 1975). and, practically simultaneously, by Gnedenko, Belyaev, and Solovyev (1969). The method of fictitious states (stages) for "Markovization" of non-Markov models as applied to queuing systems takes its origin from Erlang's work. A comprehensive exposition of existent mathematical results related to the problem can be found in Ushakov (1985, 1994), where results concerning the various models of duplicated renewal systems are presented. A detailed review of asymptotic methods is given by Gertsbakh (1984).

REFERENCES

Belyaev, Yu. K. (1962). Striping Markov processes and their applications to reliability problems (in Russian). *Proc. Fourth All-Union Meeting on Probab. Theory Math. Statist.*, Vilnius (Lithuania), pp. 309–323.

Epstein, B., and T. Hosford (1960). Reliability of some two unit redundant systems. *Proc. Sixth Nat. Symp. on RQC*, pp. 466–476.

Gaver, D. P. (1963). Time to failure and availability of parallel systems with repair. *IEEE Trans. RQC*, vol. R-12, pp. 30–38.

Gertsbakh, I. B. (1984). Asymptotic methods in reliability theory: a review. *Adv. in Appl. Probab.* vol. 16.

Gnedenko, B. V. (1964a). On duplication with renewal. *Engrg. Cybernet.*, no. 5, pp. 111–118.

Gnedenko, B. V. (1964b). On spare duplication. *Engrg. Cybernet.*, no. 4, pp. 3–12.

Gnedenko, B. V., and I. N. Kovalenko (1987). *Introduction to Queuing Theory*, 2nd ed. (in Russian). Moscow: Nauka.

Gnedenko, B. V., Yu. K. Belyaev, and A. D. Solovyev (1969). *Mathematical Methods of Reliability Theory*. New York: Academic.

Gnedenko, D. B., and A. D. Solovyev (1974). A general model for standby with renewal. *Engrg. Cybernet.*, vol. 12, no. 6.

Gnedenko, D. B., and A. D. Solovyev (1975). Estimation of the reliability of complex renewable systems. *Engrg. Cybernet.*, vol. 13, no. 3.

Solovyev, A. D. (1970). Standby with rapid repair. *Engrg. Cybernet.*, vol. 4, no. 1.

Solovyev, A. D. (1971). Asymptotic behavior of the time of first occurrence of a rare event. *Engrg. Cybernet.*, vol. 9, no. 3.

Solovyev, A. D. (1972). Asymptotic distribution of the moment of first crossing of a high level by birth and death process. *Proc. Sixth Berkeley Symp. Math. Statist. Probab.*, Issue 3.

Ushakov, I. A., ed. (1985). *Reliability of Technical Systems: Handbook* (in Russian). Moscow: Radio i Sviaz.

Ushakov, I. A., ed. (1994). *Handbook of Reliability Engineering*. New York: Wiley.

EXERCISES

7.1 There are two transition graphs (see Figures E7.1*a* and *b*). State 2 is a failed state in both cases.

  (a) Give a verbal description of the two systems to which these graphs correspond.

  (b) Which system has a larger MTTF?

(c) Which system has a larger MTBF?

(d) What is the difference between the MTTF and MTBF of the first system?

(e) Which system has a larger mean repair time?

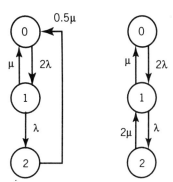

Figure E7.1.

7.2 Depict a transition graph for a unit with an exponentially distributed time to failure and a repair time having an Erlang distribution of the third order.

7.3 Depict a transition graph for a renewable unit with an exponentially distributed time to failure. $P(t) = e^{-\lambda t}$, and with a repair time distributed as

$$G(t) = p_1 e^{-\mu_1 t} + p_2 e^{-\mu_2 t} + p_3 e^{-\mu_3 t}$$

7.4 Given an interpretation of the transition graph depicted in Figure E7.4.

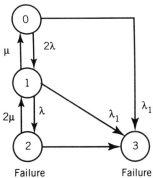

Figure E7.4.

SOLUTIONS

7.1 (a) The first system is a duplicate system with a loaded redundant unit where after the system has failed it is renewed as a whole. The second system is an ordinary duplicate system of two independent identical units operating in a loaded regime.

(b) Both systems have the same MTTF.

(c) The first system has a larger MTBF than the second one because after each failure this system starts from state 0. After a system failure the second system starts from state 1 where there is a possibility of entering a failed state immediately.

(d) There is no difference at all.

(e) The first system has twice as large a repair time.

7.2 The solution is depicted in Figure E7.2.

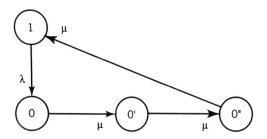

Figure E7.2.

7.3 The transition graph is depicted in Figure E7.3.

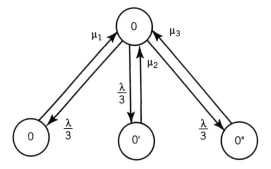

Figure E7.3.

7.4 The system is a series connection of an unrepairable unit with an exponentially distributed time to failure with parameter λ_1 and a repairable duplicated group with failure rate λ and intensity of repair μ. The redundant group consists of units in a loaded regime; there are two repair facilities. The structure of the system is depicted in Figure E7.5.

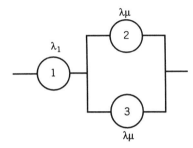

Figure E7.5.

CHAPTER 8

ANALYSIS OF PERFORMANCE EFFECTIVENESS

8.1 CLASSIFICATION OF SYSTEMS

8.1.1 General Explanation of Effectiveness Concepts

Modern large-scale systems are distinguished by their structural complexity and their requirements for sophisticated algorithms to facilitate the functioning and interacting of their subsystems. On the one hand, this allows them to fulfill many different operations and functions, while, on the other hand, it leads to stable operations with a sufficient level of effectiveness even with some failed units and subsystems and/or under extreme influences of the external environment.

The adaptation of a complex system to external influences and to internal perturbations is possible only because of the redundancy of the system's structure and its ability to readjust its functions under various circumstances. In other words, the feature of modern technical systems is not only an extreme increase in the number of interacting units but also the appearance of entirely new qualitative properties. One of these properties is the stability of operation mentioned above.

It is also very important that modern large systems, such as information systems (computer and communications networks, control systems, etc.), energy systems (electric power networks, oil and gas pipelines, etc.), and transportation systems (railroads, highways, airlines, etc.) are multifunctional. Such systems, as a result of external and internal influences, can perform some functions perfectly and, at the same time, completely interrupt the performance of other functions. This means that, according to one criterion, a system as a whole could be considered successful and, by another criterion,

298

it could be considered failed. A researcher encounters the usual difficulties associated with multicriteria analysis.

But even for a complex system predestinated for one type of operation, there is generally no strict definition of failure. Often in such systems even a significant set of failed units could lead only to a decrease in performance and not to a complete system failure. This happens because of a partial "overlapping" of different subsystem (unit) functions, the presence of different feedbacks, the means of error correction, and so forth.

We consider several simple examples. In a regional power system, a failure of some subsystem (e.g., the failure of the fuel transportation system of an electric power plant) can be compensated for by using fuel from storage. In another case a deficiency of energy can be compensated for by a partial increase in the power of neighboring plants. Sometimes clients might use another type of energy supply. Under conditions of a severe energy deficit, clients with lower levels of priority might be temporarily "turned off" from an energy system in order to decrease the total damage.

Sometimes a completely operational system might be unable to perform some of its functions because of a harmful coincidence of external circumstances. For example, consider a communications network. All equipment in the system could be in a perfect operating state, but weather may spoil the opportunity to use certain radio channels. The same effect may be observed if there is some neighboring influence of other radio transmission systems. Even network users may create excessively heavy communication traffic which can lead to system performance failures.

Of course, from a client's viewpoint, he or she is quite indifferent to the reason for a breakdown in communication: either it happens because of a system failure or because of an overloading of the communications network.

For all such systems it is natural to speak about performance effectiveness. In each concrete case the index (or indexes) of performance effectiveness should be chosen with respect to the type of system under consideration, its destination, conditions of operation, and so forth. The physical characteristics of the *performance effectiveness index* (PEI) are usually completely defined by the nature of the system's outcome and can be evaluated by the same measures. In most practical cases we can measure a system's effectiveness in relative units. We might take into account the nominal (specified) value of a system's outcome as the normalizing factor. In other words, the PEI is a measure of the quality and/or volume of the system's performed functions or operations; that is, it is a measure of the system's expediency.

Of course, a system's efficiency is not an absolute measure. It depends on the type of functions and tasks being performed and the operating environment. A system which is very efficient under some circumstances and for some operations might be quite useless and ineffective under another set of circumstances and/or operations.

In general, a PEI is dimensional. The dimension of the PEI depends on the system's outcomes, as we mentioned above. When it is possible in the following discussion, we shall consider the PEI as a ratio of the expected system outcome to its maximal outcome. In this case the PEI is nondimensional. Of course, we always assume a larger outcome is a better outcome. The use of a nondimensional PEI is very convenient in many practical evaluations.

Sometimes we encounter "pessimistic" measures which characterize a system's performance. For example, consider the acceptable error of a technological process or the permissible volume of pollution of a plant. Such indexes measure "ineffectiveness" rather than effectiveness. Usually, in such cases one can reformulate the desired outcome in "positive" terms.

If a system's outcome has an upper bound, the PEI can be expressed in a normalized form; that is, it may be considered as having a positive value lying between 0 and 1. Then we have PEI = 0 if the system has completely failed and PEI = 1 when it is completely operational. For intermediate states, $0 \leq \text{PEI} \leq 1$.

When considering a system's effectiveness, one should remember the property of monotonicity introduced earlier. In this context, an increase in the reliability of any unit leads to a simultaneous increase in the system's effectiveness. Also, a failure of any unit can only decrease (not increase) a system's effectiveness.

It is convenient for system design to determine a PEI in relative units, because in this case one does not need to measure an absolute value of a system's outcome for different states. The absolute values of a PEI are very convenient if we must compare several different competitive variants of a system. They allow us to compare variants of a system with different reliability and different efficiencies of performance. It is clear that reliability alone does not completely solve the problem of engineering design.

8.1.2 Classes of Systems

Consider a system consisting of n units. As before, we suppose that any system unit has two states: an operating state and a failed state. Let x_i be the indicator of the ith unit's state: $x_i = 1$ when the unit is up and $x_i = 0$ when the unit is down. The system then has 2^n different states as determined by the states of its units. Denote a system state by $\mathbf{X} = (x_1, x_2, \ldots, x_n)$.

If we consider the process of a system's evolution in state space, then for each unit we should consider the process $x_i(t)$, and for the system as a whole, the process $\mathbf{X}(t)$. The transformation of system states $\mathbf{X}(t)$ characterizes the system's behavior. On the basis of knowledge about this process, we can analyze a system's effectiveness.

Taking into account the length of a system's performance, it is reasonable, for effectiveness analysis, to distinguish two main classes of systems: instant and enduring.

Some systems are characterized by their instant outcome at a moment of time. The current effectiveness of an *instant system* is completely determined by its state at the moment of performance. It is clear that no instant system exists in reality because any task has some duration. Strictly speaking, we consider a system whose duration of performance is negligibly short in comparison with time intervals between changing system states $\mathbf{X}(t)$. This means that

$$P\{\mathbf{X}(t) = \mathbf{X}(t + t_0)\} = 1 - \varepsilon \qquad (8.1)$$

where t_0 is the system's task duration and ε is a practically negligible value. (The size of ε depends on the required accuracy of analysis.)

From (8.1) it follows that the current effectiveness of a system is completely determined by the current system state $\mathbf{X} = \mathbf{X}(t)$. For this state the effectiveness coefficient equals $W_\mathbf{X}$, and the system's PEI can be determined as the expected value of $W_\mathbf{X}$.

Examples of practical instant systems are missiles, production lines (during production of a single item), and a communications network during an individual call.

For an enduring system condition (8.1) is not valid. The effectiveness of an enduring system depends on a trajectory of the system's transition from one state to another. In this case the fact that some particular units have failed is very important, but the moments and the order of their failures are also equally important. In other words, for these systems the effectiveness is determined by a trajectory of states changing during the system's performance of a task.

Examples of enduring systems are different technological and chemical processes, information and computer systems, aircraft, and so on.

8.2 INSTANT SYSTEMS

Let $h_{\mathbf{X}_k}(t)$ denote the probability that an instant system at moment t is in state $\mathbf{X}_k(t)$. We assume that the current effectiveness of the system being in any state can be evaluated. Let us denote this value for state $\mathbf{X}$ as $W_\mathbf{X}$. It is natural to determine W as the expected value of $W_\mathbf{X}$, that is,

$$W_{\text{syst}}(t) = \sum_{1 \leq k \leq N} h_{\mathbf{X}_k}(t) W_{\mathbf{X}_k} \qquad (8.2)$$

where $N = 2^n$ is the total number of different system states.

It is clear that an absolutely accurate calculation of a system's effectiveness when $n \gg 1$ is a difficult, if not unsolvable, computational problem. First of all, it is connected with the necessity of determining a large number of coefficients W_k. Fortunately, it is sometimes not too difficult to split all of the system's states into a relatively small number of classes with close values W_k.

If so, we need only to group appropriate states and calculate the corresponding probabilities. W_{syst} can then be calculated as

$$W_{\text{syst}}(t) = \sum_{1 \leq j \leq M} W_j \sum_{\mathbf{X}_k \in G_j} h_{\mathbf{X}_k}(t) \tag{8.3}$$

where M is the number of different levels of the values of $W_{\mathbf{X}}$ and G_j is the set of system states for which $W_{\mathbf{X}}$ belongs to the jth level.

Later we shall consider special methods for the evaluation of the effectiveness of higher-dimensional systems.

Let us evaluate a system's effectiveness for a general case. For notational simplicity, we omit the time t in the expressions below. Let h_0 denote the probability that all units of the system are successfully operating at moment t:

$$h_0 = \prod_{1 \leq i \leq n} p_i \tag{8.4}$$

Let h_i denote the probability that only the ith unit of the system is in a down state at moment t (repairable systems can be considered as well as unrepairable). Then

$$h_i = q_i \prod_{\substack{1 \leq j \leq n \\ j \neq i}} p_i = \frac{q_i}{p_i} h_0 = g_i h_0 \tag{8.5}$$

where, for brevity, we introduce $g_i = q_i/p_i$ and h_{ij} denotes the probability that only the ith and jth units of the system are in down states at moment t:

$$h_{ij} = q_i q_j \prod_{\substack{1 \leq j \leq n \\ k \neq (i,j)}} p_k = \frac{q_i q_j}{p_i p_j} h_0 = g_i g_j h_0 \tag{8.6}$$

and so on.

We can write the general form of this probability as

$$h_{\mathbf{X}} = \prod_{i \in G_{\mathbf{X}}} p_i \prod_{i \in \bar{G}_{\mathbf{X}}} q_i = h_0 \prod_{i \in \bar{G}_{\mathbf{X}}} g_i \tag{8.7}$$

where $G_{\mathbf{X}}$ is the set of subscripts of the units which are considered operational in state $\mathbf{X}$ and $\bar{G}_{\mathbf{X}}$ is the complementary set. Sometimes it is reasonable to write (8.7) for any $\mathbf{X}$ as

$$h_{\mathbf{X}} = \prod_{\substack{1 \leq i \leq n \\ x_i \in \mathbf{X}}} p_i^{x_i} q_i^{(1-x_i)} \tag{8.8}$$

It is clear that (8.7) and (8.8) are equivalent. Using (8.4) to (8.8), we can rewrite (8.3)

$$W_{\text{syst}} = W_0 h_0 \left[1 + \sum_{1 \leq i \leq n} \tilde{W}_i g_i + \sum_{1 \leq i < j \leq n} \tilde{W}_{ij} g_i g_j + \cdots \right] \quad (8.9)$$

where W_0 is the system effectiveness for state $\mathbf{X}_0$ and $\tilde{W}_i, \tilde{W}_{ij}, \ldots$ are normalized effectiveness coefficients for states $\mathbf{X}_i, \mathbf{X}_{ij}, \ldots$. In other words, $\tilde{W}_i = W_i / W_0$, $\tilde{W}_{ij} = W_{ij} / W_0, \ldots$.

For a system consisting of highly reliable units, that is,

$$\max_{1 \leq i \leq n} q_i \ll \frac{1}{n} \quad (8.10)$$

expression (8.9) can be approximated as

$$W_{\text{syst}} \approx W_0 \left(1 - \sum_{1 \leq i \leq n} q_i \right) \left(1 + \sum_{1 \leq i \leq n} \tilde{W}_i q_i \right) = W_0 \left(1 - \sum_{1 \leq i \leq n} q_i \tilde{w}_i \right) \quad (8.11)$$

Here $w_i = 1 - \tilde{W}_i$ has the meaning of a "unit's significance."

REMARK. It is necessary to note that, strictly speaking, it is wrong to speak of a "unit's significance." The significance of a unit depends on the specific system state. For example, in a simple redundant system of two units, the significance of any unit equals 0 if both units are successfully operating, but if a single unit is operating at the time, then its significance equals 1. Other examples are considered below.

Consider some particular cases of (8.11). If $p_i(t) = \exp(-\lambda_i t)$ is close to 1 and, consequently, $q_i(t) \approx \lambda_i t$, then (8.11) can be approximated by

$$W_{\text{syst}} \approx W_0 \left(1 - \sum_{1 \leq i < +n} q_i(t) w_i \right) \approx W_0 \exp \left[-t \sum_{1 \leq i \leq n} \lambda_i (1 - W_i) \right] \quad (8.12)$$

We can see that "the significance of unit" is reflected, in this case, in the factual failure rate. (See the previous remark.)

If p_i is a stationary availability coefficient, that is, $p_i = T_i / (T_i + \tau_i)$ where T_i is the MTBF of the ith unit and τ_i is its idle time and $T_i \gg \tau_i$, then it is possible to write the approximation

$$W_{\text{syst}} \approx W_0 \left[1 - \sum_{1 \leq i \leq n} \frac{\tau_i}{T_i} w_i \right] \quad (8.13)$$

Again, we can consider "the significance of a unit" in a new form keeping in mind the same precautions as above.

We should once more emphasize that all approximate expressions (8.9) to (8.13) are valid only for highly reliable systems.

Example 8.1   To demonstrate the main ideas, we first use a simple system consisting of two redundant units. Let the system's units have corresponding probabilities of successful operation equal to p_1 and p_2. The problem is to find the probability of the system's successful operation.

Solution.   By the definition of a duplicate system, $W_0 = W_1 = W_2 = 1$ and $W_{12} = 0$. Thus, for this particular case

$$W_{syst} = 1 \cdot p_1 p_2 + 1 \cdot q_1 p_2 + 1 \cdot p_1 q_2 = 1 - q_1 q_2$$

which completely coincides with the corresponding expression for the probability of failure-free operation of a duplicate system.

It is to be understood that W is a generalization of a common reliability index. Everything depends on the chosen coefficients W_X.

Now we consider more interesting cases which cannot be put into the framework of a standard reliability scheme.

Example 8.2   An airport traffic control system consists of two stationary radars each with an effective zone of 180° (see the schematic plot of the system in Figure 8.1). For this example let us assume that the effectiveness of the system in a zone with active radar coverage equals 0.7. The availability coefficient for each radar is equal to 0.9. (Of course, nobody would use such an ineffective system in practice!) We will assume that if only one radar is operating, it means that the system PEI = 0.5. It is necessary to evaluate the PEI for the system.

Solution.

$$W_{syst} = (0.7)p_1 p_2 + (1/2)(0.7)q_1 p_2 + (1/2)(0.7)p_1 q_2$$
$$= (1/2)(0.7)p_1 + (1/2)(0.7)p_2 = (0.7)(0.9) = 0.63$$

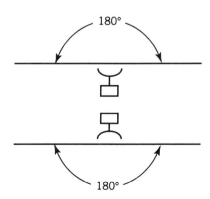

Figure 8.1. Schematic representation of an airport radar system.

Example 8.3   Consider the same airport traffic control system as in Example 8.2. To increase the effectiveness of the system, the operating zones of the radars overlap. In addition, we assume that within the overlapped zone, the effectiveness of service is higher. Let us say that the coefficient of effectiveness in an overlapping zone is practically equal to 1, while the same coefficient of effectiveness in an ordinary zone is 0.7.

The system's effectiveness is determined as the average probability of success weighted by the size of the zones with their corresponding effectiveness coefficients. There are two possibilities to design a system with overlapping zones. These two cases are depicted in Figure 8.2. The availability coefficient of each radar again equals 0.9. The problem is to compare the effectiveness of both variants and to choose the best one.

Solution.   Consider the first variant, A, with two radars in the north zone and two radars in the south zone (see Figure 8.2a). It is clear that we can consider two independent subsystems, each delivering its own outcome to the control system as a whole. The outcome of one subsystem is equal to one-half of the system's total outcome. Denote the effectiveness indexes of these two subsystems and of the whole system by W_1, W_2, and W_{syst}, respectively. Because of the identity $W_1 = W_2$, $W_{syst} = 2W_1 = 2W_2$.

Each subsystem can be in one of two useful states:

- Both radars are operating, and the probability of this is $(0.9)(0.9) = 0.81$; the coefficient of effectiveness in the zone is equal to 1.
- Only one radar is operating, and the probability of this is $(0.9)(0.1) = 0.09$; the coefficient of effectiveness in the zone is equal to 0.7.

(Recall that each subsystem covers only one-half of the zone of the operating system.)

$$W_{syst} = 2[(0.81)(1)(0.5) + (0.09)(0.7)0.5] = 0.873$$

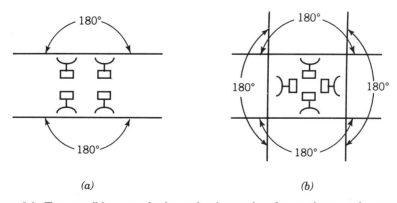

(a)   (b)

Figure 8.2. Two possible ways of using redundant radars for an airport radar system.

TABLE 8.1 Analysis of Variant B of Example 8.3

Type of State	Number	Probability	Effectiveness Coefficient	Product
NSEW	1	0.9^4	1	0.6561
N'SEW	4	$4(0.9)^3(0.1)$	$(1/2)(1 + 0.7)$	0.3060
N'S'EW	2	$2(0.9)^2(0.1)^2$	0.7	0.01134
N'SE'W	4	$4(0.9)^2(0.1)^2$	$(1/4)1 + (1/2)(0.7)$	0.01944
N'S'E'W	4	$4(0.9)(0.1)^3$	$(1/2)(0.7)$	0.00126

Now let us consider the second variant, B (see Figure 8.2*b*). In this case we have to analyze $2^3 - 1 = 7$ different states. The results of this analysis are presented in Table 8.1. Here we denote the corresponding radars by N, S, E, and W and use the symbols N', S', E', and W' to denote their idle states. The final result can be found by summing all values in the last column of Table 8.1:

$$W = 0.99314$$

Thus, variant B is the preferable one.

Example 8.4 As Russian nonmilitary authors, we never had access to information about former Soviet military systems, even the out-of-date systems. So for illustration we are forced to use an illustrative narrative from the proceedings of one of the early IEEE Reliability Conferences.

Just after World War II there were antiaircraft missile systems of the following simple type. There was a radar searching for a target with information displayed on a screen. After locating the target, a conveying radar was switched in and information about the target was processed by a computer and displayed by the same monitor. If the searching radar failed, the conveying radar was used for searching (with a lower efficiency). The last step

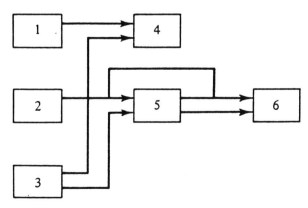

Figure 8.3. Simplified block diagram of an aircraft radar system. 1 = searching radar; 2 = optical device; 3 = conveying radar; 4 = display; 5 = computer; 6 = control equipment.

TABLE 8.2   Effectiveness of Different Modes of the System in Example 8.4

Mode Number	Stage of Operation			W_X
	Searching	Finding	Guiding	
1	Searching radar	Display	Watching radar and computer	1.0
2	Watching radar	Display	Watching radar and computer	0.60
3	Searching radar	Display	Optical equipment and computer	0.30
4	Optical equipment	Optical equipment	Optical equipment and computer	0.15
5	Optical equipment	Optical equipment	Optical equipment	0.10

connected with the destruction of the target was fulfilled by means of control equipment and a controlled missile. In case of a failure of the electronic equipment (which was so unreliable at that time!), a pilot could use an optical system for pursuing the target (see Figure 8.3).

Thus, the system could be in different states because of the failures of the equipment. Different modes of the system under consideration are presented in Table 8.2. The probabilities of a successful operation at some given moment of time are

For the searching radar, $p_1 = 0.80$.
For the optical equipment, $p_2 = 0.99$.
For the watching radar, $p_3 = 0.80$.
For the display, $p_4 = 0.95$.
For the computer, $p_5 = 0.90$.
For the control system, $p_6 = 0.95$.

Solution.   Let the probability of the kth mode be denoted by h_k. Then

$$h_1 = p_1 p_3 p_4 p_5 p_6 \approx 0.52$$
$$h_2 = q_1 p_3 p_4 p_5 p_6 \approx 0.13$$
$$h_3 = q_3 p_1 p_4 p_5 p_6 \approx 0.13$$
$$h_4 = p_2 p_5 p_6 (q_4 p_1 p_3 + q_4 q_1 p_3 + q_4 q_1 q_3 + p_4 q_1 q_3) \approx 0.08$$

(in this case one should take into account the impossibility of performing the operation by means of previous modes)

$$h_5 = p_2 q_5 p_6 (q_4 + p_4 q_1 q_3) \approx 0.01$$

The final result is that the probability of success W_{syst} is equal to

$$W_{\text{syst}} = (0.51) \cdot 1 + (0.13)(0.6) + (0.13)(0.3)$$
$$+ (0.08)(0.15) + (0.01)(0.1) = 0.66$$

8.3 ENDURING SYSTEMS

If the period of time it takes to perform a system task is sufficiently long, that is, during this period a number of different state changes can occur, then one needs to investigate an *enduring system*. In this case a probabilistic measure is distributed over a continuous space of trajectories of the changing system states. Let $Z(t, t + t_0)$ denote some fixed trajectory. In the continuous trajectory space we can determine a density f_Z for each such trajectory. At the same time, if a system moves from one state to another in correspondence to such a trajectory, one can characterize it by some predetermined outcome (effectiveness), say W_Z.

Now we can write an expression similar to (8.2)

$$W_{\text{syst}} = \int_{G_Z} W_Z \, dF(Z) \tag{8.14}$$

where G_Z is the space of all possible system state trajectories in the interval $(t, t + t_0)$. The simplicity of (8.14) is deceptive. In general, it is very difficult to compute the densities of trajectories and to find analytical expressions for the outcomes of a system for each particular case. (We will return to this topic later.) To illustrate this statement, consider a simple duplicated system consisting of two unrepairable units. Initially, both units are in an operating state. The expression for this case is

$$W_{\text{syst}}(t, t + t_0) = W_0 h_0 \left[1 + \frac{1}{p_1} \int_{t_0}^{t+t_0} \tilde{W}_1(t_1) \, dF(t_1) + \frac{1}{p_2} \int_{t_0}^{t+t_0} \tilde{W}_2(t_2) \, dF(t_2) \right.$$
$$\left. + \frac{1}{p_1 p_2} \int_{t}^{t+t_0} \int_{t}^{t+t_0} \tilde{W}_{12}(t_1, t_2) \, dF(t_1) \, dF(t_2) \right]$$

Here $\tilde{W}$ again is a normalized effectiveness coefficient relative to the nominal trajectory with no failures.

Thus, even for a very simple enduring system, the expression for the evaluation of effectiveness is quite complex. But the complexity of the expression is not all that makes this problem difficult. One also needs very detailed information about the reliability of the system's units as well as some knowledge about the effectiveness coefficients for different trajectories. In this case one is interested in finding an approximate solution.

Let us denote $q_i(t) = 1 - p_i(t)$. If the following condition is valid:

$$\max_{1 \le i \le n} q_i(t, t + t_0) \ll \frac{1}{n} \tag{8.15}$$

it is possible to write an approximate formula for unrepairable systems:

$$W_{syst} \approx W_0 \left[1 - \sum_{1 \le i \le n} \left(q_i(t, t + t_0) - \int_t^{t+t_0} \tilde{W}_i(x_i) \, dF_i(x_i) \right) \right]$$

For repairable systems such an analysis becomes extremely difficult and boring. For numerical calculations one can introduce a discrete lattice to describe the system's trajectory in the space of system states. But in this case one encounters a complex factorial problem. Of course, the largest practical difficulties arise in the determination of the effectiveness coefficients for different state trajectories in both cases: continuous and discrete.

For enduring systems W_{syst} is also a generalized index in comparison with the standard reliability indexes. As usual, a generalized (or more or less universal) method permits one to obtain any different particular solutions but with more effort. So, for simple reliability problems, one need not use this general approach. At the same time we should mention that, in general, a system performance effectiveness analysis cannot be done via common reliability methods.

We first consider two simple examples which can be solved by the use of standard reliability methods.

Example 8.5 Consider a unit operating in the time interval $[0, Z]$. An outcome of the unit is proportional to the operating time; that is, if a random TTF is more than Z, then the outcome of the unit is proportional to Z. Let $p(t)$ be the probability of a failure-free operation during time t and let $q(t) = 1 - p(t)$. Find an effectiveness index W_{syst}.

Solution. Simply reformulating the verbal description gives the result

$$W_{syst} = Zp(Z) + \int_0^Z t \, dq(t)$$

and, after integrating by parts,

$$W_{syst} = Zp(Z) + Zq(Z) - \int_0^Z q(t) \, dt = Z - \int_0^Z q(t) \, dt = \int_0^Z p(t) \, dt$$

As one can see (and as would be expected), the result coincides with the conditional MTTF inside the interval $[0, Z]$.

Example 8.6 Consider a system consisting of n identical and independent units. The system's behavior can be described with the help of the birth and death process (BDP). The effectiveness of the system during the time interval $[t, t + t_0]$ is completely determined by the lowest state which the system attains. If the system's lowest state is k, denote the effectiveness coefficient by W_k (see Figure 8.4). The problem is to determine W_{syst}.

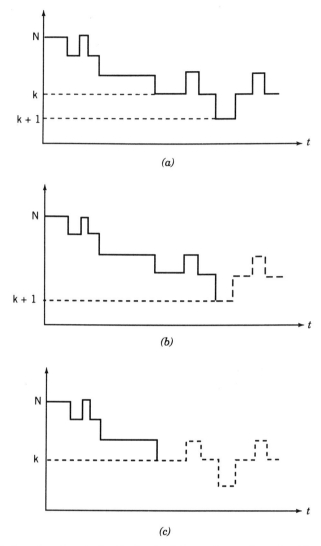

Figure 8.4. Sample of a stochastic track: (a) an observed track without absorbing states; (b) a track absorbed at state $k + 1$; (c) a track absorbed at state k.

Solution. To solve the problem, one should write the BDP equations (see Chapter 1). At moment $t = 0$, assume all system units are operating; that is, the system is in state 0. Let α_k be the transition intensity from state k to state $k + 1$ and let β_k be the transition intensity from state k to state $k - 1$. If we consider a process without an absorbing state, then the system of linear differential equations is

$$\frac{dR_k(t)}{dt} = -(\alpha_k + \beta_k)R_k(t) + \alpha_{k-1}R_{k-1}(t) + \beta_{k+1}R_{k+1}(t)$$

$$\text{for} \quad 0 \le k \le n, \, \alpha_{n+1} = \beta_{-1} = 0 \quad (8.16)$$

and the initial condition is $R_0(0) = 1$.

To solve the problem, we should solve (8.16) n times for different absorbing states. Namely, we should solve n subproblems of type (8.16) for absorbing states $n, n-1, \ldots, 1$. Let $R_k^*(t)$ denote the probability that the process is in the absorbing state when the process is "cut" up to the absorbing state k (see Figure 8.4). From (8.16) it is clear that $R_k^*(t)$ differs from $R_k(t)$. Moreover, the sum of the $R_k^*(t)$'s over all k does not equal 1.

We can use the methods described in Chapter 1. But our purpose is not to actually find the above-mentioned probabilities. For further consideration of this particular example, let us assume that we know the probabilities $R_k^*(t)$ which are the probabilities of reaching an absorbing state k in the corresponding subproblem (8.16).

It is clear that if state k is absorbing, $R_k^*(t)$ is the probability that in the original state space the process would also reach states with larger subscripts. This means that

$$R_k^*(t) = \sum_{k \le j \le n} S_k(t)$$

where $S_k(t)$ is the probability that the worst state that the initial process reached in $[0, t]$ is k. Hence, $S_k(t) = R_k^*(t) - R_{k+1}^*(t)$. The final result is the following:

$$W_{\text{syst}} = \sum_{0 \le k \le n} W_k S_k(t)$$

Example 8.7   Consider a system that involves the collection and transmission of information. The system consists of two identical and independent communication channels. If a channel fails, the system capacity decreases to 0.3 times a nominal value. For simplicity, assume that each channel is characterized by an exponentially distributed TTF with parameter λ. Let the duration of a given information collection equal $(0.1)/\lambda$. The volume of the collected information is proportional to the operating time, that is,

$$W_0 = t$$
$$W_1(x) = W_2(x) = x + (0.3)(t - x) = (0.3)t + (0.7)x$$
$$W_{12}(x_1, x_2) = \min(x_1, x_2) + 0.3[\max(x_1, x_2) - \min(x_1, x_2)]$$
$$= 0.3\max(x_1, x_2) + (0.7)\min(x_1, x_2)$$

The task is to calculate W_{syst} expressed via the absolute amount of collected information.

Solution.   In this case (8.14) can be written in the form

$$W_{\text{syst}}(t) = p^2 t + 2p \int_0^t (0.3)t + (0.7)x\lambda e^{-\lambda x}\, dx$$

$$+ \int_0^t \lambda e^{-\lambda x} \int_0^x [(0.3)x + (0.7)y]\, dx\, dy$$

where we have denoted $p = p(t) = e^{-\lambda t}$. After substituting the input data $p = 0.905$ and $t = 1/\lambda$, one obtains the final result

$$W_{\text{syst}} = (0.819 + 0.109 + 0.002)(1/\lambda) = 0.94(1/\lambda)$$

This value is the amount of information collected by the system as a whole during a time equal to the MTTF of a single channel.

8.4   PARTICULAR CASES

Below we consider several particular cases for which one can obtain simple results. Such kinds of structures are often encountered in practice.

8.4.1   Additive Type of a System Unit's Outcome

We first consider an instantaneous system containing n independent units. Each of them performs its own task, which implements a determined portion W_i of the total system outcome. Therefore, the system's outcome W_{syst} can be represented as the sum of the W_i's. Each unit i can be in one of two states: successfully operating or failure, with probabilities p_i and q_i, respectively.

For such a system W_{syst} can be written as

$$W_{\text{syst}} = \sum_{1 \le i \le n} W_i p_i \tag{8.17}$$

Expression (8.17) can also be written for dependent units. This follows from the fact that the expected value of a sum of random values equals the sum of its expected values, regardless of their dependence.

For concreteness, let us consider a system with two types of units. Let us call a unit an *executive unit* if it produces a portion of the system's outcome. All of the remaining units will be called *administrative units*. The system's outcome again consists of the sum of the individual outcomes of its executive units. The coefficient of effectiveness of the ith executive unit, $i = 1, \ldots, N$, depends on $\mathbf{X}$, the state of both the structural and executive units of the system, that is, $W_i(\mathbf{X})$, $1 \leq i \leq n$, $N \leq n$.

In this case a unit's outcome depends on two factors: the operating state of the unit itself and the state of the system. Finally, we can write

$$W_{\text{syst}} = \sum_{1 \leq i \leq n} p_i \, \mathrm{E}\{W_i(\mathbf{X})\} \tag{8.18}$$

where $\mathrm{E}\{W_i(\mathbf{X})\}$ is the unit's average coefficient of effectiveness. (In other words, it is a W_i of the separate ith unit.)

$$\mathrm{E}\{W_i(\mathbf{X})\} = \sum_{\text{all } \mathbf{X}} \mathrm{Pr}\{\mathbf{X}\} W_i(\mathbf{X}) \tag{8.19}$$

A clear practical example of such a system can be represented by the so-called nonsymmetrical branching system with a simple treelike hierarchical structure. This system consists of N executive units controlled by "structural" units at higher hierarchy levels (see Figure 8.5). The total number of hierarchy levels is M. Each executive unit of the system can produce its outcome if it is operating itself, and if all of its controlling units are also operating. Each executive unit i has its own outcome W_i being a portion of

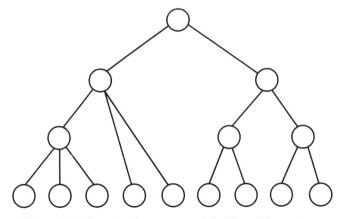

Figure 8.5. Sample of a nonsymmetrical branching system.

the total system outcome, that is,

$$W_{\text{syst}} = \sum_{1 \le i \le N} E\{W_i\}$$

Denote the probability of a successful operation of the highest unit in the system hierarchy by p_1; the corresponding probability of the units of the second level controlling the ith executive unit by p_{2i}; the same for the third level, p_{3i}; and so on. Thus, the successful operation of the ith executive unit can occur with probability

$$R_i = \prod_{1 \le j \le M-1} p_{ij} \tag{8.20}$$

Now it is easy to calculate the system's effectiveness

$$W_{\text{syst}} = \sum_{1 \le i \le n} W_i R_i \tag{8.21}$$

Again, we use the fact that the mean for the sum of dependent random variables equals the sum of their means.

Example 8.8 Consider a power supply system whose structure is presented in Figure 8.6. Units 0, 1, and 2 are structural and units 3 to 10 are executive. The outcome of each of them equals the power distributed to consumers (in conditional units). All absolute outcomes of the system units and their availability coefficients are presented in Table 8.3. Find W_{syst} with the condition of independence of the system's units.

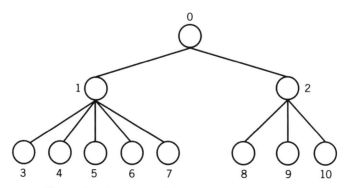

Figure 8.6. Structure of the system in Example 8.8.

TABLE 8.3   Parameters of the System's Units for Example 8.8

Unit	p_i	W_i	$p_i W_i$
0	0.99	—	
1	0.98	—	
2	0.97	—	
3	0.9	5	4.5
4	0.95	10	9.5
5	0.9	10	9.0
6	0.9	5	4.5
7	0.95	15	14.25
8	0.98	10	9.8
9	0.9	5	4.5
10	0.95	15	14.25

Solution.   Any executive unit performs its function if it is successfully operating itself, along with the common units and the corresponding units of the second level. Thus, using (8.21), one obtains

$$W_{syst} = p_0 \left(p_1 \sum_{3 \leq i \leq 7} p_i W_i + p_2 \sum_{8 \leq i \leq 10} p_i W_i \right)$$

$$= (0.99)[(0.98)(41.75) + (0.97)(28.55)] = 67.92$$

For an enduring system operating in a time interval $[t, t + t_0]$, the coefficient of effectiveness for the ith unit will depend on the moment of its failure: $W_i(x)$, $t \leq x \leq t + t_0$. In this case an expression for W can also be written in a very simple form

$$W_{syst}(t, t + t_0) = \sum_{1 \leq i \leq n} \left[p_i(t, t + t_0) W_i(t + t_0) + \int_t^{t+t_0} W_i(x) \, dF_i(x) \right]$$

where $F(x)$ is the distribution of a random time to failure of the ith unit.

Example 8.9   Let us consider a spy satellite designed for the collection and transmission of information. This system is unrepairable and can be considered as enduring. The system consists of three communication channels. Their capacities and failure rates correspondingly are: $V_1 = 100$ Mbps, $V_2 = 200$ Mbps, $V_3 = 250$ Mbps, and $\lambda_1 = 0.0001$ 1/hr, $\lambda_2 = 0.0003$ 1/hr, and $\lambda_3 = 0.0004$ 1/hr. Find W_{syst} (in absolute value) in two forms: (a) the mean capacity of the system as a whole at moment $t = 1000$, and (b) the mean volume of transmitted information during 5000 hours.

Solution. (a) For an arbitrary moment t one can write

$$W_{\text{syst}} = \sum_{1 \le i \le 3} W_i e^{-\lambda_i t}$$

After the substitution of numerical input data

$$W_{\text{syst}}(1000) = 100 e^{-0.1} + 200 e^{-0.3} + 250 e^{-0.4} = 405.5 \text{ Mbps}$$

(b) If, during the period $[0, t_0]$ in this example, there was no failure, a channel has collected $W_i t_0$ bits of information. If a failure has occurred at the moment $t < t_0$, then a channel has collected $W_i t$ bits of information. Taking this into account, one can write

$$W_{\text{syst}}(0, t_0) = \sum_{1 \le i \le 3} \left[W_i t_0 e^{-\lambda_i t_0} + \int_0^{t_0} \lambda_i W_i x e^{-\lambda_i x}\, dx \right] = \frac{W_i}{\lambda_i} \left[1 - e^{-\lambda_i t_0} \right]$$

Note that, since the amount of transmitted information is proportional to t_0, the total operating time is

$$W_{\text{syst}}(0, t_0) = q \int_0^{t_0} W_{\text{syst}}(x)\, dx$$

Substituting the input data (in the same time dimension), one obtains

$$W_{\text{syst}}(0, 5000)$$
$$= \left[\frac{100}{0.00001} 0.39 + \frac{200}{0.00003} 0.78 + \frac{250}{0.00004} 0.86 \right] \cdot 3600 = 5.2 \cdot 10^9 \text{ Mbits}$$

8.4.2 Systems with a Symmetrical Branching Structure

Now we consider a system whose structure presents a particular case of the system structure discussed in Example 8.8. The branching structure has a symmetry, which means that each controlling unit controls the same number of units in the lower level. Also, all units of the same hierarchical level have the same reliability characteristics; that is, the system is homogeneous.

Now we will consider a more complex measure of system effectiveness: one that depends in a nonlinear way on the number of executive units performing their functions successfully.

The successful performance of a unit of any hierarchy level means that the unit is in an operating state and all of its controlling units are also in operating states. It is clear that an executive unit will not operate successfully if at least one "structural" unit which controls it has failed.

Note that the executive units of the system are dependent through their common controlling units. Indeed, a failure of any controlling unit leads to the failure of all controlled units at lower levels, including the corresponding executive units. Therefore, a failure of some "structural" unit leads to the stopping of successful operations of the corresponding branch as a whole; that is, the corresponding set of executive units does not produce its outcome. A failure of the highest-level controlling unit leads to an interruption of successful operations at all executive units.

It is understandable that the problem of effectiveness evaluation for dependent executive units is not trivial. We introduce the following notation:

p_j is the probability of a successful operation of a unit in the jth hierarchy level, $0 \leq j \leq n$.

a_j is the "branching power" of the $(j - 1)$th-level unit which shows how many units of the jth level are controlled by this unit.

x_j is the random number of successfully performing units in the jth hierarchy level.

N_j is the total number of units in the jth level.

$P_j\{x_j\}$ is the distribution of x_j.

$W(x_n)$ is the coefficient of the system's effectiveness if x_n executive units are successfully performing.

(See the explanations in Figure 8.7.)

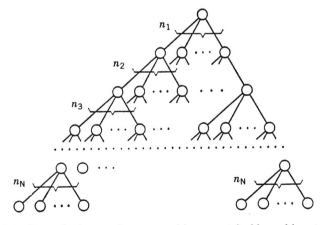

Figure 8.7. General scheme of a system with symmetrical branching structure.

REMARK. Here we use the two expressions: "successfully operating" and "successfully performing." Really these expressions have a very slight difference. In this text we will understand that "operating" means that a unit is in an up state itself, independent of the states of any other unit in the system, and "performing" means that the unit is operating itself and, at the same time, all of its controlling units are successfully operating. (See the explanations in Figure 8.7.)

For the system under consideration,

$$W_{\text{syst}} = \sum_{0 \le x_n \le N_n} P_n(x_n)W(x_n) = E\{W(x_n)\} \tag{8.22}$$

where N_n is the total number of the executive units:

$$N_n = \prod_{1 \le i \le n} a_i \tag{8.23}$$

In general, the function $W(x_n)$ can be arbitrary. For simplicity, let us suppose that $W(x)$ is a continuous differentiable function of x. It is known that any such function can be represented in the form of a Taylor series. In the case under consideration,

$$W(x_n) = \sum_{k \ge 1} x_n^k \frac{d^k W(x_n)}{d^k x_n} \tag{8.24}$$

For practical purposes, one can use an approximation taking (8.24) with relatively small k. Using (8.24), we can easily write

$$W_{\text{syst}} = E\left\{ \sum_{k \ge 1} B_k x_n^k \right\} = \sum_{k \ge 1} B_k E\{x_n^k\} = \sum_{k \ge 1} B_k M_k \tag{8.25}$$

where M_k is the moment of the distribution of the number of successfully performing executive units.

To find M_k, we write the moment generating function. First, consider a group of executive units depending on a single unit in the $(n-1)$th level. We have N_{n-1} such groups. A random number of successfully operating executive units in a group, x, has a binomial distribution $B(a_n, p_n)$. The moment generating function for the distribution of successfully operating executive units of the above-mentioned group is denoted by

$$g(e^{x_n}) = \left[p_n e^{x_n} + q_n \right]^{a_n} \tag{8.26}$$

Now consider all executive units which depend on N_{n-1} controlling units at the $(n-1)$th level. (At this stage of consideration we are not interested in all of the remaining units in the system.) The random number of successfully operating units at this level, x_n, also has a binomial distribution $B(N_{n-1}, p_{n-1})$.

Note that if no units at the $(n-1)$th level are successfully operating, no units at the nth level are successfully performing, even though all executive units are operating. This event occurs with probability:

$$P_{n-1}(0) = q_{n-1}^{N_{n-1}}$$

If only one unit at the $(n-1)$th level is operating successfully, then not more than a_n executive units can perform successfully. The random number of successfully performing executive units will have a binomial distribution with moment generating function (8.26). This event occurs with probability

$$P_{n-1}(1) = \binom{N_{n-1}}{1} p_{n-1} q_{n-1}^{N_{n-1}-1}$$

If two units at the $(n-1)$th level are operating successfully then not more than 2^{a_n} executive units can perform successfully. The probability of this event equals $P_{n-1}(2)$ where

$$P_{n-1}(2) = \binom{N_{n-1}}{2} p_{n-1}^2 q_{n-1}^{N_{n-1}-2}$$

Arguing in the same manner, we obtain the moment generating function of the distribution of the random number of all successfully performing executive units, x_n, taking into account the random number of successfully operating units at the $(n-1)$th level as a whole:

$$G_n(e^{x_n}) = \sum_{0 \le x_{n-1} \le N_{n-1}} P_{n-1}(x_{n-1}) \left[g(e^{x_n}) \right]^{x_{n-1}} \tag{8.27}$$

If we let

$$e(Y) = G(e^{x_n}) = \left[p_n e^{x_n} + q_n \right]^{a_n} \tag{8.28}$$

then (8.27) can be rewritten as

$$G_n(e^{x_n}) = \sum_{0 \le x_{n-1} \le N_{n-1}} P_{n-1}(x_{n-1}) \left[e^Y \right]^{x_{n-1}} = G_{n-1}(e^Y) \tag{8.29}$$

From (8.28) it follows that

$$Y = a_n \ln(p_n e^{x_n} + q_n) \tag{8.30}$$

We can continue similar arguments for the units at the remaining upper levels of the system's hierarchy.

Thus, we have the recurrent expression

$$G_n(e^{x_n}) = G_{n-1}(e^Y) = G_{n-1}\left[(p_n e^{x_n} + q_n)^{a_n}\right] \tag{8.31}$$

Using the chain rule, we obtain recurrent expressions for the desired initial moments M_k.

Indeed, the first moment can be found in the following way:

$$M_n^1 = \frac{d[G_n(e^{x_n})]}{dx_n}\bigg|_{x_n=0} = \frac{d[G_{n-1}(e^Y)]}{dY}\frac{dY}{dx_n}\bigg|_{x_n=0}$$

$$= M_{n-1}^1 \frac{d[a_n \ln(p_n e^{x_n} + q_n)]}{dx_n}\bigg|_{x_n=0} \tag{8.32}$$

Finally,

$$M_n^1 = M_{n-1}^1 a_n p_n \tag{8.33}$$

Continuing this recurrent procedure, we obtain the result in closed form

$$M_n^1 = p_0 \prod_{1 \le i \le M} p_i a_i \tag{8.34}$$

We mention that (8.34) could be obtained in a simpler way. Indeed, it directly follows from (8.21) that

$$M_n^1 = N_n \prod_{0 \le i \le n} p_i = p_0 \prod_{1 \le i \le n} p_i a_i$$

The second moment of the distribution of the random value x_n can be found in a similar way.

$$M_n^2 = \frac{d^2[G_n(e^{x_n})]}{dx_n^2}\bigg|_{x_n=0} = \frac{d^2[G_{n-1}(e^Y)]}{dY^2}\frac{dY}{dx_n}\bigg|_{x_n=0} + \frac{dG(Y)}{dY}\frac{dY^2}{dx_n^2}\bigg|_{x_n=0}$$

The recurrent equation for M_n^2 is

$$M_n^2 = M_{n-1}^2 a_n p_n + M_{n-1}^1 a_n p_n q_n \tag{8.35}$$

and the final result in closed form is

$$M_n^2 = p_0 \prod_{1 \leq i \leq n} a_i p_i \left[\prod_{1 \leq i \leq n} a_i p_i + \prod_{1 \leq i \leq n} q_i \prod_{1 \leq k \leq n} a_k p_k \right] \qquad (8.36)$$

Closed-form expressions for higher-order moments are enormously complicated. One is advised to use the above-obtained recurrent expressions for
computer calculations.

Example 8.10 Consider different variants of a branching system (see
Figure 8.8). Each system has six executive units. The problem of interest is to
choose the best structure for two cases: (a) the system outcome is a given
linear function of the number of operating executive units; and (b) the system
outcome is a given quadratic function of the number of operating executive
units.

Solution.   For the first case $W(x_n) = Ax_n$. Then

$$W_{syst} = AM_n^1 = Ap_0(p_1 a_1)(p_2 a_2) = A(p_0 p_1 p_2)$$

that is, according to the chosen effectiveness measure, all four variants are
equivalent.
   If $W(x_n) = Bx_n^2$, the effectiveness of any state of the system is proportional to the square of the number of successfully performing executive units.
Then

$$W_{syst} = E\{W(x_n)\} = BM_n^2 = 6Bp_0 p_1 p_2 (6p_1 p_2 + a_2 q_1 q_2 p_2)$$

In this case the value of the system effectiveness increases as a_i increases.
This means that the variant a is best for the second type of system effectiveness. For example, such situations appear when one considers Lanchester's
models of the second order when the effectiveness of an army division is
proportional to the square of the number of its combat personnel.

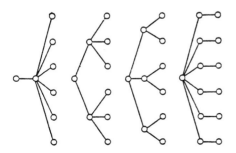

Figure 8.8. Variants of a system structure with six executive units.

The system with the largest a_2 is the most effective. Thus, the higher the level of centralized control from the center, the better is the result.

8.4.3 Systems with Redundant Executive Units

Many instant systems are used for fulfilling a given task. For example, an antiaircraft or antimissile defense system is designed to destruct a target. To improve the system's effectiveness, N redundant executive units can be used. If the system is in state X, each executive unit fulfills its task with probability $W_i(X)$. For example, the efficiency of an antiaircraft missile system depends on the state of its subsystems which are used for searching, controlling, and so on. There are two main cases: (1) when all executive units perform their common task simultaneously (2) when units perform the same task sequentially.

Case 1 All units are dependent through the system state $\mathbf{X}$. Then

$$W_{\text{syst}} = \sum_{\text{all } \mathbf{X}} P(\mathbf{X}) \left[1 - \prod_{1 \leq i \leq N} (1 - W_i(\mathbf{X})) \right] \tag{8.37}$$

In particular, for the branching system considered in the previous section, the problem can be solved in the following elegant way. Let D be the probability of success of a single executive unit (e.g., the kill probability of an enemy's aircraft). Then, if x_n executive units are acting simultaneously, the total probability of success is

$$W(x_n) = 1 - (1 - D)^{x_n} \tag{8.38}$$

An interesting particular case arises if we consider a symmetrical branching system. As discussed above, the effectiveness of the branching system is completely determined by the number of successfully performing executive units, x_n. Thus, using (8.2), we can write

$$W_{\text{syst}} = \sum_{0 \leq x_n \leq N} P_n(x_n) W(x_n)$$

$$= 1 - \sum_{0 \leq x_n \leq N} P(x_n)(1 - D)^{x_n} = 1 - G_n(1 - D) \tag{8.39}$$

The second term in (8.39) is a moment generating function with the substitution of $1 - D$ as a variable. Thus, (8.39) can be rewritten as

$$W_{\text{syst}} = 1 - G_n(1 - D) = 1 - G_{n-1}\left(\left[p_n(1 - D) + q_n \right]^{a_n} \right) \tag{8.40}$$

Using a recurrent procedure, we finally obtain

$$W_{syst} = 1 - \left(p_0 \left(p_1 \ldots \left(p_{n-1} \left(p_n (1 - D) + q_n \right)^{a_n} + q_{n-1} \right)^{a_{n-1}} \right. \right.$$
$$\left. \left. + \cdots + q_1 \right)^{a_1} + q_0 \right) \quad (8.41)$$

Case 2   Assume that the system's executive units are operating sequentially. The system states are supposed to change between the use of two consequent executive units. If the time interval between the two system performances is large enough, the result of their operations might be independent. The same result will also be valid if one considers the simultaneous operation of several executive units controlled by identical and independent controlling systems. For example, one can consider the destruction of an enemy aircraft in the overlapping zone of action of several antiaircraft systems. In this case

$$W_{syst} = 1 - \prod_{1 \leq i \leq N} \left[1 - \sum_{all\ \mathbf{X}} P(\mathbf{X}) W_i(\mathbf{X}) \right] \quad (8.42)$$

8.5   SYSTEMS WITH INTERSECTING ZONES OF ACTION

8.5.1   General Description

Suppose that a system consists of n executive units. Unit i has its own zone Z_i of action. Each unit is characterized by its own effectiveness of action W_i in the zone Z_i. These zones can be overlapping (see Figure 8.9).

The joint effectiveness of several executive units in such an overlapping zone depends on the types of systems and their tasks. Such systems appear in satellite intelligence systems, radio communication networks, power systems, and antiaircraft and antimissile systems (overlapping zones of destruction).

In general, in the entire zone in which a system as a whole is operating, 2^n different overlapping subzones may be created. Then the problem of computing the system's effectiveness cannot be reduced and we need to use a general expression:

$$W_{syst} = \sum_{1 \leq i \leq 2^n} H_i W_i$$

where H_i is the probability that the system is in state i, W_i is the conditional effectiveness performance index for this system state, and n is the total number of system units. Of course, the number 2^n is huge if $n > 10$. Moreover, in a computational sense, for some hundreds of units the problem cannot be solved in general: there is not sufficient memory to store the data and there is not sufficient time to perform the computations!

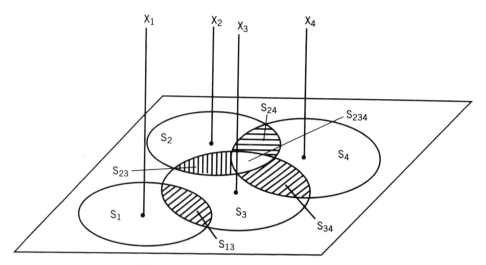

Figure 8.9. Sample of overlapped zones.

Fortunately, in practice, if one considers a territorial system, the number of overlapping zones is usually small enough. On the other hand, if there is a strong overlapping of different zones, the case can be significantly reduced: if all zones are totally overlapping, the system becomes a common redundant system, and its analysis involves no special analytic difficulties.

If we consider a territorial system, the number of units acting in the same zone is not usually large. A zone of the whole system action can be represented as

$$Z = \bigcup_{1 \le i \le n} Z_i$$

For further purposes, let us introduce zones Z_{a_i} which are disjoint. Within each zone Z_{a_j} the same set of serving units is acting. The subscript a_j represents the set of subscripts of the executive units acting in zone Z_{a_j}. Thus, $i \in a_j$ means that the ith unit serves in zone Z_{a_j}. Let the number of different zones Z_{a_j} be M, that is, $1 \le j \le M$. It is clear that zones Z_{a_j} are disjoint and we can write

$$Z = \bigcup_{1 \le j \le M} Z_{a_j}$$

As we mentioned above, $M \ll 2^n$ in practice.

Because of failures the actual set of units operating in zone Z_{a_j} is random. In general, if a_j includes m_j subscripts, zone Z_{a_j} can be characterized by 2^{m_j} different possible levels of effectiveness. For each possible set of acting units, say a_{jk_j}, where $k_j \geq 2$, we observe some specified coefficient of effectiveness $W_{a_{jk_j}}$. As a result, for such a system we can write

$$W_{\text{syst}} = \sum_{1 \leq j \leq M} Z_{a_j} \sum_{1 \leq k_j \leq 2^{m_j}} P_{k_j} W_{a_{jk_j}} \tag{8.43}$$

Another, more compact representation of (8.43) is

$$W_{\text{syst}} = \sum_{1 \leq j \leq M} Z_{a_j} E\{W_{a_j}\} \tag{8.44}$$

Such a simple and obvious modification of the general expression (8.2) sometimes allows us to obtain constructive results for some important and interesting practical cases.

8.5.2 Additive Coefficient of Effectiveness

In this case for any set of acting units in the zone we have

$$W_{\text{syst}_{a_j}} = \sum_{i \in a_j} W_j \tag{8.45}$$

As an example, we can consider pollution in some region when each of several polluting companies makes its own "investment" in the total level of pollution. Pollution is assumed to be additive. (Of course, in this case it would be more reasonable to speak of loss rather than effectiveness.) It is clear that in this case for the system as a whole

$$W_{\text{syst}} = \sum_{1 \leq j \leq M} W_{a_j} = \sum_{1 \leq i \leq n} p_i Z_i W_i \tag{8.46}$$

Let us illustrate this by a simple example.

Example 8.11 Consider a system consisting of two units and three acting zones (see Figure 8.10). Let us denote

$Z_1' = Z_1 + Z_3 = $ the acting zone of the first unit.
$Z_2' = Z_2 + Z_3 = $ the acting zone of the second unit.
$Z_3 = $ the acting zone of both units.

The effectiveness coefficient of the first unit is W_1, and the effectiveness coefficient of the second unit is W_2. By assumption of the additive character of the joint effect of the units, $W_3 = W_1 + W_2$ for zone Z_3.

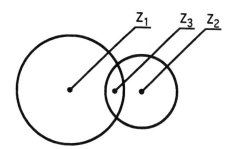

Figure 8.10. Two overlapped zones.

Because all zones Z_1, Z_2, and Z_3 are independent, we can write for the whole system

$$
\begin{aligned}
W_{\text{syst}} &= W_1 + W_2 + W_3 \\
&= p_1 Z_1 W_1 + p_2 Z_2 W_2 + Z_3 \big[p_1 p_2 (W_1 + W_2) + q_1 p_2 W_2 + p_1 q_2 W_1 \big] \\
&= p_1 W_1 (Z_1 + Z_3) + p_2 W_2 (Z_2 + Z_3) \\
&= p_1 W_1 Z_1' + p_2 W_2 Z_2'
\end{aligned}
$$

8.5.3 Multiplicative Coefficient of Effectiveness

In this case for any set of acting units in the zone we have

$$
W_{\text{syst}_{a_j}} = \sum_{i \in a_j} W_j
$$

It is more natural to consider a loss rather than a "positive" outcome. For example, $1 - W_i$ is the kill probability of a target in the acting zone of the ith unit of the system. Thus, if the ith unit does not act, this probability is 0. In other words, the probability of the enemy's survival (or, more accurately, the loss of the attacker) is $W_1 = 1$. If a unit acts successfully, the enemy's damage is larger: the probability of the enemy's survival equals some $W_i < 1$. If in zone Z_{a_i} units $i_1, i_2, \ldots, i_{k_i}$ act together successfully, then the probability of the enemy's survival is

$$
W_{a_j} = W_{i_1} W_{i_2} \cdots W_{i_{k_j}}
$$

Taking into account the probability of success of the units, we can write

$$
W_{\text{syst}} = \sum_{1 \le j \le M} Z_{a_j} \prod_{i \in a_j} (W_i p_i + q_i) \tag{8.47}
$$

If we take into account that $p_i W_i + q_i = 1 - p_i w_i$, (8.47) can be rewritten as

$$W_{\text{syst}} = \sum_{1 \le j \le M} Z_{a_j} \prod_{i \in a_j} (1 - p_i w_i) \qquad (8.48)$$

The final results (8.47) and (8.48) are illustrated by a simple example.

Example 8.12 The system under consideration is the same as in Example 8.11 (see Figure 8.10 for an explanation). We use the same notation. To facilitate understanding, keep in mind the case of a target's destruction in the zone of defense. Thus, q_i is the probability of failure of the ith executive unit, $p_i = 1 - q_i$, and W_i is the probability of a target's destruction by the ith executive unit. Assume that the probability of a target's appearance in a zone is proportional to its size. Then the probability of the target passing through the defense zone equals

$$
\begin{aligned}
W_{\text{syst}} &= Z_1(W_1 p_1 q_1) + Z_2(W_2 p_2 + q_2) \\
&\quad + Z_3(p_1 p_2 W_1 W_2 + p_1 q_2 W_1 + p_2 q_1 W_2 + q_1 q_2) \\
&= Z_1(W_1 p_1 + q_1) + Z_2(W_2 p_2 + q_2) + Z_3(p_1 W_1 + q_1)(p_2 W_2 + q_2)
\end{aligned}
$$

or, equivalently

$$W_{\text{syst}} = Z_1(1 - w_1 p_1) + Z_2(1 - w_2 p_2) + Z_3(1 - p_1 w_1)(1 - p_2 w_2)$$

8.5.4 Redundant Coefficient of Effectiveness

Now we consider a "positive" outcome in a zone. This type of effectiveness coefficient is, in effect, complementary to the one considered in the previous section. For any set of acting units in the zone,

$$W_{\text{syst}_{a_j}} = 1 - \prod_{i \in a_j} w_i \qquad (8.49)$$

where $w_i = 1 - W_i$ and W_i has the previous meaning, that is, the probability of success. For this system we obtain

$$W_{\text{syst}} = \sum_{1 \le j \le M} Z_{a_j} \left[1 - \prod_{i \in a_j} (p_i w_i + q_i) \right] \qquad (8.50)$$

If we again take into account that $p_i w_i + q_i = 1 - p_i W_i$, (8.50) can be rewritten as

$$W_{\text{syst}} = \sum_{1 \le j \le M} Z_{a_j} \left[1 - \prod_{i \in a_j} (1 - p_i W_i) \right] \qquad (8.51)$$

Again, let us illustrate the final result by a simple example.

Example 8.13 Consider again the system represented in Figure 8.10. Using the previous arguments, we have

$$
\begin{aligned}
W_{\text{syst}} &= Z_1 p_1 W_1 + Z_2 p_2 W_2 \\
&\quad + Z_3 \left[p_1 p_2 (1 - w_1 w_2) + p_1 q_2 W_1 + p_2 q_1 W_2 \right] \\
&= Z_1 \left[1 - (1 - p_1 w_1) \right] + Z_2 \left[1 - (1 - p_2 w_2) \right] \\
&\quad + Z_3 \left[1 - (p_1 w_1 + q_1)(p_2 w_2 + q_2) \right]
\end{aligned}
$$

8.5.5 Boolean Coefficient of Effectiveness

This case is very close to ordinary redundancy applied to each zone. In other words, at least one executive unit must act in a zone to fulfill the operation within that zone. Thus, if a_j is a set of units acting in the jth zone and unit i delivers the outcome W_j in this zone, then any subset—a_j^*, $a_j^* \subseteq a_j$ and $a_j^* \neq \varnothing$—delivers the same effectiveness W_j. For example, if we consider a communication with a zone, it is sufficient to have at least one path of connection with this zone. For the system as a whole, we have

$$
W_{\text{syst}} = \sum_{1 \leq j \leq M} W_j Z_{a_j} \left[1 - \prod_{i \in a_j} q_j \right] \tag{8.52}
$$

We assume that (8.52) does not demand any additional comments.

8.5.6 Preferable Maximal Coefficient of Effectiveness

If a set of units a_j^* might act in the jth zone, then for the actual operation the unit with the maximal possible effectiveness coefficient is chosen

$$
W_{a_j^*} = \max_{i \in a_j^*} W_j \tag{8.53}
$$

Enumerate all of the system units in decreasing order of their effectiveness indexes W_i: $W_1 \geq W_2 \geq \cdots \geq W_n$. Then in the jth zone the kth one uses the unit which is characterized by the effectiveness coefficient W_k (of course, $k \in a_j$) if and only if the kth unit itself is operational and there are no other operational units with $i < k$. This means that all units belonging to the set a_j and having smaller numbers have failed at the moment of use.

After this argument, it is simple to write the following expression:

$$
W_{\text{syst}} = \sum_{1 \leq j \leq M} Z_{a_j} \sum_{k \in a_j} W_k p_k \prod_{\substack{i < k \\ i \in a_j}} q_i \tag{8.54}
$$

This follows directly from the formulation of the problem. We again think that there is no special need to explain it in more detail.

8.5.7 Preferable Minimal Coefficient of Effectiveness

In this case, if a set of units a_j^* might act in the jth zone, then for the actual operation the unit with the minimal possible effectiveness coefficient is chosen

$$W_{a_j} = \min_{i \in a_j^*} W_i \qquad (8.55)$$

This kind of effectiveness coefficient can be chosen if one investigates damage rather than a "positive" outcome.

Actually, this case is not distinguished from the previous one. One can even keep formula (8.54) with only one very essential difference: the enumeration of the system's executive units must be done in an increasing order of effectiveness indexes W_i: $W_1 \le W_2 \le \cdots \le W_n$.

8.6 ASPECTS OF COMPLEX SYSTEMS DECOMPOSITION

As mentioned above, the problem of effectiveness analysis arises in connection with the analysis of complex systems. Thus, the more complex a system is, the more important and, at the same time, the more difficult is the evaluation of its effectiveness. Thus, the problem of simplifying the evaluation of effectiveness, in particular, the methods of decomposition, seems very important.

Above we considered systems consisting of units with two states: an operating state and an idle state. But one sometimes deals with complex systems consisting of many such subsystems which themselves can be considered as complex systems. This is equivalent to the consideration of a system consisting of units with more than two states.

Let n be the total number of system units. Suppose the system is divided into M subsystems by some rule (it can be a functional principle or a constructive one). Each jth subsystem includes n_i units and, consequently, has

$$m_j = 2^{n_i}$$

different states. Now the system consists of M new units, each with $m_j \ge 2$ states. Of course, such a system representation does not lead to a decrease in the total number of system states m; that is, it does not follow that

$$m = \sum_{1 \le i \le M} 2^{n_i} \le 2^n$$

But such a new system representation may still help to generate new ideas.

First, it may be possible to characterize subsystems via some simpler description. For example, we can find one main characterization parameter for the entire system. In this case the dimension of the problem could be essentially decreased.

Second, it may be possible to use a simpler description of the states of the subsystems in comparison with complete enumeration. In this case the number of subsystems M is usually not very large. The second case leads to the construction of upper and lower bounds on the system effectiveness index W_{syst}. We will consider it in the next section.

8.6.1 Simplest Cases of Decomposition

It would be very constructive to represent a system's effectiveness index as a function of the W's of its subsystems. Is this ever possible? If so, when? The problem is to present the system's effectiveness as a function of the subsystem's effectiveness:

$$W_{syst} = f(W_1, W_2, \ldots, W_M) \tag{8.56}$$

Assume that for any system state a^*, which is expressed as a composition of subsystem states a_j^*, that is, $a^* = (a_1^*, \ldots, a_M^*)$, the condition

$$W_{a^*} = \sum_{1 \le j \le M} W_{a_j^*} \tag{8.57}$$

is true. Then, for such an additive system, (8.56) can be written as

$$W_{syst} = \sum_{1 \le j \le M} W_j \tag{8.58}$$

The statement is clear if one remembers that the mean of a linear function equals the function of the mean values of its variables. Thus, if it is possible to choose subsystems in such a way that (8.58) holds, we can use the simple expression (8.57).

The next results can be formulated for multiplicative systems. Note that for multiplicative systems (8.56) can be written as

$$W_{syst} = \prod_{1 \le j \le M} W_j \tag{8.59}$$

if and only if for any system state a^* which is expressed as a composition of subsystem states a_j^*, that is, $a^* = (a_1^*, \ldots, a_M^*)$, the following condition is valid:

$$W_{a^*} \prod_{1 \le j \le M} W_{a_j^*} \tag{8.60}$$

Expression (8.60) means that the W of any subsystem does not depend on the states of other subsystems.

Statement (8.59) becomes clear if one remembers that the mean of the product of independent random variables equals the product of the means of its variables. Thus, if subsystems are chosen in such a way that (8.60) holds, we can use the simple expression (8.59).

Unfortunately, the number of practical examples where we may obtain such a fantastic bargain in the evaluation of performance effectiveness is exhausted by these two trivial cases. Also, unfortunately, such systems are quite rare in engineering practice.

Fortunately, however, a similar approach can be used to obtain bounds of a system's effectiveness index in the case of regional systems.

8.6.2 Bounds for Regional Systems

1. Consider a regional system with a multiplicative effectiveness coefficient in a zone. Let us consider a zone with a set of executive units A. Assume that the system is divided into M subsystems. In this case the units of the set A can belong to several different subsystems. This means that the set A can be divided into several nonintersecting subsets A_j, $1 \le j \le M$. (Some of the A_j can be empty.) If the W_i's are normalized effectiveness coefficients, that is, if $0 \le (W_i/W_0) \le 1$, then, for any A,

$$\prod_{i \in A} (W_i p_i + q_i) \le \sum_{1 \le j \le M} \prod_{i \in A_j} (W_i p_i + q_i) \qquad (8.61)$$

From (8.61) it immediately follows that for these systems

$$W_{\text{syst}} \le \sum_{1 \le j \le M} W_j \qquad (8.62)$$

2. For systems with a redundancy type of effectiveness coefficient, we have

$$W_{\text{syst}} \le \sum_{1 \le j \le M} W_j \qquad (8.63)$$

To confirm (8.63), we show that this is correct for a zone with two acting units belonging to different subsystems. Keeping in mind that $0 \le W_i \le 1$, $i = 1, 2$, we can easily write

$$1 - (p_1 w_1 + q_1)(p_2 w_2 + q_2) \le p_1 W_1 + p_2 W_2 \qquad (8.64)$$

3. For systems with a Boolean type of effectiveness coefficient, one can obviously write

$$W_{\text{syst}} \le \sum_{1 \le j \le M} W_j \qquad (8.65)$$

4. For systems in which one chooses for operation the unit with the maximal effectiveness coefficient in a zone, we have

$$\sum_{k \in a} p_k W_k \prod_{i < k, i \in a} q_i \le \sum_{a_j \in a} \sum_{k \in a_j} p_k W_k \qquad (8.66)$$

Expression (8.66) is clear as any product of q_i's is always less than 1.
5. For systems in which one chooses for operation the unit with the minimal effectiveness coefficient in a zone, we have absolutely the same result (8.66).

Unfortunately, we have obtained only one-sided bounds. From a practical point of view, a lower bound of any "positive" effectiveness index (the larger, the better) is reasonable: one has a guaranteed result. But all bounds considered here yield a restriction on the upper side.

8.6.3 Hierarchical Decomposition and Bounds

Using Subsystem W_i's   Let the system be represented as a composition of M subsystems. For each subsystem one can calculate its own effectiveness W_i. Let the jth subsystem include n_j units. This subsystem has, in general,

$$m_j = 2^{n_j}$$

different states. Thus, we must analyze m_j different states for each subsystem.

If it is possible to express a system's effectiveness index W_{syst} as a function of the W_i's of the subsystems in the form

$$W_{\text{syst}} = f(W_1, W_2, \ldots, W_M) \qquad (8.67)$$

then it is enough to calculate W_j, as we did before,

$$W_j = \sum_{1 \le i \le m_j} P\{X_{ji}\} W_{ji} \qquad (8.68)$$

and, after this, use (8.67).

The total number of computations to obtain the desired result is proportional to

$$\sum_{1 \le j \le M} 2^{m_j} \ll 2^n \tag{8.69}$$

Unfortunately, such a procedure cannot be used too frequently as functions such as (8.67) are seldom known. The two simplest examples were shown above. At any rate, this method allows one to obtain at least some rough estimates of the unknown value of W_{syst}.

Let us give several examples of the effectiveness of such a decomposition. For a system consisting of $n = 400$ units, a strict evaluation of the system's effectiveness is practically impossible because the number of all possible system states exceeds the so-called googol (10^{100}). As the reader may know, the googol is sometimes jokingly called "the greatest number in the universe." Indeed, everything in the universe—its maximal diameter, its time of existence since the Big Bang, its total number of smallest elementary particles—measured by the smallest physical units (length or time, respectively) is smaller than this definitely restricted number. Thus, any attempt just to enumerate all the states of the above-mentioned system is unrealistic.

But if the system is divided into 20 subsystems, each consisting of 20 units, the number of calculations will still be large—$20 \cdot 20^{20} \approx 2 \cdot 10^7$—but at least it is a realistic number. If it is possible to divide the system into 40 subsystems, the corresponding number equals $40 \cdot 2^{10} \approx 4 \cdot 10^4$, which is unconditionally acceptable.

REMARK. We mention that very complex systems are usually considered in engineering practice in a hierarchical way with more than two levels. This permits one to independently analyze first the system as a whole; then each subsystem as a part of the system, but performing its own functions; then some more-or-less autonomous parts of these subsystems; and so on. Such a mode is very effective in the evaluation of a system's effectiveness.

Let us again consider a system consisting of $n = 400$ units. Suppose the system is divided into 5 subsystems, each subsystem is divided into 5 autonomous parts, where each part consists of 16 units. Thus, the total number of calculations can be evaluated as the total number of parts in the system, multiplied by the number of calculations for one such part: $5 \cdot 5 \cdot (2^{16}) \approx 200,000$. It is significantly less than in the initial case.

It is interesting to note that, for a system consisting of $n = 1000$ units, one can obtain an even smaller number of calculations if the system is represented by a three-level hierarchy: 5 subsystems, each of 5 parts, each of 5 complex units, each of 8 units of the lowest level. The number of calculations required is equal to $5 \cdot 5 \cdot 5 \cdot (2^8) = 32,000$.

Distributions of Subsystem Levels of W   A more accurate method than the previous one is described next. From the viewpoint of a system's user, all subsystem states can be divided into a very restricted number of groups. The states of each such group are characterized by a value close to the value of the subsystem's effectiveness coefficient W_j. It is clear that the number of such groups could be very small, say 10. This number does not depend on the initial number of subsystem states. (One has also to take into account that there is no necessity to consider groups with levels of effectiveness which appear with an infinitesimally small probability and/or with levels of one essentially negligible effectiveness.)

At any rate, the first step in the analysis of a system's effectiveness consists in a detailed analysis of each subsystem. For each subsystem j, we need to analyze all possible states X_{ji}, $1 \leq i \leq 2^{m_i}$. Also, for each such subsystem we need to choose a reasonable lattice of the effectiveness coefficient values. Assume this lattice has K_j different cells:

- The first cell includes those states whose effectiveness coefficients W_{ij} satisfy the condition $1 = W'_{0j} \leq W'_j < B_1$, where B_1 is the first threshold of the lattice; for all states belonging to this cell of the lattice, one computes the total probability R_1 as the sum of the probabilities of all states whose effectiveness values are included in this cell.
- The second cell includes those states whose effectiveness coefficients W_{ij} satisfy the condition $B_1 \leq W'_j < B_2$, where B_2 is the second threshold of the lattice; the corresponding probability computed is R_2.
  $\vdots$
- The K_jth cell includes those states whose effectiveness coefficients W_{ij} satisfy the condition $B(K_j - 1) \leq W'_j < 0$, where B_{K_i} is the last threshold of the lattice; the corresponding probability computed is R_{K_i}.

We may now analyze the system as a whole. In each cell of the lattice, we choose a "middle" state which corresponds to the average value of W'_j. For future analysis, this state now becomes a "representative" of all of the remaining states related to this cell. Thus, we choose K_j representatives for each subsystem. We should choose an appropriate number of representatives, say X^*_{ji}. Each of them appears with probability R_{ji}. The number of representatives is determined with respect to the required accuracy of the analysis and the available computer capacity.

After these preliminary steps we consider

$$K = \prod_{1 \leq j \leq M} K_j \tag{8.70}$$

different system states and, for each of them, evaluate the effectiveness coefficient. We then consider all K system states

$$\mathbf{X} = \left(X^*_{ji}; 1 \leq j \leq M \right) = \left(X^*_{1i_1}, X^*_{2i_i}, \ldots, X^*_{Mi_M} \right)$$

and write the expression for W_{syst} with the use of (8.62)

$$\text{PEI} = \sum_{\text{all } X} W\big(X^*_{1i_1}, \ldots, X^*_{Mi_M}\big) \prod_{\substack{1 \leq j \leq M \\ X^*_{ik} \in \mathbf{X}}} R^{X^*_{jk}}_{jk}\big(1 - R_{jk}\big)^{(1 - X^*_{jk})} \quad (8.71)$$

This expression is not too pleasant in visual form because of the notation used. Neither is it easy to compute. But its nature is simple and completely coincides with (8.61).

Of course, if we decide to distinguish several levels of a system's hierarchy, the methodology would be the same but the corresponding description, in a general form, would be even longer than (8.71). We would like to emphasize that a hierarchical model needs less computation.

This method of representative selection can be successfully used for obtaining lower and upper bounds.

1. Let us choose from among the states of the lattice cell a state with a minimal effectiveness coefficient and consider this state as a representative of this cell. Denote this aggregate state by X_j^{min}. If we substitute X_j^{min} instead of $X^*_{ji_j}$ in (8.71), we will obtain a lower bound for the system index, W_{syst}^{min}.
2. If a state X_j^{max} with a maximal effectiveness coefficient is chosen as the representative of the cells, then the same procedure gives us an upper bound for W_{syst}^{max}.

Thus, we obtain two-sided bounds for W_{syst}:

$$W_{syst}^{min} \leq W_{syst} \leq W_{syst}^{max} \quad (8.72)$$

In general, for practical purposes, it is enough to have an approximate expression (8.71). We should emphasize that reliability (and also effectiveness performance) computations are usually provided not for a precise evaluation of different indexes, but usually for a comparison of competitive variants at some design stage. For such purposes, we may use an approximate solution as a direction for design.

8.7 PRACTICAL RECOMMENDATION

An analysis of the performance effectiveness of a system must be carried out by a researcher who deeply comprehends the system as a whole, knows its operation, and understands all demands on the system. It is a necessary condition of successful analysis. Of course, the systems analyst should also be acquainted with operations research methods. As with any operations research problem, the task is concrete and its solution is more of an art than a science.

For simplicity of discussion, we demonstrate the effectiveness analysis methodology referring to an instant system. The procedure of a system's effectiveness evaluation, roughly speaking, consists of the following tasks:

- A formulation of an understandable and clear goal of the system.
- A determination of all possible system's tasks (operations, functions).
- A choice of the most appropriate measure of system effectiveness.
- A division of a complex system into subsystems.
- A compilation of a structural–functional scheme of the system which reflects the interaction of the system's subsystems.
- A collection of reliability data.
- A computation of the probabilities of the different states in the system and its subsystems.
- An estimation of the effectiveness coefficients of different states.
- A performance of the final computations of the system's effectiveness.

Of course, the effectiveness analysis methodology of enduring systems is quite similar, with the exception of some terms.

We need to remark that the most difficult part of an effectiveness analysis is the evaluation of the coefficients of effectiveness for different system states. In only extremely rare cases is it possible to find these coefficients by means of analytical approaches. At any rate, in the initial stages of a system's design there is no other way. The most common method is to simulate the system with the help of a computerized model or a physical analogue of the system. In the latter case, the analyst introduces different failures at appropriate moments into the system and analyzes the consequences. The last and the most reliable method is to perform experiments with the real system or, at least, with a prototype of the system.

Of course, one has to realize that usually all of these experiments set up to evaluate effectiveness coefficients are very difficult and they demand much time, money, and other resources. Consequently, one has to consider how to perform only really necessary experiments. This means that a prior evaluation of different state probabilities is essential: there is no need to analyze extremely rare events.

One can see that the analysis of a system's effectiveness performance is not routine. Designing a mathematical model of a complex system is, in some sense, a problem similar to the problem of designing a system itself. Of course, there are no technological difficulties—no time or expense for engineering design and production.

CONCLUSION

It seems that the first paper devoted to the problem discussed in this chapter was the paper by Kolmogorov (1945). This work focused on an effectiveness measure of antiaircraft fire. The total kill probability of an enemy's aircraft was investigated. The random nature of the destruction of different parts of an aircraft and the importance of these parts was assumed. It is clear that from a methodological viewpoint the problem of system effectiveness analysis is quite similar: one has only to change slightly the terminology.

The first papers concerning a system's effectiveness evaluation appeared in the early 1960s [see, e.g., Ushakov (1960, 1966, 1967)]. Some special cases of system effectiveness evaluation were considered in Ushakov (1985, 1994) and Netes (1980, 1984).

One can find an analysis of the effectiveness of symmetrical branching systems in Ushakov (1985, 1994) and Ushakov and Konyonkov (1964). Territorial (regional) systems with intersecting zones of action were studied in Ushakov (1985, 1994). Here one can also find an analysis of decomposition methods. The general methodology and methods of system effectiveness analysis are described in Ushakov (1985, 1994).

REFERENCES

Kolmogorov, A. N. (1945). A number of target hits by several shots and general principles of effectiveness of gun-fire (in Russian). *Proc. Moscow Inst. Math.*, Issue 12.

Netes, V. A. (1980). Expected value of effectiveness of discrete system (in Russian). *Automat. Comput. Sci.*, no. 11.

Netes, V. A. (1984). Decomposition of complex systems for effectiveness evaluation. *Engrg. Cybernet.* (USA), vol. 22, no. 4.

Ushakov, I. A. (1960). An estimate of effectiveness of complex systems. In *Reliability of Radioelectronic Equipment* (in Russian). Moscow: Sovietskoye Radio.

Ushakov, I. A. (1966). Performance effectiveness of complex systems. In *On Reliability of Complex Technical Systems* (in Russian). Moscow: Sovietskoe Radio.

Ushakov, I. A. (1967). On reliability of performance of hierarchical branching systems with different executive units. *Engrg. Cybernet.* (USA), vol. 5, no. 5.

Ushakov, I. A., ed. (1985). *Reliability of Technical Systems: Handbook* (in Russian). Moscow: Radio i Sviaz.

Ushakov, I. A., ed. (1994). *Handbook of Reliability Engineering*. New York: Wiley.

Ushakov, I. A., and Yu. K. Konyonkov (1964). Evaluation of effectiveness of complex branching systems with respect to their reliability. In *Cybernetics in Service for Communism* (in Russian), A. Berg, N. Bruevich, and B. Gnedenko, eds. Moscow: Nauka.

EXERCISES

8.1   A conveyor system consists of two lines, each producing N items per hour. Each of the lines has an availability coefficient $K = 0.8$. When one of the lines has failed, the other decreases its productivity to $0.7N$ because of some technological demands. There is a suggestion to replace this system with a new one consisting of one line with a productivity of $1.7N$ items per hour and an availability coefficient $K_1 = 0.9$. Is this replacement reasonable from an effective productivity viewpoint or not?

8.2   A branching system has one main unit and three executive ones. There are two possibilities: (1) to use a main unit with PFFO $p_0 = 0.9$ and an executive unit with PFFO $p_1 = 0.8$ or (2) to use a main unit with PFFO $p_0 = 0.8$ and an executive unit with PFFO $p_1 = 0.9$. Is there is a difference between these two variants if the system's effectiveness depends on (a) the average number of successfully operating executive units, (b) a successful operation at least one executive unit, and (c) a successful operation of all executive units?

SOLUTIONS

8.1   The old system of two lines has the following states:

· Both lines operate successfully. In this case the effective productivity of the system is $2N$. This state occurs with probability $P = (0.8)(0.8) = 0.64$.

· One line has failed and the other is operating. This state occurs with probability $P = 2(0.8)(0.2) = 0.32$. During these periods the system productivity equals $0.7N$.

· Both lines have failed. This state occurs with probability $P = (0.2)(0.2) = 0.04$. The system productivity obviously equals 0. Thus, the total average productivity of the old system can be evaluated as

$$W_{old} = (0.64)(2.0) + (0.32)(0.7) \approx 1.5$$

The new system has an average effective productivity equal to

$$W_{new} = (1.7)(0.9) = 1.53$$

Thus, the average productivity of both systems is very close. The increase in productivity is about 1.5%. One has to solve this problem taking into account expenses for installation of the new system, on the one hand, and the potential decrease in the cost of repair, on the other hand (the new system will fail less often).

8.2 (a) The average number of successfully operating executive units depends only on the product $p_0 p_1$, so both systems are equivalent.

(b) For (1) one has $W_{syst} = (0.9)(1 - 0.8)^3 \approx 0.898$ and for (2) $W_{syst} = (0.8)(1 - 0.9)^3 \approx 0.799$. Thus, the first variant is more effective.

(c) For (1) one has $W_{syst} = (0.9)(0.8)^3 \approx 0.460$ and for (2) $W_{syst} = (0.8)(0.9)^3 \approx 0.584$. In this case the second variant is more effective.

CHAPTER 9

TWO-POLE NETWORKS

Above we considered systems with a so-called "reducible structure." These are series, parallel, and various kinds of mixtures of series and parallel connections. As mentioned, they are two-pole structures which can be reduced, with the help of a simple routine, into a single equivalent unit. However, not all systems can be described in such a simple way.

We would like to emphasize that most existing networks, for example, communication and computer networks, transportation systems, gas and oil pipelines, electric power systems, and others, have a structure which cannot be described in terms of reducible structures, even if they are considered as two-pole networks.

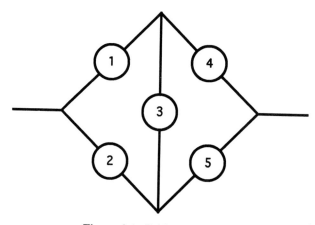

Figure 9.1. Bridge structure.

The simplest example of a system with a nonreducible structure is the so-called *bridge structure* (see Figure 9.1). This particular structure is probably not of great practical importance, but it is reasonable to consider it in order to demonstrate the main methods of analysis of such kinds of structures.

9.1 RIGID COMPUTATIONAL METHODS

9.1.1 Method of Direct Enumeration

The bridge structure cannot be represented as a connection of parallel–series or series–parallel subsystems of independent units (links). For this system the structure function $\varphi(X)$, where $X = (x_1, x_2, x_3, x_4, x_5)$, can be written in tabular form (see Table 9.1) where all possible system states and corresponding structure function values are presented.

Because each Boolean variable has two possible different values, 0 or 1, the system can be characterized by $2^5 = 32$ different states. In Table 9.1 we enumerate all possible values of the variables $x_1, x_2, \ldots, x_5$ and denote them as $X_1, X_2, \ldots, X_{32}$. Some X_k's are states of successful operation of the bridge system (the set G) and some of them are not (the set $\overline{G}$). In this notation the structure function of the bridge system can be written as

$$\varphi(x_1, \ldots, x_5) = \varphi(X_1) \cup \varphi(X_2) \cup \cdots \cup \varphi(X_{32}) = \bigcup_{X_k \in G} \varphi(X_k) \quad (9.1)$$

The probability of a system's successful operation is

$$\Pr\{\varphi(x_1, \ldots, x_5) = 1\} = E\left\{ \bigcup_{X_k \in G} \varphi(X_k) \right\} = \sum_{X_k \in G} E\{\varphi(X_k)\} \quad (9.2)$$

Each vector X_k can be expressed through its component x's and $\bar{x}$'s. For example (see Table 9.1),

$$X_8 = (\bar{x}_1, x_2, \bar{x}_3, x_4, x_5)$$

From Table 9.1 it follows that the vector X_8 belongs to G, and so it will be taken into account in (9.2). Then

$$\varphi(X_8) = \bar{x}_1 x_2 \bar{x}_3 x_4 x_5$$

and

$$\begin{aligned} E\{\varphi(X_8)\} &= E\{\bar{x}_1 x_2 \bar{x}_3 x_4 x_5\} \\ &= E\{\bar{x}_1\} E\{x_2\} E\{\bar{x}_3\} E\{x_5\} = q_1 p_2 q_3 p_4 p_5 \end{aligned}$$

We do not write the detailed expression for $\varphi(X)$ here. This can be easily obtained from Table 9.1 by taking into account that the corresponding term

TABLE 9.1 Description of the Structure Function of the Bridge Structure

States of units					Vector	Value
x_1	x_2	x_3	x_4	x_5	X_k	$f(X_k)$
1	1	1	1	1	X_1	1
0	1	1	1	1	X_2	1
1	0	1	1	1	X_3	1
1	1	0	1	1	X_4	1
1	1	1	0	1	X_5	1
1	1	1	1	0	X_6	1
0	0	1	1	1	X_7	0
0	1	0	1	1	X_8	1
0	1	1	0	1	X_9	1
0	1	1	1	0	X_{10}	1
1	0	0	1	1	X_{11}	1
1	0	1	0	1	X_{12}	1
1	0	1	1	0	X_{13}	1
1	1	0	0	1	X_{14}	1
1	1	0	1	0	X_{15}	1
1	1	1	0	0	X_{16}	0
0	0	0	1	1	X_{17}	0
0	0	1	0	1	X_{18}	0
0	0	1	1	0	X_{19}	0
0	1	0	0	1	X_{20}	1
0	1	0	1	0	X_{21}	0
0	1	1	0	0	X_{22}	0
1	0	0	0	1	X_{23}	0
1	0	0	1	0	X_{24}	1
1	0	1	0	0	X_{25}	0
1	1	0	0	0	X_{26}	0
0	0	0	0	1	X_{27}	0
0	0	0	1	0	X_{28}	0
0	0	1	0	0	X_{29}	0
0	1	0	0	0	X_{30}	0
1	0	0	0	0	X_{31}	0
0	0	0	0	0	X_{32}	0

of type $E\{\varphi(X_k)\}$ has p_i for $x_i = 1$ and q_i for $\bar{x}_i = 1$. Based on Table 9.1, the following equation can be written:

$$E\{\varphi(X)\} = E\{\varphi(X_1)\} + E\{\varphi(X_2)\} + \cdots + E\{\varphi(X_{32})\}$$

Omitting intermediate results, we give the final formula for the connectivity probability (in the case of identical units) in two equivalent forms

$$E\{\varphi(X)\} = p^5 - 5p^4 + 2p^3 + 2p^2 \tag{9.3}$$

$$E\{\varphi(X)\} = 1 - 2q^2 - 2q^3 + 5q^4 - 2q^5 \tag{9.4}$$

Expression (9.4) is useful for the calculation of the reliability of a highly reliable system where $q \ll 1$. In this case we have the approximation $E\{\varphi(X)\} \approx 1 - 2q^2$.

Of course, such a method of direct enumeration allows one to compute the probability of the connectivity of a nonreducible two-pole network only in principle. This method is inadequate for problems one meets in practice.

9.1.2 Method of Boolean Function Decomposition

Sometimes the method of decomposition of a Boolean function $\varphi(X)$ is very effective. Any Boolean function can be represented as

$$\varphi(x_1, \ldots, x_k, \ldots, x_n) = x_k \varphi(x_1, \ldots, 1_k, \ldots, x_n) \cup \bar{x}_k \varphi(x_1, \ldots, 0_k, \ldots, x_n) \tag{9.5}$$

where we use 1_k (or 0_k) to show that 1 (or 0) is placed at the kth position. If we interpret the terms of the Boolean function as events, we can say that these two events are mutually exclusive because the first term includes x_k, and the second one includes $\bar{x}_k$. In this case we can write

$$E\{\varphi(x_1, \ldots, x_k, \ldots, x_n)\} = E\{x_k \varphi(x_1, \ldots, 1_k, \ldots, x_n)\}$$
$$+ E\{\bar{x}_k \varphi(x_1, \ldots, 0_k, \ldots, x_n)\} \tag{9.6}$$

Now we can see that x_k and $\varphi(\ldots 1_k \ldots)$ are independent as well as $\bar{x}_k$ and $\varphi(\ldots 0_k \ldots)$, and thus (9.6) can finally be rewritten as

$$E\{\varphi(x_1, \ldots, x_k, \ldots, x_n)\} = E\{x_k\} E\{\varphi(x_1, \ldots, 1_k, \ldots, x_n)\}$$
$$+ E\{\bar{x}_k\} E\{\varphi(x_1, \ldots, 0_k, \ldots, x_n)\} \tag{9.7}$$

Now consider the bridge structure. Let us choose unit x_3 for decomposition. Then (9.7) can be rewritten as

$$E\{\varphi(x_1, x_2, x_3, x_4, x_5)\} = E\{x_3\} E\{\varphi(x_1, x_2, 1, x_4, x_5)\}$$
$$+ E\{\bar{x}_3\} E\{\varphi(x_1, x_2, 0, x_4, x_5)\}$$
$$= p_3 E\{(x_1 \cup x_2)(x_4 \cup x_5)\} + q_3 E\{x_1 x_4 \cup x_2 x_5\} \tag{9.8}$$

It is clear that the new structure functions under the operators E in the last row of (9.8) are functions of the reducible series–parallel and parallel–series structures. We can easily calculate the reliability index for each of these cases.

This procedure should become clear with the following explanations. What does it mean that $x_3 = 1$? It means that in the initial bridge structure, unit

x_3 is absolutely reliable (it is always in the operational state). Thus, the bridge structure becomes a simple series–parallel structure. Similarly, $x_3 = 0$ (or, equivalently, $\bar{x}_3 = 1$) means that unit x_3 is eliminated (it is always in a failed state). This means that the structure becomes a parallel–series structure.

Finally, we may write

$$E\{\varphi(x_1, x_2, x_3, x_4, x_5)\} = p_3[(1 - q_1 q_2)(1 - q_4 q_5)]$$
$$+ q_3[1 - (1 - p_1 p_4)(1 + p_2 p_5)] \quad (9.9)$$

We note that the same result can be obtained with the use of direct probabilistic arguments, namely with the use of the formula:

$$\Pr\{\varphi(x_1, x_2, x_3, x_4, x_5) = 1\}$$
$$= \Pr\{\varphi(x_1, x_2, x_3, x_4, x_5)|x_3 = 1\} \Pr\{x_3 = 1\}$$
$$+ \Pr\{\varphi(x_1, x_2, x_3, x_4, x_5)|x_3 = 0\} \Pr\{x_3 = 0\} \quad (9.10)$$

Obviously, (9.9) and (9.10) are equivalent.

We should mention that a Boolean function can be decomposed by any variable. In this particular example such a decomposition can be done with respect to any x_k. (The only advantage of using the unit x_3 was to produce a clearer explanation.) In Exercise 9.1 we obtain the result by decomposing the bridge structure with respect to another unit.

Sometimes it is reasonable to decompose a Boolean function with respect to several variables. For example, if we choose two variables, the expression takes the form

$$\varphi(\ldots x_k \ldots x_j \ldots) = x_k x_j f(\ldots 1_k \ldots 1_j \ldots) \cup \bar{x}_k x_i \varphi(\ldots 0_k \ldots 1_j \ldots)$$
$$\cup x_k \bar{x}_j \varphi(\ldots 1_k \ldots 0_j \ldots) \cup \bar{x}_k \bar{x}_j \varphi(\ldots 0_k \ldots 0_j \ldots)$$

In general, this decomposition method is not practically effective. If a nonreducible structure is complex, the problem of choosing a unit or units with respect to which such a decomposition could be reasonably done becomes very difficult. Moreover, for a complex Boolean function, an analysis of reduced functions still remains very difficult in general cases.

In short, this idea of network decomposition, in fact, nearly always represents only a nice illustrative example and not an effective tool for engineers. All of the difficulties connected with the numerical analysis of nonreducible structures leads to a need to find other methods. One effective analytical method is obtaining the lower and upper bounds of the unknown value of $E\{\varphi(X)\}$.

9.2 METHOD OF PATHS AND CUTS

9.2.1 Esary – Proschan Bounds

Consider an arbitrary two-pole network. *Network connectivity* for some state is defined as the possibility of connecting two network poles with at least one chain of its links (units). (We assume that all network vertexes, or nodes, are absolutely reliable.) We call this set of units a *path*. In general, there may be several such paths in the network. Each path represents a series connection of corresponding units. In other words, the path A_k^* is defined as a subset of network units such that, if all units of A_k^* are up, the network poles are connected in spite of the failure of all remaining units. In general, a path A_k^* might include "hanging" (or even isolated) units and loops. But these units are not critical at all. They may fail with no influence whatsoever on the network's connectivity; they can be deleted without changing the network connectivity. In this connection it is reasonable to consider a *simple path*. A simple path A_k is the minimal set of units x_j, $x_j \in A_k$, whose up states guarantee that the network poles are connected in spite of the states of the remaining network units. If at least one unit of a path has failed, the path is disconnected.

Denote the structure function of path A_k by

$$\alpha_k = \bigcap_{j \in A_k} X_j$$

It is clear that a network with only one path is a series system. If we enumerate all of the possible N paths in a two-pole network, the structure function can be represented as

$$\varphi(\mathbf{X}) = \bigcup_{1 \leq k \leq N} \alpha_k(\mathbf{X}) \tag{9.11}$$

In other words, we present the network as a parallel connection of all possible different simple paths. An example of such a representation for the bridge structure is depicted in Figure 9.2*a* We see that some units belong to different simple paths. This means that the paths in a nonreducible structure, in general, are not independent, and, therefore, $E\{\varphi(\mathbf{X})\}$ cannot be expressed in any simple form.

We can formulate a similar natural condition of a connectivity violation by the following equivalent statement expressed via a network's *cut*. We call a cut a set of network units such that the elimination of all of them leads to a disconnection of the network poles. Let B_j^* denote the subset of network units forming the cut. The cut, in general, may include some extra units. The restoration of such units does not lead to the restoration of the network's connectivity. We will consider only *simple cuts* B_k. A simple cut is a minimal

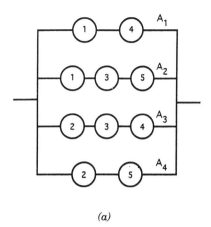

(a)

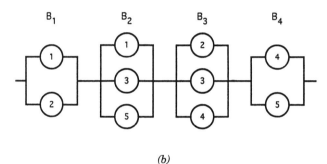

(b)

Figure 9.2. Presentation of a bridge structure as (a) a parallel–series structure and (b) as a series–parallel structure with replicated links.

Legend: A_k = path; B_k = cut.

set of units whose failure leads to the disruption of the network's connectivity. It is clear that the cut's structure function B_k is

$$\beta_k(\mathbf{X}) = \bigcup_{j \in B_k} x_j$$

This is the structure function of a parallel system. Of course, a purely parallel system has only one cut.

If we enumerate all possible M different simple cuts of a two-pole network, the network can be represented as a series connection of simple cuts. An example of a bridge structure is displayed in Figure 9.2b. The structure function of the network in this case can be written as

$$\varphi(\mathbf{X}) = \bigcap_{1 \le j \le M} \beta_j(\mathbf{X}) \tag{9.12}$$

It is clear that (9.11) and (9.12) are equivalent Boolean equations represented in conjunctive and disjunctive forms.

Notice that a network's simple cuts may be interdependent because they may contain the same units. Thus, because of this dependence, simple direct probabilistic computations with the use of (9.12) by means of substitution of x_i's for p_i's are impossible.

For convenience, we introduce the so-called *connectivity function*

$$h(\mathbf{p}) = E\{\varphi(\mathbf{X})\} = h(p_1, p_2, \ldots, p_n) \tag{9.13}$$

where $\mathbf{p} = (p_1, p_2, \ldots, p_n)$. In fact, this function is simply the probability that a two-pole network is connected.

Applying the above results for parallel–series and series–parallel connections of dependent units, we can immediately write

$$h = E\{\varphi(\mathbf{X})\} \le 1 - \prod_{1 \le k \le N} \left[1 - E\{\alpha_k(\mathbf{X})\}\right] = U \tag{9.14}$$

and

$$h = E\{\varphi(\mathbf{X})\} \ge \prod_{1 \le j \le M} E\{\beta_j(\mathbf{X})\} = L \tag{9.15}$$

Here h is a brief notation for $h(\mathbf{p})$, U is an upper bound, and L is a lower bound.

We now point out that units belonging to a simple path (or to a simple cut) are themselves independent (these sets consist of different units). Thus, it is possible to use the formula for a series connection to compute $E\{A_k(\mathbf{X})\}$ and the formula for a parallel connection to compute $E\{B_j(\mathbf{X})\}$.

In the detailed form the bounds can be written as follows

$$U = 1 - \prod_{1 \le k \le N}\left[1 - \prod_{i \in A_k} p_i\right]$$

and

$$L = \prod_{1 \le j \le M}\left(1 - \prod_{i \in B_k} q_i\right)$$

This method of obtaining network reliability bounds was introduced by Esary and Proschan (1962). We illustrate this method by the example of a bridge structure (see Figure 9.1). All simple cuts and paths of the structure are enumerated in Figure 9.2.

Using (9.14) and (9.15), we can write

$$(1 - q_1 q_2)(1 - q_1 q_3 q_5)(1 - q_4 q_5)(1 - q_2 q_3 q_4) \le E\{\varphi(\mathbf{X})\}$$
$$= h \le 1 - (1 - p_1 p_4)(1 - p_1 p_3 p_5)(1 - p_2 p_5)(1 - p_2 p_3 p_4)$$

In this particular case, the set of units forming paths A_2 and A_3 coincide with the sets forming cuts B_2 and B_3, respectively. For more complex networks, this fact is a rare and unusual exclusion.

9.2.2   Litvak – Ushakov Bounds

These bounds are based on sets of nonintersecting simple paths and cuts. We first point out that, for complex networks, the enumeration of all of the different simple paths and cuts is a very difficult problem demanding a large computer memory and an enormous computational time. For systems of any practical dimension, this enumeration problem is essentially impossible.

For this reason, one sometimes attempts to make the computations shorter. On a heuristic level, an explanation of the main idea of such an attempt follows. If we consider a very complex multiunit network, we often find that a lower bound includes some very "thick" cuts, that is, cuts with a large number of units. It is clear that such a parallel connection is characterized by an extremely high reliability. There is a temptation to exclude such "thick" cuts from consideration: they are very reliable in comparison with the remaining cuts! In other words, the value $1 - q_1 q_2 \ldots q_k$ is so close to 1 that it seems reasonable to replace it with 1. This leads to an increase in the reliability index. Thus, after this, the new lower bound should be even higher than the initial strong lower bound. The higher the lower bound, the better.

Analogously, the strong upper bound includes some "very long" series connections which may be very unreliable. Again, the question arises: why should one take into account such a practically absolutely unreliable series connection of units for the computation of a reliability index of a parallel connection? Indeed, for very large m, $p_1 p_2 \ldots p_m \approx 0$. If one neglects such "very long" paths, the new upper bound becomes lower. This again produces a better upper bound than we have initially.

We must emphasize that such a "heuristic heuristic" leads to very rough mistakes. Indeed, the higher the lower bound, the better, but only *if* the lower bound remains a lower bound.

We may obtain strange results using these "simplifications" and "improvements" of the bounds: the obtained "improved" bounds may not even bound the unknown value $h = E\{\varphi(\mathbf{X})\}$. In fact, an "improved" lower bound, obtained in such manner, may be even larger than an "improved" upper bound. Thus, we may obtain new "bounds" which lose all mathematical meaning at all.

Once more, we would like to emphasize that a heuristic is not an arbitrary guess on an "intuitive level." In our opinion, a heuristic must usually be an "almost proven" simplification of an existing strong solution. Sometimes, instead of a proof, one may deliver a set of numerical examples, covering the parametrical area of domain, as a confirmation of the heuristic's validity. Such "experimental mathematics" occupies more and more room in computational methods and very often replaces exact proofs.

Let us illustrate some possible mistakes of using the above-mentioned "simplification" on an example of a bridge structure consisting of identical

TABLE 9.2

p	$\tilde{U}$	$\tilde{L}$
0.9	0.9639	0.9801
0.5	0.4375	0.5625
0.1	0.0199	0.0361

units. Represent approximations of an upper bound in the form

$$U = 1 - \left(1 - p^2\right)^2\left(1 - p^3\right)^2 \approx \tilde{U} = 1 - \left(1 - p^2\right)^2$$

and of a lower bound in the form

$$L = \left(1 - q^2\right)^2\left(1 - q^3\right)^2 \approx \tilde{L} = \left(1 - q^2\right)^2$$

We prefer not to use straight dull transformations to prove that $\tilde{U} < \tilde{L}$. We show this on numerical examples for $p = 0.9$, $p = 0.1$, and $p = 0.5$ (see Table 9.2). Within the entire area of domain p, the "improved" upper bound is smaller than the "improved" lower bound! Thus, these new "bounds" have no meaning.

We give these extended explanations with the purpose of showing that one should be very cautious in relying on so-called "common sense" and "intuition." They do not always work. At the same time, we will show below that an incorrect idea, which led to an incorrect result, after a *correct understanding* and explanation, can be used to obtain the correct results.

We now give a precise definition of the new bounds. To make the discussion easier, we introduce a useful and natural definition. In reliability theory a monotone structure is important. In terms of the structure function, the characterization properties of monotone structures are the following:

1. $\varphi(1, 1, \ldots, 1) = 1$; that is, a system with all units successfully operating is operating.
2. $\varphi(0, 0, \ldots, 0) = 0$; that is, a system with all units in a failed state is always in a failed state.
3. If $\mathbf{X} > \mathbf{Y}$, where $\mathbf{X} = (x_1, x_2, \ldots, x_n)$ and $\mathbf{Y} = (y_1, y_2, \ldots, y_n)$, then $\varphi(\mathbf{X}) \geq \varphi(\mathbf{Y})$; that is, a failure of any unit can only worsen the system's reliability index.

The connectedness function introduced above is monotone with respect to any of its variable p_k: $h(\mathbf{p}) > h(\mathbf{p}')$ if $\mathbf{p} > \mathbf{p}'$. In other words, if any unit's reliability is increased, the system's reliability is increased. All reducible systems and nonreducible systems of a network type belong to this class of monotone structures.

We are now ready to obtain new lower and upper bounds for nonreducible structures. We accompany our arguments with illustrations on the simple example of the bridge structure. We say that two paths *intersect* if they have at least one common unit. In the opposite case, we say that the paths are disjoint. In any two-pole network, we can distinguish a set of disjoint paths. Obviously, the maximal number of disjoint paths equals the number of units in a "minimal" cut (a cut with a minimal number of units). Consider the so-called *complete set of disjoint, or nonintersecting paths*. This means that each such set includes all possible nonintersecting simple paths. Of course, a network might have several different sets of such a type. In general, there might be very many different complete sets of paths. To enumerate all of these sets is an extremely difficult computational problem but, fortunately, we do not need to do it.

We introduce the following notation:

a_i is the ith complete set of nonintersected paths.

$X(i)$ is the set of units included in the ith complete set of paths (this set of network units is determined by the concrete complete set of disjoint simple paths).

V is the number of different complete sets of nonintersecting simple paths.

L_i^* is the connectivity function for $X(i)$.

For the bridge structure, for example, we can draw Figure 9.3 and construct Table 9.3. In Table 9.3, L_i^* denotes the corresponding Litvak–Ushakov bound. In this case $V = 3$, and the maximal number of disjoint paths equals 2. This corresponds to the number of units in the "thinnest" cut.

For any complete set of paths, we have $X(i) \leq X$. In other words, we consider a new system consisting of disjoint sets of units. One can construct this system using an initial network by eliminating all units which are not included in $X(i)$. Sometimes, this procedure needs to split the nodes, that is, to use the same node in different paths. Consequently, $\varphi(X(i)) \leq \varphi(X)$.

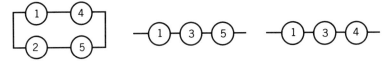

Figure 9.3. All possible representations of a bridge structure by nonintersected paths.

TABLE 9.3   Various Complete Sets of Nonintersecting Simple Paths for a Bridge Structure

a_i	$X(i)$	L_i^*
a_1	$\{x_1, x_4\}, \{x_2, x_5\}$	$1 - (1 - p_1 p_4)(1 - p_2 p_5)$
a_2	$\{x_1, x_3, x_5\}$	$p_1 p_3 p_5$
a_3	$\{x_2, x_3, x_4\}$	$p_2 p_3 p_4$

Thus,

$$L_i^* = E\{\varphi(\mathbf{X}(i))\} \le E\{\varphi(\mathbf{X})\} = h \qquad (9.16)$$

Thus, the probability of connectivity of the initial network is not smaller than a parallel connection of simple paths belonging to the above-mentioned complete set. In other words, we obtained a lower bound for h. In general, one has several different complete sets of nonintersecting simple paths. Each set generates its own lower bound. Choosing the highest lower bound among all of the above-obtained L_i^*, we find the best (nonimproved) lower bound in the class of bounds of this type. (We would like to emphasize that within the described procedure we deal *only* with lower bounds.) For the bridge structure, a lower bound of this type can be written as

$$L^* = \max\{[1 - (1 - p_1 p_4)(1 - p_2 p_5)], p_1 p_3 p_5, p_2 p_3 p_4\}$$

In general, the best lower bound of this type is

$$L^* = \max_{1 \le i \le V} L_i^* = \max_{1 \le i \le V} \left\{ \left[1 - \prod_{k \in a_i} \left(1 - \prod_{j \in A_k^i} p_j \right) \right] \right\} \qquad (9.17)$$

where A_k^i is a set of indexes of units of the **k**th path of the **i**th complete set $\mathbf{a}_i$.

In a similar manner, we can distinguish sets of disjoint cuts in the two-pole network. The maximal number of cuts in each such set equals the number of units in a shortest path. Let the number of units in a shortest path be d. Consider sets which include the maximal number of disjoint cuts. By analogy with the previous case, call each such set a *complete set of nonintersecting simple cuts*. Again, we mention that, in general, there are many different complete sets.

We introduce the following notation:

b_i is the ith complete set of disjoint cuts.

$\mathbf{Y}(i)$ is the set of units included in the ith complete set of cuts.

W is the number of such different sets.

U_i^* is the connectivity function for $\mathbf{Y}(i)$.

For the bridge structure, the detailed description in these terms is depicted in Figure 9.4 and Table 9.4. In this case, $W = 3$, and the maximal number of

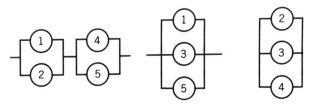

Figure 9.4. All possible representations of a bridge structure by nonintersected cuts.

TABLE 9.4   Complete Sets of Nonintersecting Simple Cuts for the Bridge Structure

a_i	$Y(i)$	U_i^*
b_1	$\{x_1, x_2\}, \{x_4, x_5\}$	$(1 - q_1 q_2)(1 - q_4 q_5)$
b_2	$\{x_1, x_3, x_5\}$	$1 - q_1 q_3 q_5$
b_3	$\{x_2, x_3, x_4\}$	$1 - q_2 q_3 q_4$

disjoint cuts equals 2. This corresponds to the number of units in the "shortest" path.

Any complete set of nonintersecting simple cuts $Y(i)$ can be constructed on the basis of the initial network by the insertion of some absolutely reliable units. In certain particular cases, it is enough to make some existing units absolutely reliable, and the initial network is transformed into a structure with disjoint cuts. (Below we illustrate this on the example of a bridge structure.) By the monotonicity property, this new structure has a reliability index higher than the initial one. Indeed, the system reliability can only increase if one increases the unit reliability or if one inserts into the structure an absolutely reliable unit. In other words, $Y(i) \leq X$ and, consequently, $\varphi(Y(i)) \leq \varphi(X)$:

$$U_i^* = E\{\varphi(Y(i))\} \geq E\{\varphi(X)\} = h \qquad (9.18)$$

Thus, the probability of the connectivity of the initial network is not larger than a series connection of simple cuts belonging to the above-mentioned complete set. Thus, we have obtained a lower bound for h. Each complete set of nonintersecting simple paths (if a network is characterized by several such sets) has its own upper bound. To find the best upper bound, we need to choose the smallest value of U_i^*. In this case we may improve the upper bound because we choose the smallest among several upper bounds. (Let us emphasize once more that we deal only with upper bounds.) For the bridge structure, an upper bound of this type is

$$U^* = \min\{(1 - q_1 q_2)(1 - q_4 q_5), 1 - q_1 q_3 q_5, 1 - q_2 q_3 q_4\}$$

In general, the best upper bound of this type is

$$U^* = \min_{1 \leq i \leq W} U_i^* = \min_{1 \leq i \leq W} \prod_{k \in b_i} \left[1 - \prod_{j \in B_k^i} q_j \right] \qquad (9.19)$$

where $\mathbf{B_k^i}$ is a set of indexes of units of the **k**th cut of the **i**th complete set $\mathbf{b_i}$. These bounds were obtained by Ushakov and Litvak (1977).

9.2.3 Comparison of the Two Methods

We emphasize that the bounds (9.17) and (9.19) can be used as a correct approximation for the calculation of network bounds. Indeed, if there are V different sets of nonintersecting simple paths and W analogous sets of cuts, it is not necessary to take them into account. Consideration of only part of the entire number of sets of simple paths gives us a lower bound smaller than in (9.17), and consideration of only part of the entire number of sets of simple cuts gives us an upper bound larger than in (9.19). Thus, both bounds become worse, but at any rate, in both cases the lower bound remains lower and the upper bound remains higher.

One of the main advantages of obtaining bounds with the use of disjoint paths and cuts is the simplicity of obtaining a good estimation by using the following simple method.

First, consider the lower bound. Find in the initial two-pole network a shortest path, say X_1. Put this path into a special file containing a set of nonintersecting simple paths, a_1. Exclude all the units belonging to path X_1 from the initial network. Then from this remaining part of the network, extract a new shortest path, say X_2. Add this path to the file a_1. Delete all units of this new path from the network. Continue the procedure. As a result of such a procedure, a smaller subnetwork remains on each step. If the minimal cut of the initial network is of size M, then such a procedure continues at most M steps. As a result, we obtain a complete set of paths. The parallel connection of these paths usually gives one of the best lower bounds in this class.

For obtaining a shortest path in each case, we may use any known algorithm. In practice, even the first subset of nonintersecting simple paths gives us a good approximation of the lower bound (if not the best one).

Of course, we should mention that the "length" of the path can be measured not only by the number of units in this path. In reliability this is determined by the value $\prod_{i \in A_j} p_j$, where A_j is the set of all sets a_i. We call a path "shorter" if this value is larger.

A similar idea can be used to obtain an upper bound. Find a "minimal" cut, say Y_1, from the initial two-pole network and form a file for the set B_1. Each unit of a cut might be considered as a link with left and right ends, one of which belongs to the "left" and another belongs to the "right" poles of the cut. This corresponds to the case where one introduces absolutely reliable links to connect all left sides of all units to one node; the same procedure is applied to the right ends of the links of this cut. The procedure continues to the next simple cut.

The cut might be considered as "minimal" not only if it has a smallest number of units. In reliability the value of $\prod_{i \in B_k} q_i$ is taken into account (here B_k is the set of all units in the kth cut).

The more such variants of different subsets of nonintersecting simple paths and cuts are taken into account, the better are the bounds.

TABLE 9.5   Comparison of Bounds for Two Methods

P_i	L	L^*	h	U	U^*
0.01	0.0^86	0.0002	0.0002	0.0002	0.0004
0.1	0.0026	0.0199	0.0202	0.0219	0.0361
0.3	0.1123	0.1719	0.1981	0.2160	0.2601
0.5	0.4301	0.4375	0.5000	0.5699	0.5625
0.7	0.7840	0.7399	0.8016	0.8877	0.8281
0.9	0.9781	0.9639	0.9798	0.9974	0.9801
0.99	0.9998	0.9996	0.9998	0.9^84	0.9998

Remark: $0.0^86 = 0.000000006$ and $0.9^84 = 0.999999994$

The method introduced by Esary and Proschan becomes inefficient for networks of large dimension. At the same time, it does not allow one the opportunity to obtain approximations. The Litvak–Ushakov method allows one to obtain simple approximate bounds in a very effective way.

Let us numerically compare both methods (see Table 9.5), considering a bridge structure with identical units. In principle, one can combine both bounds and write

$$\max(L, L^*) \le h \le \min(U, U^*)$$

The only difficulty in using this expression for complex networks is in obtaining L and U by the Esary–Proschan method.

9.2.4   Method of Set Truncation

As we showed above, any structure function can be written in one of two equivalent forms: (9.11) and (9.12). First of all, recall the expressions for the union of intersecting (joint) events, each of which represents the existence of a corresponding path. From an elementary course in probability theory, one knows that

$$E\{\varphi(\mathbf{X})\} = E\left\{ \bigcup_{1 \le j \le N} \alpha_j(\mathbf{X}) \right\} = \sum_{1 \le i \le N} E\{\alpha_i\} - \sum_{1 \le i < j \le N} E\{\alpha_i(\mathbf{X}) \cap \alpha_j(\mathbf{X})\}$$
$$+ \sum_{1 \le i < j < k \le N} E\{\alpha_i(\mathbf{X}) \cap \alpha_j(\mathbf{X}) \cap \alpha_k(\mathbf{X})\} - \cdots$$
$$+ (-1)^{N+1} E\left\{ \bigcap_{1 \le i \le N} \alpha_i(\mathbf{X}) \right\} \qquad (9.20)$$

If one truncates this set using an odd s number of terms, an upper bound can be obtained. A lower bound can be obtained if the set is truncated using an even s number of terms. The more terms of the series taken into account, the more accurate is the result. Both these bounds are good when the p_i's are close to 0. In this case one can use a small number of terms in the sum.

An analogous expression can be obtained for highly reliable networks. In this case one evaluates unreliability rather than reliability. We briefly repeat the above arguments, but we now omit detailed explanations:

$$
\begin{aligned}
E\{\overline{\varphi(\mathbf{X})}\} = E\left\{\bigcup_{1\le j\le M} \overline{\beta_j(\mathbf{X})}\right\} = {}& \sum_{1\le i\le M} E\{\overline{\beta_j(\mathbf{X})}\} \\
& - \sum_{1\le i<j\le M} E\{\overline{\beta_i(\mathbf{X})} \cap \overline{\beta_j(\mathbf{X})}\} \\
& + \sum_{1\le i<j<k\le M} E\{\overline{\beta_i(\mathbf{X})} \cap \overline{\beta_j(\mathbf{X})} \cap \overline{\beta_k(\mathbf{X})}\} - \cdots \\
& + (-1)^{N+1} E\left\{\overline{\bigcap_{1\le i\le M} \beta_i(\mathbf{X})}\right\} \qquad (9.21)
\end{aligned}
$$

We obtain upper and lower bounds for unreliability in the same way as above: truncation of this set by an odd number of terms leads to an upper bound for unreliability, and truncation of this set by an even number of terms leads to a lower bound for unreliability. The more terms of the sum taken into account, the more accurate is the result.

We note that the method just described coincides with the so-called method of *absorption of the powers*. Corresponding to this method, a special operator $\mathscr{L}$ is used for the multiplication of polynomials. The meaning of this operator becomes clear from an example composed of two terms

$$
\begin{aligned}
\Pr\{\alpha_i(\mathbf{X}) \wedge \alpha_j(\mathbf{X}) = 1\} = \Pr\left\{\bigcap_{i\in A_j} x_i \wedge \bigcap_{i\in A_k} x_i = 1\right\} = {}& \prod_{i\in(A_j\cup A_k)} p_i \\
= \Pr\left\{\mathscr{L}\left(\prod_{i\in A_j} p_i \cdot \prod_{i\in A_k} P_i\right)\right\} = {}& \prod_{i\in A_j} p_i \cdot \prod_{i\in(A_k\setminus A_j)} p_i \\
= {}& \prod_{i\in A_k} p_i \cdot \prod_{i\in(A_j\setminus A_k)} p_i
\end{aligned}
$$

We introduce the notation

$$
a_j = \Pr\{\alpha_j(\mathbf{X}) = 1\} = \prod_{i\in A_j} p_i
$$

$$
b_k = \Pr\{\beta_k(\mathbf{X}) = 0\} = \prod_{i\in B_k} q_i
$$

We easily rewrite exact expressions represented above as

$$E\{\varphi(\mathbf{X})\} = \sum_{1 \leq j \leq N} a_j - \sum_{1 \leq j < k \leq N} \mathscr{L}(a_j \cdot a_k) + \cdots$$

$$+ (-1)^s \sum_{1 \leq j_1 < \cdots < j_s \leq N} \mathscr{L}(a_{j_1} \cdot a_{j_2} \cdot \cdots \cdot a_{j_s}) - \cdots$$

$$+ (-1)^N \mathscr{L}\left(\prod_{1 \leq j \leq N} a_i \right) \qquad (9.22)$$

or

$$E\{\varphi(\mathbf{X})\} = 1 - \sum_{1 \leq j \leq M} b_i + \sum_{1 \leq i < j \leq M} \mathscr{L}(b_j \cdot b_k) - \cdots$$

$$+ (-1)^s \sum_{1 \leq j_1 < \cdots < j_s \leq M} \mathscr{L}(b_{j_1} \cdot b_{j_2} \cdot \cdots \cdot b_{j_s}) - \cdots$$

$$+ (-1)^N \mathscr{L}\left(\prod_{1 \leq i \leq M} b_i \right) \qquad (9.23)$$

Naturally, (9.22) and (9.23) are totally equivalent to (9.20) and (9.21).

9.2.5 Generalization of Cut-and-Path Bounds

Both lower and upper bounds, either Esary–Proschan or Litvak–Ushakov, are valid not only for the analysis of network connectivity, but they are also valid for network units with various physical parameters. The reader can find details in Litvak and Ushakov (1984).

Let us assume that we observe some units which are characterized by a parameter w. This parameter may be a resistance, a capacitance, the time of traveling, the cost of transportation, and so on. For a complex network with such units, a function $\varphi(\mathbf{w})$ exists which is analogous to the above-considered structure function. This function is measured by a system parameter which coincides with that of a single system unit. For example, a network of resistances has its main parameter measured in ohms; a road network, with a known traveling time associated with each road, might be characterized by the traveling time from origin to destination, and so on. For such networks bounds analogous to the above-mentioned can be used.

Let us consider a parameter w and an analogue of a structure function: $\alpha(\mathbf{w}) = \alpha(w_1, \ldots, w_n)$ for a series connection of n units and $\beta(\mathbf{w}) = \beta(w_1, \ldots, w_m)$ for a parallel connection of m units. Assume the functions $\alpha(\mathbf{w})$ and $\beta(\mathbf{w})$ satisfy the following properties:

1. $\partial\alpha(w_i, w_j)/\partial w_i$ does not increase in w_i and does not decrease in w_j;
2. $\alpha(w_1, \ldots, w_n) \geq \alpha(w_1, \ldots, w_n, w_{n+1})$;
3. $\alpha(w_1, \ldots, w_j, 0, w_{j+1}, \ldots, w_n) = 0$;
4. $\alpha(w_1, \ldots, w_j, w_{max}, w_{j+1}, \ldots, w_n) = \alpha(w_1, \ldots, w_j, w_{j+1}, \ldots, w_n)$;
5. $\partial\beta(w_i, w_j)\partial w_i$ does not decrease in w_i and does not increase in w_j;
6. $\beta(w_1, \ldots, w_m) \leq \beta(w_1, \ldots, w_m, w_{m+1})$;
7. $\beta(w_1, \ldots, w_j, 0, w_{j+1}, \ldots, w_m) = \beta(w_1, \ldots, w_j, w_{j+1}, \ldots, w_m)$;
8. $\beta(w_1, \ldots, w_j, w_{max}, w_{j+1}, \ldots, w_m) = w_{max}$.

[Some additional restrictions on the first and second partial derivatives of functions $\alpha(\mathbf{w})$ and $\beta(\mathbf{w})$ are formulated in Litvak and Ushakov (1984).]

Let A_j and B_k be the corresponding sets of units in the jth simple path and the kth simple cut, respectively. Then, for a network, we can write the following lower and upper bounds of the Esary–Proschan type:

$$\alpha\big[\beta(\mathbf{w} \in B_1), \beta(\mathbf{w} \in B_2), \ldots, \beta(\mathbf{w} \in B_M)\big]$$

$$= L_{E} \leq \varphi(\mathbf{w}) \leq U_{EP}$$

$$= \beta\big[\alpha(\mathbf{w} \in A_1), \alpha(\mathbf{w} \in A_2), \ldots, \alpha(\mathbf{w} \in A_N)\big] \quad (9.24)$$

where N and M are the total number of network simple paths and simple cuts, respectively.

In the same case, the following lower and upper bounds of the Litvak–Ushakov type can be written:

$$\max_{1 \leq k \leq K^\alpha} \Big\{\beta\big[\alpha(\mathbf{w} \in A_1^k), \alpha(\mathbf{w} \in A_2^k), \ldots, \alpha(\mathbf{w} \in A_{N_k}^k)\big]\Big\} = L_{LU} \leq \varphi(\mathbf{w}) \leq U_{LU}$$

$$= \min_{1 \leq j \leq K^\beta} \Big\{\alpha\big[\beta(\mathbf{w} \in B_1^j), \beta(\mathbf{w} \in B_2^j), \ldots, \beta(\mathbf{w} \in B_{M_j}^j)\big]\Big\} \quad (9.25)$$

where K^α and K^β are different possible groups of nonintersecting simple path and cut sets, respectively; N_k and M_j are the total number of sets of disjoint paths and cuts in the corresponding sets.

There are also the "mirror"-like parameters. They are described in the above-given terms as

1. $\partial\alpha^*(w_i, w_j)/\partial w_i$ does not decrease in w_i and does not increase in w_j;
2. $\alpha^*(w_1, \ldots, w_n) \leq \alpha^*(w_1, \ldots, w_n, w_{n+1})$;
3. $\alpha^*(w_1, \ldots, w_j, 0, w_{j+1}, \ldots, w_n) = \alpha^*(w_1, \ldots, w_j, w_{j+1}, \ldots, w_n)$;
4. $\alpha^*(w_1, \ldots, w_j, w_{max}, w_{j+1}, \ldots, w_n) = w_{max}$;
5. $\partial\beta^*(w_i, w_j)/\partial w_i$ does not increase in w_i and does not decrease in w_j;
6. $\beta^*(w_1, \ldots, w_m) \geq \beta^*(w_1, \ldots, w_m, w_{m+1})$;
7. $\beta^*(w_1, \ldots, w_j, 0, w_{j+1}, \ldots, w_m) = 0$;
8. $\beta^*(w_1, \ldots, w_j, w_{max}, w_{j+1}, \ldots, w_m) = \beta^*(w_1, \ldots, w_j, w_{j+1}, \ldots, w_m)$.

For a network, the following lower and upper bounds of the Esary–Proschan type are valid:

$$\beta^*\left[\alpha^*(\mathbf{w} \in A_1), \alpha^*(\mathbf{w} \in A_2), \ldots, \alpha^*(\mathbf{w} \in A_N)\right]$$

$$= L_{\mathrm{EP}} \leq \varphi(\mathbf{w}) \leq U_{\mathrm{EP}}$$

$$= \alpha^*\left[\beta^*(\mathbf{w} \in B_1), \beta^*(\mathbf{w} \in B_2), \ldots, \beta^*(\mathbf{w} \in B_M)\right] \quad (9.26)$$

where N and M are the total number of network simple paths and simple cuts, respectively.

The following lower and upper bounds of the Litvak–Ushakov type are valid:

$$\max_{1 \leq j \leq K^\beta}\left\{\alpha\left[\beta^*(\mathbf{w} \in B_1^j), \beta^*(\mathbf{w} \in B_2^j), \ldots, \beta^*\left(\mathbf{w} \in B_{M_j}^j\right)\right]\right\}$$

$$= L_{\mathrm{LU}} \leq \varphi(\mathbf{w}) \leq U_{\mathrm{LU}}$$

$$= \min_{1 \leq k \leq K^\alpha}\left\{\beta^*\left[\alpha^*(\mathbf{w} \in A_1^k), \alpha^*(\mathbf{w} \in A_2^k), \ldots, \alpha^*\left(\mathbf{w} \in A_{N_k}^k\right)\right]\right\} \quad (9.27)$$

where K^α, K^β, N_k, and M_j were defined above.

We see that (9.24) and (9.25) are dual with respect to (9.26) and (9.27).

Examples of the parameters w and the corresponding functions α, β, α^*, and β^* are given in Table 9.6.

TABLE 9.6   Networks with Different Physical Nature

Number	Parameter W_c	$f\{w\}$	$F\{w\}$
1	Operation without "cutoff" failures	$f_1\{w\} = \bigcap_i w_i$	$F_1\{w\} = \bigcup_i w_i$
2	Operation without "short-circuit" failures	$f_2\{w\} = \bigcup_i w_i$	$F_2\{w\} = \bigcap_i w_i$
3	Probability of operation without "cutoff" failures	$f_1\{w\} = \prod_i w_i$	$F_1\{w\} = \coprod_i w_i$
4	Probability of operation without "short-circuit" failures	$F_2\{w\} = \coprod_i w_i$	$F_2\{w\} = \prod_i w_i$
5	Capacitance	$f_1\{w\} = \left(\sum_i w_i^{-1}\right)^{-1}$	$f_1\{w\} = \sum_i w_i$
6	Resistance	$f_3\{w\} = \sum_i w_i$	$F_2\{w\} = \left(\sum_i w_1^{-1}\right)^{-1}$
7	Random operating time to "cutoff" failure	$f_1\{w\} = \min_i w_i$	$F_1\{w\} = \max_i w_i$
8	Random operating time to a "short-circuit" failure	$f_2\{w\} = \max_i w_i$	$F_2\{w\} = \min_i w_i$
9	Carrying capacity of a transportation network	$f_1\{w\} = \min_i w_i$	$F_1 = \sum_1 w_i$
10	Cost of hauls in a transportation network	$f_2\{w\} = \sum_i w_i$	$F_2\{w\} = \min_i w_i$

Note: Here $\coprod w_i$ is a formula for the determination of the probability of faultless operation for a parallel connection of elements: $\coprod w_i = 1 - \prod(1 - w_i)$.

9.3   METHODS OF DECOMPOSITION

9.3.1   Moore – Shannon Method

Moore and Shannon (1956) analyzed the reliability of networks of a special type: a structure of these systems possesses a recurrent property. This means that a system with such a type of a structure consists of units whose structure repeats the structure of the system as a whole; in turn, each unit consists of units of the next level which repeat the same structure, and so on until one comes to the lowest level when units are "elementary," that is, cannot be decomposed into a next level of units. We explain this method by means of examples of a two-pole network, namely, a bridge structure. For an explanation, see Figure 9.5.

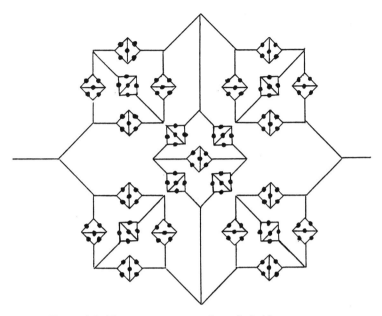

Figure 9.5. Recurrent construction of a bridge structure.

We will call a number of layers of "embedded" bridges within the structure by the *system order*. Consider a system of order n. The problem is to compute the system reliability if the "elementary" unit's reliability is known and equals p. (All units are assumed to be identical.) Let us consider a system of the first order, that is, a system whose units are "elementary." Its reliability is determined by the known connectivity function $p_1 = h(\mathbf{p}) = h(p)$ where $\mathbf{p} = (p_1, \ldots, p_n)$ is a vector whose components are the probabilities for respective units (see Section 9.1). Note that we can write $h(\mathbf{p}) = h(p)$ because all of the units are identical, and so the function h can be considered as a function of the argument p. For a second-order system, we write $p_2 = h(p_1) = h(h(p))$, and so on. Finally, for a system of the nth order, we have $p_n = h^{(1)}(h^{(2)}(\ldots h^{(n)}(p)\ldots))$ where the subscript denotes the iteration step [functions $h(p)$ are the same]. The iterations are represented by arrows in a plot of the function $h(p)$ (see Figure 9.6). The diagonal is the "structure function" of the unit and the S-shaped curve is the structure function of the bridge system (this is the probability of connectedness for the system).

If a unit has a probability of a failure-free state p, the system has the probability of connectivity p_1 [see the intersection of the first vertical dashed line with the curve $h(p)$]. We then consider a unit with a probability of a failure-free state equal to p_1 (the first horizontal dashed line). From this point, we build the second vertical dashed line until this intersects the curve $h(p)$. The new value $p_2 = h(p_1)$ is found. The procedure continues until the final result.

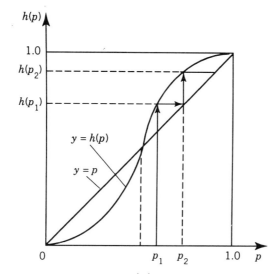

Figure 9.6. Function $h(p)$ for a bridge structure.

Incidentally, for a bridge structure, the function $h(p)$ is symmetric. Thus, if p equals 0.5, the system of any order has the same probability of connectedness equal to 0.5. If $p < 0.5$, then the higher the system order, the smaller is the system reliability. If $p > 0.5$, such a recurrent procedure allows one, in principle, to reach any desirable level of system reliability.

Of course, there are other examples of recurrent structures (see Figure 9.7). We also present the function $h(p)$ in the same figure.

Note that a parallel structure generates, under a recurrent procedure, a parallel system with a larger number of parallel units. Obviously, such a procedure improves the system reliability for any value of a unit reliability. A

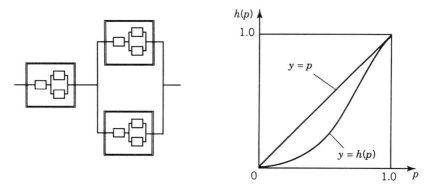

Figure 9.7. Example of another type of recurrent structure and its function $h(p)$.

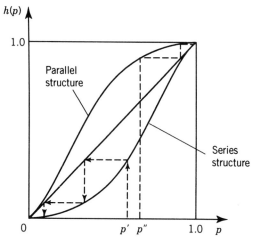

Figure 9.8. Functions $h(p)$ for series and parallel structures.

series structure generates, under a recurrent procedure, a series system with a larger number of units. In this case, on the contrary, the system reliability decreases with each step. [The corresponding functions $h(p)$ are presented in Figure 9.8.]

9.3.2 Bodin Method

Consider a network with a structure φ consisting of n units. Assume that the network can be represented as N two-pole modules, each of which might be considered as a "macro unit." We call this procedure a *modular decomposition*. Such a representation is typical in engineering design because of the natural decomposition of the process of design: different engineers, or groups of engineers, are responsible for designing different subsystems and parts of the system. Each such part of a system can be considered as a module. A modular representation of the system may be not unique (see Figure 9.9).

Using such a modular decomposition, a system with a structure function $\varphi(\mathbf{X})$ can be represented as a new structure $\zeta(\chi_1, \chi_2, \ldots, \chi_N)$, where χ_j is the structure function of the jth module: $\chi_j(\mathbf{X}_j)$. Here $\mathbf{X}_j$ is the set of initial units of the system such that

$$\chi_i \cap \chi_j = \varnothing \qquad \text{and} \qquad \bigcup_{1 \le j \le N} \mathbf{X}_j = \mathbf{X}$$

From the equivalency of the two forms of the system's representation, the following structure functions are equivalent:

$$\varphi(\mathbf{X}) = \psi(\chi_1(\mathbf{X}_1), \chi_2(\mathbf{X}_2), \ldots, \chi_N(\mathbf{X}_N))$$

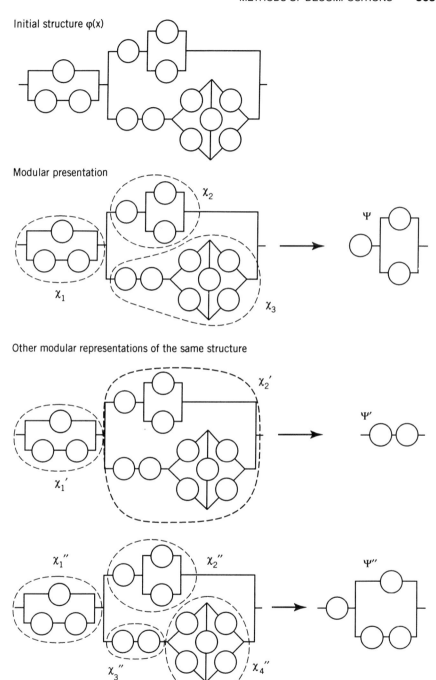

Figure 9.9. Example of a modular presentation of a particular structure.

Thus,

$$h_\varphi(\mathbf{p}) = h_\psi\big[h_{\chi_1}(\mathbf{p}_1), \ldots, h_{\chi_N}(\mathbf{p}_N)\big]$$

Recall that the structure is monotone and, consequently, if any p_i, $i = 1, \ldots, n$, is substituted by some $p_i' < p_i$, the value of the function $h(\mathbf{p})$ cannot increase. This rule can be applied to the modular function h's. Thus, we can immediately write

$$h_\varphi(\mathbf{p}) = \geq \left\{\begin{array}{l} L_\psi\big[h_{\chi_1}(\mathbf{p}_1), \ldots, h_{\chi_N}(\mathbf{p}_N)\big] \\ h_\psi\big[L_{\chi_1}(\mathbf{p}_1), \ldots, L_{\chi_N}(\mathbf{p}_N)\big] \end{array}\right\} \geq L_\psi\big[L_{\chi_1}(\mathbf{p}_1), \ldots, L_{\chi_N}(\mathbf{p}_N)\big]$$

$$(9.28)$$

Moreover,

$$L_\psi\big[L_{\chi_1}(\mathbf{p}_1), \ldots, L_{\chi_N}(\mathbf{p}_N)\big] \geq L_\varphi(\mathbf{p}) \tag{9.29}$$

Thus, the initial network might be represented as a new network with a smaller number of units (each unit is a module). For this new aggregated structure, one can apply lower and upper bounds of the Esary–Proschan or Litvak–Ushakov type. In this case one substitutes lower bounds for modules to obtain the lower bound of the network, and one substitutes upper bounds for modules to obtain the upper bound of the system structure function. Of course, these bounds are worse than the direct bounds obtained by both of the above-described methods, but they are obtained with much less computational difficulty. The reader can find a more detailed and careful discussion in Barlow and Proschan (1975).

Of course, analogous results will be obtained if lower and upper bounds of the Litvak–Ushakov type are used in the same manner. We present the final result without explanation:

$$h_\varphi(\mathbf{p}) \geq \left\{\begin{array}{l} L_\psi^*\big[h_{\chi_1}(\mathbf{p}_1), \ldots, h_{\chi_M}(\mathbf{p}_M)\big] \\ h_\psi\big[L_{\chi_1}^*(\mathbf{p}_1), \ldots, L_{\chi_M}^*(\mathbf{p}_M)\big] \end{array}\right\}$$

$$\geq L_\psi^*\big[L_{\chi_1}^*(\mathbf{p}_1), \ldots, L_{\chi_M}^*(\mathbf{p}_M)\big] \geq L_\varphi(\mathbf{p}) \tag{9.30}$$

9.3.3 Ushakov Method

The Bodin method requires special conditions for its use: a network must be represented in modular form. Below we suggest a method which is free of this restriction. This method is based on a network decomposition by simple paths and cuts.

Decomposition by Simple Paths Consider an initial two-pole network $G = (\varphi, \mathbf{X})$ with structure function φ. This structure includes n units: $\mathbf{X} = (x_1, \ldots, x_n)$. Extract from G any simple path, say A_k. Delete all units of the path A_k from G. As a result, a new subnetwork G_k^* appears. This subnetwork, in particular, may have "hanging" units (links); that is, some links may be connected to a remaining subnetwork only by one end and another end connects to nowhere. (It is even possible that some units may by totally separated from the remaining subnetwork.) In the result of path extraction, an initial network may also split into two separate parts. We delete all hanging units from further consideration and denote the remaining network as the subnetwork G_k. If some unit (link) of an extracted link connects to a node which belongs to A_k and G_k simultaneously, the corresponding node is divided into identical nodes: one belonging to A_k and the other to G_k. Thus, instead of the initial network G, depicted on Figure 9.9 we obtain a new network consisting of G_k and A_k connected in parallel. (see Figure 9.10). The set of units remaining after the first step may be smaller than the initial set because some hanging or isolated links can be omitted completely:

$$\mathbf{X'} = \left\{ \left(\bigcup_{j \in A_k} x_j \right) \vee \left(\bigcup_{j \in G_k} x_j \right) \right\} \subseteq \bigcup_{j \in G} x_j = \mathbf{X}$$

or, in other words, $|G_k| + |A_k| \leq |G| = n$.

For the new structure, we can write

$$\varphi(\mathbf{X}) \geq \varphi'(\mathbf{X'}) = \alpha_k(\mathbf{X}) \cup \varphi_k(\mathbf{X}) \tag{9.31}$$

It is clear that (9.31) follows in a simple way from the monotonicity property. Indeed, imagine that we restore all excluded, separated, and/or hanging units in the structure $G_k \cup A_k$ and add absolutely reliable connecting units between all divided nodes to join them. By this act we transform the structure $G_k \cup A_k$ again into the structure G. Adding absolutely reliable units can never decrease the structure function. From this (9.31) follows.

Of course, from (9.31) with no special proof, it follows that

$$h_k(\mathbf{p}) = 1 - \left[1 - E\{\varphi_k(\mathbf{X})\} \right] \left[1 - E\{\alpha_k(\mathbf{X})\} \right]$$
$$= 1 - \left[1 - h_k(\mathbf{p}) \right] \left[1 - h_\alpha(\mathbf{p}) \right] \leq E\{\varphi(\mathbf{X})\} = h(\mathbf{p}) \tag{9.32}$$

where h_α is the connectivity function of the path.

Thus, a laborious computation of the probability of a network's connectivity is reduced to a simple, easily solvable problem. For example, the structure in Figure 9.9 reduced to the connection of two substructures in parallel (see Figure 9.11).

Initial network with appointed units

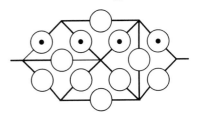

State of the network units after the path selection

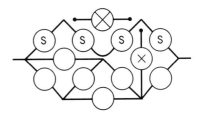

Final structure without superfluous units

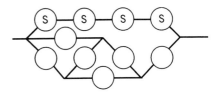

Legend: (•) = appointed unit

(S) = selected unit

(×) = hanging unit

Figure 9.10. Example of a path extraction from a network.

(⊗) = isolated unit

Obviously, there are many different ways to choose a path subjected to extraction from the initial network. Depending on the choice, we obtain different qualities of lower bounds: better or worse. (We remark that the same nonuniqueness of a solution exists for Bodin's method.)

In general, decomposition problems relate to an art more than to a routine. At any rate, we advise one to try to choose a shortest path, or a very short path for A_k.

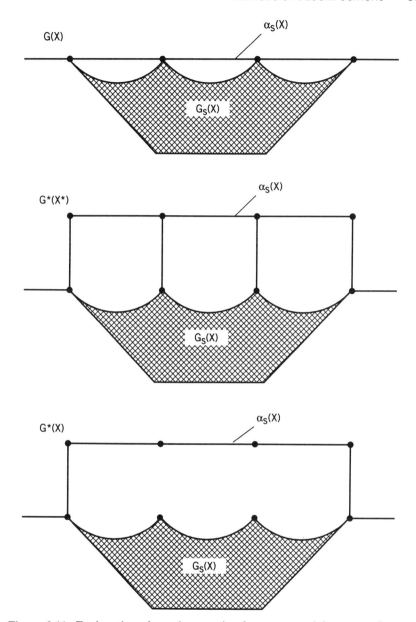

Figure 9.11. Explanation of a path extraction from a network in a general case.

If a network is very complex, the procedure can be continued. Incidentally, the systematic application of the described procedure leads one to the Litvak–Ushakov lower bound.

Decomposition by Simple Cuts   Consider again the network $G = (\varphi, \mathbf{X})$. Extract from G any simple cut, say B_k. Thus cut divides a network into three parts: a "left" subnetwork, the dividing cut itself, and a "right" subnetwork.

Here we call a subnetwork "left" if it includes the "input" network pole, and we call a subnetwork "right" if it includes the "output" pole. (In particular cases each of these subnetworks may consist only of a pole itself, if the dividing cut connects with a pole.) This cut has a set of "left" nodes which are closer to the "input pole" and a set of "right" nodes which are closer to the "output pole" of the network. Connect all "left" nodes of the cut by perfectly reliable units. This tightens them into one node. Do the same with the "right" nodes. The initial network G is now transformed into three fragments: the "left" two-pole subnetwork G'_k (its poles are the input pole of

Initial network with appointed units

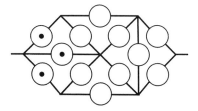

State of the network units after the cut selection

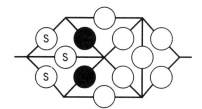

Final structure without superfluous units

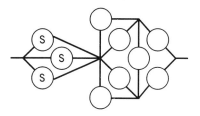

Legend:　(•) = appointed unit

　　　　　(S) = selected unit

 = unit tightened into a node

Figure 9.12. Example of a cut extraction from a network.

the initial network and the "left" pole of the cut B_k), the cut B_k, and the "right" two-pole subnetwork G_k'' (its poles are the "right" pole of the cut and the output pole of the initial network). An example of such a decomposition is presented in Figure 9.12 where the initial network corresponds to the network in Figure 9.9.

These subnetworks—G_k' and G_k''—may have a "short connection" of units after introducing absolutely reliable units. This happens, for example, if two

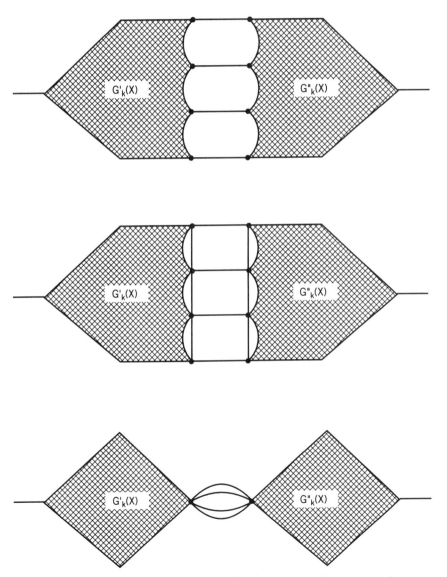

Figure 9.13. Explanation of a cut extraction from a network in a general case.

"tightened" nodes have a unit between them. Such units are excluded from further consideration. In the example in Figure 9.12, x_4, x_6, and x_8 are such units. Thus, the new structure is built from the initial one by introducing several perfectly reliable units. Using the monotonicity property, we immediately obtain

$$\varphi(\mathbf{X}) \le \varphi''(\mathbf{X}'') = \varphi''_k(\mathbf{X}'') \cup \beta_k(\mathbf{X}'') \cup \varphi''_k(\mathbf{X}'') \tag{9.33}$$

Of course, from (9.33) it immediately follows that

$$h_k(\mathbf{p}) = \mathrm{E}\{\varphi'_k(\mathbf{X})\}\,\mathrm{E}\{\beta_k(\mathbf{X})\}\,\mathrm{E}\{\varphi''_k(\mathbf{X})\} = h'_k(\mathbf{p})h_\beta(\mathbf{p})h''_k(\mathbf{p})$$

$$\ge \mathrm{E}\{\varphi(\mathbf{X})\} = h(\mathbf{p}) \tag{9.34}$$

where h_β is the connectivity function of the cut.

Thus, the computational problem relating to the probability of a network's connectedness is again reduced to a simple solvable problem. For example, the structure in Figure 9.13*a* turns into the structure displayed in Figure 9.13*c*.

As in the above-mentioned case, there are many different ways to choose the cut B_k. Depending on the choice, we obtain better or worse lower bounds. Again the same advice might be given : try to choose the minimal, or a very "thin" cut, for B_j.

For very complex networks the procedure may be continued. Again, we remark that the systematic application of this procedure leads to the Litvak–Ushakov upper bound.

9.4   MONTE CARLO SIMULATION

We intend to address Monte Carlo simulation in more detail in the forthcoming second volume of this book. But this method is so essential in the numerical analysis of network reliability that a very brief review is placed here. We consider only the case of highly reliable networks for which a direct use of Monte Carlo simulation takes too much computer time and requires a large computer memory. There are some special methods of accelerating a Monte Carlo simulation but we will not discuss them here.

9.4.1   Modeling with Fixed Failed States

To simplify our explanation, we analyze the case where all network units are identical and independent. Of course, this is a rough picture of reality. But all of these ideas could be easily extended or adapted to a more general case. Furthermore, for many practical purposes, we can approximate a real network with the help of such a simplified version.

Consider a network consisting of n units. If the units are highly reliable and if the network cuts are "thick" enough, a direct application of a Monte Carlo simulation cannot be considered as a reasonable approach because failures appear very seldom. Indeed, excessive computer time is needed for simulating a significant sample: we need at least several tens of network failures, that is, events in which we are interested for a confident statistical analysis. An effective method of investigation of such networks involve mixing both analytical and simulation approaches.

First, we analytically compute the probabilities of different system states with a fixed number of failed units. These computations can be based on simple combinatorial methods. Then a simulation can be used to determine the conditional probabilities of a system's failure for different states with a fixed number of failed units. The description of the general procedure follows.

1. Analytically calculate the probabilities of states $H_{(k)}$, which are characterized by exactly k failed units

$$P_{(k)} = \binom{n}{k} q^k p^{n-k} \tag{9.35}$$

2. Use a Monte Carlo simulation to estimate the conditional probabilities $\hat{\Phi}_{(k)}$ that the system state $H_{(k)}$ is operational.
3. Compute the total probability of a successful system operation:

$$P = P_{(0)} + \sum_{1 \le k \le n} \hat{\Phi}_{(k)} P_{(k)} \tag{9.36}$$

If the smallest simple cut of the network has j units, we can use the simplified formula

$$P = \sum_{0 \le k \le j} P_{(k)} + \sum_{j+1 < +k < +n} \hat{\Phi}_{(k)} P_{(k)}$$

This mixture of an analytical method with simulation becomes more effective with increasing j.

In considering networks with identical units, we can use a binomial distribution for the calculation of the probabilities H_k. A binomial distribution is unimodal which, for highly reliable systems with a large number of units n and a probability of unit failure $q \ll 1/n$, allows one to find the left and right margins outside of which a calculation of the conditional probabilities $\hat{\Phi}_{(k)}$ is unreasonable because of its negligible influence on the final result.

The conditional probability $\hat{\Phi}_{(k)}$ is estimated by

$$\hat{\Phi}_{(k)} = \frac{\nu_k(N_k)}{N_k} \tag{9.37}$$

where N_k is the total number of experiments with k failed units and $\nu(N_k)$ is the number of successful trials.

We remark that in this case it is possible to build an optimal statistical experiment of a restricted size to obtain each $\hat{\Phi}_{(k)}$ with a high degree of confidence in order to obtain the best result for the network as a whole.

9.4.2   Simulation Until System Failure

This method is more effective because there are no "noninformative" trials at all. In this case units are extracted from the network one by one until the system fails. The failure criterion may be chosen as the loss of connectedness between two poles or in another way (e.g., a measure of the total connectedness of all nodes might be chosen). When the system fails, the number of failed units is put in a special file in the computer memory. After a sufficient number of experiments N, we use the estimating function:

$$\hat{\Phi}_{(k)} = \frac{1}{N} \sum_{1 \leq s \leq N} \delta(k_s) \tag{9.38}$$

where

$$\delta(k_s) = \begin{cases} 1 & \text{if } k_s \leq k \\ 0 & \text{otherwise} \end{cases}$$

Here the monotonicity of the network structure is used very directly: the failure of a new unit might only deteriorate the operating system and cannot restore the failed system. The final calculation is performed with the help of (9.36).

Obviously, the number of trials needed to obtain the conditional probabilities $\hat{\Phi}_{(k)}$ does not depend on the reliability level of the network units. The only dependence is on the "depth" of each experiment until a network failure. The number of trials in this case depends almost linearly on the dimension of the network.

Both described methods for finding $\hat{\Phi}_{(k)}$ can be applied when system units are different. In this case, instead of computing the binomial probabilities in (9.35), we compute

$$P_{(k)} = \sum_{\mathbf{d}} \prod_{1 \leq j \leq n} p_j^{d_j} q_j^{1-d_j}$$

where $\mathbf{d} = (d_1, d_2, \ldots, d_n)$ is a vector with components d_i equal to 1 when the ith unit is functioning and 0 when it has failed.

The conditional probability is calculated as

$$\hat{\Phi}_{(k)} = \frac{1}{N} \sum_{1 \le s \le N} \delta[k_s(\mathbf{d}_s)]$$

where k_s is the number of unit failures which have led to a system failure when a random sequence of failures forms $\mathbf{d}_s$. In this general case, $\mathbf{d}_s$ is formed by the random exclusion of the network units with respect to their probabilities of failure. If the probability of the ith network unit failure is q_i, then at the first step this unit is excluded with probability

$$P_i^* = \frac{q_i}{\sum_{1 \le i \le n} q_i}$$

Assume that at the first step the unit i is chosen and deleted from the network. Then at the second step a new unit, say j, is excluded with probability

$$P_j^* = \frac{q_j}{\sum_{1 \le k \le n} q_k - q_i}$$

and so on. In this case the procedure of statistical evaluation is not so simple, but it is simpler and more effective than a direct simulation.

CONCLUSION

As we mentioned above, the first work dedicated to the problem of two-pole network connectivity was published by Moore and Shannon (1956). They analyzed the connectivity of so-called "hammock-like" networks.

Some results about the connectivity of complete graphs with "random links" can be found in Kelmans (1970) and Stepanov (1970).

A series of works by Esary and Proschan (1962, 1963, 1970) is dedicated to obtaining network bounds based on the enumeration of all simple paths and cuts. A decomposition method with the use of this same idea was proposed by Bodin (1970). A detailed discussion of this subject is presented in Barlow and Proschan (1975).

The bounds based on disjoint simple paths and cuts were obtained by Ushakov and Litvak (1977). The same bounds can be found in Colbourn (1987). Computational aspects of the Litvak–Ushakov bounds are considered in Kaustov et al. (1986).

Litvak and Ushakov (1984) generalized both the Esary–Proschan and Litvak–Ushakov bounds on networks of units with a different physical nature. Some additional properties of analogous bounds are also discussed there.

Some details concerning Monte Carlo simulations are found in Ushakov (1985, 1994).

We did not consider multiterminal (multipole) networks. This subject lies outside the scope of this book. Bounds on the total connectivity of multiterminal networks in terms of disjoint trees were found by Lomonosov and Polessky [Lomonosov and Polessky (1971, 1972) and Polessky (1971)]. The reader can find related results in Satynarayana (1980), Colbourn (1987) and Reinschke and Ushakov (1988).

We also avoided the consideration of complex networks with regular structures which are characterized by a permanent systematic rule of system recurrent enlarging. These systems, in principle, remind one of recurrent constructing as in the case considered by Moore and Shannon but more general. These results are interesting but cumbersome. A reader interested in this problem is advised to see Gadasin and Ushakov (1975), Kozlov and Ushakov (1979), and Ushakov (1994).

An in-depth review on some special aspects of the problem, can be found in Agraval and Barlow (1984).

REFERENCES

Agraval, A., and R. E. Barlow (1984). A survey on network reliability and domination theory. *Oper. Res.*, vol. 32, no. 2.

Barlow, R. E., and F. Proschan (1975). *Statistical Theory of Reliability and Life Testing: Probabilistic Models*. New York: Holt, Rinehart, and Winston.

Barlow, R. E., and A. S. Wu (1978). Coherent systems with multistate components. *Math. Oper. Res.*, no. 4.

Birnbaum, Z. W., and J. D. Esary (1965). Modules of coherent binary systems. *SIAM J. Appl. Math.*, no. 13.

Birnbaum, Z. W., J. D. Esary, and S. C. Saunders (1961). Multicomponent systems and structures, and their reliability. *Technometrics*, no. 3.

Bodin, L. D. (1970). Approximation to system reliability using a modular decomposition. *Technometrics*, vol. 12, pp. 335–344.

Colbourn, C. (1987). *The Combinatorics of Network Reliability*. Oxford: Oxford Univ. Press.

Esary, J. D., and F. Proschan (1962). The reliability of coherent systems. In *Redundancy Techniques for Computing Systems*, R. H. Wilcox and C. W. Mann, eds., pp. 47–61. Washington, DC: Spartan Books.

Esary, J. D., and F. Proschan (1963). Coherent structures of nonidentical components. *Technometrics*, vol. 5, pp. 183–189.

Esary, J. D., and F. Proschan (1970). A reliability bound for systems of maintained interdependent components. *J. Amer. Statist. Assoc.*, vol. 65, pp. 329–338.

Gadasin, V. A., and I. A. Ushakov (1975). *Reliability of Complex Information and Control Systems* (in Russian). Moscow: Sovietskoe Radio.

Kaustov, V. E., E. I. Litvak, and I. A. Ushakov (1986). The computational effectiveness of reliability estimates by method of non-link-intersecting chains and cuts. *Soviet J. Comput. Systems Sci.*, vol. 24, no. 4, pp. 59–62.

Kelmans, A. K. (1970). On estimation of probabilistic characteristics of random graphs. *Automat. Remote Control*, vol. 32, no. 11.

Kozlov, B. A., and I. A. Ushakov (1979). *Handbuch zur Berechnung der Zuverlassigkeit* (in German). Munich: Springer.

Litvak, E. I., and I. A. Ushakov (1984). Estimation of parameters of structurally complex systems. *Engrg. Cybernet.* (USA), vol. 22, no. 4, pp. 35–49.

Lomonosov, M. V., and V. Polessky (1971). An upper bound for the reliability of information networks. *Problems Inform. Transmission*, vol. 7, pp. 337–339.

Lomonosov, M. V., and V. Polessky (1972). Lower bound of network reliability. *Problems Inform. Transmission*, vol. 8, pp. 118–123.

Moore, M. E., and C. E. Shannon (1956). Reliable circuits using less reliable relays. *J. Franklin Inst.*, vol. 262, part 1, pp. 191–208, and vol. 262, part 2, pp. 281–297.

Polessky, V. (1971). A lower bound for the reliability of information networks. *Problems Inform. Transmission*, vol. 7, pp. 165–171.

Reinschke, K., and I. A. Ushakov (1988). *Zuverlassigkeitsstructuren* (in German). Munich: Oldenberger.

Satyanarayana, A. (1980). *Multi-Terminal Network Reliability*. Berkeley: Operations Research Center.

Stepanov, V. E. (1970). On the probability of a graph connectivity. *Probab. Theory Appl.*, vol. 15, no. 1.

Ushakov, I. A., ed. (1985). *Reliability of Technical Systems*: *Handbook* (in Russian). Moscow: Radio i Svyaz.

Ushakov, I. A., ed. (1994). *Handbook of Reliability Engineering*. New York: Wiley.

Ushakov, I. A., and E. I. Litvak (1977). An upper and lower estimate of the parameters of two-terminal networks. *Engrg. Cybernet.* (USA), vol. 15, no.1.

EXERCISES

9.1 Write an expression for the probability of connectedness of a bridge structure if the decomposition is made by unit 5.

9.2 How many simple paths does a series connection of n units have? A parallel connection of n units?

9.3 How many simple cuts does a series connection of n units have? A parallel connection of n units?

9.4   How do you interpret the following:
   (a) A simple path in a system of type "k of n";
   (b) A simple cut in a system of type "k of n";
   (c) How many simple paths does such a system have?
   (d) How many simple cuts does such a system have?

9.5   For the structure depicted in Figure E9.2, construct a structure for obtaining the upper bound of the probability of connectedness by "shortening" links 5 and 9.

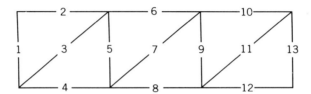

Figure E9.2.

SOLUTIONS

9.1   The solution can be easily found with the help of Figure E9.1. The resulting expression is

$$P = q_5\{p_4[1 - q_3(1 - p_1 p_2)]\} + p_5\{1 - [q_2(p_1(1 - p_3 p_4))]\}$$

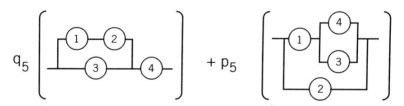

Figure E9.1.

9.2   A series connection itself represents a single path. A parallel connection of n units has n paths. Each path consists of one unit.

9.3 A series connection of n units has n cuts. Each cut consists of one unit. A parallel connection itself represents a single cut.

9.4 (a) Any group of k units in this system represents a simple path.
(b) Any group of $n - k + 1$ units in this system represents a simple cut.
(c) There are

$$\binom{n}{k} = \frac{n!}{(n-k)!\,k!}$$

simple paths in this system.
(d) There are

$$\binom{n}{n-k+1} = \frac{n!}{(n-k+1)!\,(k-1)!}$$

simple paths in this system.

9.5 The result is depicted in Figure E9.3.

Figure E9.3.

CHAPTER 10

OPTIMAL REDUNDANCY

10.1 FORMULATION OF THE PROBLEM

One very effective method for the improvement of reliability is redundancy including the use of additional units, circuits, and blocks. This method is especially convenient when the principal solution of the system design is found: the use of redundancy usually does not lead to a serious change in the system's structure. But the use of extra units entails additional expense. Naturally, one should try to find the least expensive method of reliability improvement; that is, a designer faces two dual problems:

1. Find an optimal allocation of redundant units among different subsystems to fulfill some given request according to some specified reliability index and make the total cost of the system as low as possible.
2. Find an optimal allocation of redundant units among different subsystems which maximizes the chosen reliability index with some specified restriction on the total cost of the system.

The choice of restrictions depends on the character of the specific engineering problem. Sometimes one encounters several different restrictions: cost, volume, weight, and so forth.

Consider a series system composed of n independent redundant groups which we will sometimes call subsystems. A redundant group is not necessarily a separate part of a system. In this context, it may be a group of units of the same type which uses the same redundant units. For instance, in spare-parts problems a redundant group might be a set of identical units allocated throughout the entire system in quite different places.

10.1.1 Simplest Problems

In most cases we deal with a system reliability index in multiplicative form: as the probability of a successful operation, an availability coefficient, and so on. We denote this index by $R(\mathbf{X})$:

$$R(\mathbf{X}) = R(x_1, \ldots, x_n) = \prod_{1 \leq i \leq n} R_i(x_i) \qquad (10.1)$$

where $\mathbf{X} = (x_1, \ldots, x_n)$ is the set of the system's redundant units x_i's of the ith type, $1 \leq i \leq n$, and the $R_i(x_i)$ are the reliability indexes of the ith redundant group (subsystem).

As it follows from (10.1), this reliability index is the probability of success for a series system consisting of independent subsystems.

Sometimes it is more convenient to present this index in an additive form

$$L(\mathbf{X}) = L(x_1, \ldots, x_n) = \sum_{1 \leq i \leq n} L_i(x_i) \qquad (10.2)$$

where $L(\mathbf{X}) = -\ln R(\mathbf{X})$ and $L_i(x_i) = -\ln R_i(x_i)$.

If a system is highly reliable, that is,

$$Q_i(x_i) = 1 - R_i(x_i) \ll \frac{1}{n} \qquad (10.3)$$

or, equivalently,

$$Q(\mathbf{X}) = Q(x_1, \ldots, x_n) \ll 1 \qquad (10.4)$$

one can use the approximation

$$Q(\mathbf{X}) = Q(x_i, \ldots, x_n) \approx \sum_{1 \leq i \leq n} Q_i(x_i) \qquad (10.5)$$

The cost function may be arbitrary but usually it is an additive function of the form

$$C(\mathbf{X}) = C(x_1, \ldots, x_n) = \sum_{1 \leq i \leq n} C_i(x_i) \qquad (10.6)$$

and, moreover, for most practical problems

$$C_i(x_i) = c_i x_i \qquad (10.7)$$

Considering the optimal redundancy problem for one restricted factor, it is possible to formulate the two following problems. The so-called direct problem of optimal redundancy can be written as

$$\min_{\mathbf{X}} \left[C(\mathbf{X}) | R(\mathbf{X}) \geq R_0 \right] \tag{10.8}$$

and the inverse problem as

$$\max_{\mathbf{X}} \left[R(\mathbf{X}) | C(\mathbf{X}) \leq C_0 \right] \tag{10.9}$$

where R_0 and C_0 are given restrictions for the specific problems.

Of course, similar problems can be formulated for the two previous goal functions $L(\mathbf{X})$ and $Q(\mathbf{X})$. In these cases the smaller the value of the goal function, the higher is the reliability of the system. Thus, the only difference in the formal expression is in the sign of the inequality in (10.8) and the operator "min" instead of "max" in (10.9). (We ask the reader to formulate these two problems in Exercise 10.1.)

Of course, there are many other goal functions and some of them will be investigated below.

10.1.2 Several Restrictions

In the presence of several restricting factors, one can consider the two following problems:

1. Find an optimal solution for the chosen goal function under specified restrictions on the other goal function.
2. Find "the best" (in some special sense which will be explained below) set of competitive factors with specified restrictions on the single chosen goal function.

The first problem is an ordinary optimization problem with restrictions. The second problem is more difficult in a computational sense but it contains nothing extraordinarily new in a mathematical sense. We will consider two main problems of this kind:

(a) $$\max_{\mathbf{X}} \left[R(\mathbf{X}) | C_1(\mathbf{X}) \leq C_{01}, \ldots, C_M(\mathbf{X}) \leq C_{0M} \right] \tag{10.10}$$

where the C_{0j}'s, $1 \leq j \leq M$, are given restrictions on the different kinds of expenditures for the system as a whole (we can assume that smaller C_{0j}'s are more admissible); and

(b) $$\min_{\mathbf{X}} \left[C(\mathbf{X}) | R_1(\mathbf{X}) \geq R_{01}, \ldots, R_M(\mathbf{X}) \geq R_{0M} \right] \tag{10.11}$$

where the R_{0j}'s, $1 \leq j \leq M$, are the required reliability index values for different subsystems of the system under investigation.

The second problem belongs to the class of so-called problems of multicriteria optimization. In this case a designer has several restrictions on different system's parameters such as cost, weight, volume, and so on. It is impossible, in mathematical terms, to determine "a best solution." All solutions which we obtain are, in some sense, incomparable: a cheaper system can weigh more than a more expensive one; a very compact system can be heavier and more expensive than another with a larger volume, but cheaper; and so forth. Thus, the problem lies outside the scope of mathematics: the problem of choosing a preferable object is up to the decision maker.

But then what is the mathematical problem in this case? It consists of selecting the so-called nonimprovable solutions. This means that none of the selected variants (solutions) is strictly better than another.

A set of multicriteria problem solutions is called the *Pareto set*. In mathematical terms one can write the problem in the form

$$\underset{\mathbf{X}}{\text{MIN}} \left\{ C_1(\mathbf{X}), \ldots, C_M(\mathbf{X}) | R(\mathbf{X}) \geq R_0 \right\} \tag{10.12}$$

where the symbol MIN denotes optimization in the Pareto sense.

10.1.3 Practical Problems

One might remark that redundancy is not often encountered in practice. But such a real problem as an optimal supply of technical systems with spare parts is, in fact, an optimal redundancy problem. An exact solution to this problem with respect to all specific practical restrictions and specific goal functions is very complicated.

A solution to this problem should be *practical*, it should reflect the real situation, and it should be based on a sound mathematical background. To satisfy all of these restrictions is an extremely difficult problem. Solution of such kind of problems is more art than science. This part of the problem will be considered in Chapter 13 dedicated to heuristic methods. Here we pay attention to the mathematical side of the problem.

10.2 OPTIMAL REDUNDANCY WITH ONE RESTRICTION

10.2.1 Lagrange Multiplier Method

One of the first attempts to solve the optimal redundancy problem was based on the classical *Lagrange multiplier method* (LMM). Of course, strictly speaking, this method is not appropriate in this case as a system's reliability and cost are represented as functions of discrete arguments x_i (number of redundant units), and the restrictions on accessible resources (or on required values of reliability) are fixed in the form of inequalities. The LMM can sometimes be used to obtain approximate solutions and/or to find bounds of the exact solutions in the case of the discrete optimization problem.

To solve the direct optimal redundancy problem, we construct the Lagrange function, $\Lambda(\mathbf{X})$:

$$\Lambda(\mathbf{X}) = C(\mathbf{X}) + \lambda^* L(\mathbf{X}) \tag{10.13}$$

where λ^* is called a Lagrange multiplier. The goal is to minimize $\Lambda(\mathbf{X})$, taking into account the equation $L(\mathbf{X}^*) = L^o$. Thus, the system of equations to be solved is

$$\frac{d\Lambda(\mathbf{X})}{dx_i} = \frac{dC(\mathbf{X})}{dx_i} + \lambda^* \frac{dL(\mathbf{X})}{dx_i} = 0 \tag{10.14}$$

for all $i = 1, 2, \ldots, n$, and the equation

$$L(\mathbf{X}^*) = L^o$$

The unknown arguments are: the x_i^*'s, $i = 1, 2, \ldots, n$, and λ^*.

Usually, the function $C(\mathbf{X})$ is assumed to be separable. If the function $L(\mathbf{X})$ is also separable and differentiable, the first n equations of (10.14) can be rewritten as

$$\frac{dL_1(x_1)}{c_1 \, dx_1} = \cdots = \frac{dL_n(x_n)}{c_n \, dx_n} = \lambda \tag{10.15}$$

where $\lambda = 1/\lambda^*$ and

$$L_i(x_i) = -\ln R_i(x_i). \tag{10.16}$$

On a physical level, (10.16) means that for separable functions $L(\mathbf{X})$ and $C(\mathbf{X})$, the optimal solution corresponds to an equality of relative increments of reliability of each redundant group for an equal and small resource investment.

In general, (10.16) yields no solution in closed form but only suggests the following algorithm:

1. At some arbitrary point $x_1^{(1)}$, calculate the derivative for some fixed redundant group, say the first one:

$$\frac{dL_1(x_1)}{c_1 \, dx_1} \bigg|_{x_1^{(1)}} = -\lambda^{(1)}$$

REMARK. Of course, one would like to choose a value of $x_1^{(1)}$ close to an expected optimal solution. For example, if you consider spare parts for time t and know that the unit MTTF is T, this value should be a little larger than t/T. In other words, this choice should be based on engineering experience and intuition.

2. For the remaining redundant groups, calculate the values of $x_i^{(1)}$ which satisfy the conditions

$$\left.\frac{dL_i(x_i)}{dx_i}\right|_{x_i^{(1)}} = c_i \lambda^{(1)}$$

3. Calculate

$$L\left(\mathbf{X}^{(1)}\right) = \sum_{1 \le i \le n} L_i\left(x_i^{(1)}\right) = L^{(1)}$$

4. Compare $L^{(1)}$ with L^o. If $L^{(1)} < L^o$ choose $x_i^{(2)} < x_i^{(1)}$; if $L^{(1)} > L^o$ choose $x_i^{(2)} > x_i^{(1)}$. After selecting a new value of $x_i^{(2)}$, return to step 1 of the algorithm.

A stopping point criterion corresponds to the condition

$$\left|L\left(\mathbf{X}^{(N)}\right) - L^o\right| \le \varepsilon$$

where ε is some specified possible error.

A solution to the inverse optimal redundancy problem is left as an exercise.

Unfortunately, there is only one case where the direct optimal redundancy problem can be solved in closed form. This is the case of a highly reliable system with active redundancy where

$$L_i(x_i) = -\ln R_i(X_i) = -\ln[1 - Q_i(x_i)] \approx Q_i(x_1) = q_i^{x_i}$$

But even for this case the solution is somewhat awkward:

$$x_i^o = \frac{1}{|\ln q_i|} \ln\left[Q_0 \frac{c_i}{\ln q_i} \left(\sum_{1 \le i \le n} \frac{c_i}{\ln q_i}\right)^{-1}\right] \tag{10.17}$$

The inverse optimal redundancy problem cannot be solved, even in this simplest case. One might use an iterative procedure which involves solving the direct problem at each stage. We explain this in more detail. One first solves the direct optimal redundancy problem for some fixed $Q_0^{1(1)}$ and finds $\mathbf{X}_0^{(1)} = (x_1^{(1)}, \ldots, x_n^{(1)})$. Then $\mathbf{X}^{(1)}$ is substituted into the equation $L(\mathbf{X}^*) = L^o$. If $C(\mathbf{X}^{(1)}) = C^{1(1)} < C_0$, one must choose a new value of $Q_0^{(2)} < Q_0^{(1)}$; if $C(\mathbf{X}^{(1)}) = C^{1(1)} > C_0$, one must choose a new value of $Q_0^{(2)} < Q_0^{(1)}$. This procedure continues until the result is obtained with a required accuracy.

The solutions obtained by the LMM are not necessarily discrete because we assumed that all functions are continuous themselves. But the nature of the problem is essentially discrete. An immediate question arises: Is it possible to use an integer extrapolation for each noninteger x_i? If this extrapolation is possible, is the obtained solution optimal?

Unfortunately, even if one tries all possible combinations of extrapolated solutions for the x_i's, namely, substituting $[x_i]$ and $[x_i] + 1$, it may happen that there is no optimal solution among all 2^n of these combinations!

We give a simple example. Consider a system of two units. We need to maximize the reliability index under a specified restriction on the redundant units, say limited to at most 5 cost units. Let the solutions (obtained with the help of the LMM with respect to some specified reliability parameters) be $x^1 = 0.8$ and $x_2 = 0.2$. The possible integer roundings are: $(0, 0)$, $(0, 1)$, $(1, 0)$, and $(1, 1)$. Let the unit costs be $c_1 = 10$ and $c_2 = 1$. This means that the use of a single redundant unit of the first type will violate the given restriction on the total cost of the system. Thus, the third and fourth solutions are improper because of the restriction violation. As the solution one would choose the second one because it dominates the first solution. Thus, the only possible solution among the rounded pairs is $(0, 1)$. But the global optimal solution is $(0, 5)$ which was not even considered!

10.2.2 Steepest Descent Method

The *steepest descent method* (SDM) is based on a very simple idea: to reach the extremum of a multidimensional unimodal function, one should move from an arbitrary point in the direction of the maximal gradient of the goal function.

Let us explain this statement with a simple analogy. Suppose that a skier stands on a slope (his initial position) leading to a valley. Suppose that the valley has a single lowest point and any path has no pit or ditch (no local extremum). The question is: What is the fastest way to ski down? It would seem that the skier should move in the direction of maximal possible incline at each point on his path to the bottom. This direction coincides with the gradient of the function at each point.

This analog is useful in the case of continuous functions of continuous arguments. But in the case of optimal redundancy, all arguments are discrete. If we continue the analogy with the skier, one sees that there are restrictions on the choice of the direction of the skier's movement: he can move only in the north–south or east–west directions and he can change direction only at the vertexes of a discrete lattice with specified steps. This means that at each vertex one should use the direction of the largest partial derivative. Because of this, one sometimes speaks of the *method of coordinate steepest descent*.

This idea may be applied to the optimal redundancy problem. At the first step of the optimization procedure, one determines which redundant group (subsystem) of the system produces the largest ratio of the system's reliability index to cost. Then one adds a redundant unit to the chosen group and goes to the next step. The procedure continues in accordance with the same rule.

For reliability optimization let us use the goal function

$$L(\mathbf{X}) = -\ln R(\mathbf{X}) = - \sum_{1 \le i \le n} \ln R_i(x_i) \tag{10.18}$$

In this case both functions $C(\mathbf{X})$ and $L(\mathbf{X})$ are separable, and the relative increment can be calculated in a very simple way.

For convenience of further exposition, let us introduce the following notation: $\mathbf{X}_i = \mathbf{X} + e_i$ where e_i is the unit vector whose components all equal 0, except the ith one which equals 1. In other words, if $\mathbf{X} = (x_1, \ldots, x_i, \ldots, x_n)$, then $\mathbf{X}_i$ is the vector with the ith component increased by 1: $\mathbf{X} = (x_1, \ldots, x_i + 1, \ldots, x_n)$.

Suppose at the Nth step the optimization process reaches a point $\mathbf{X}^{(N)}$. This means that during the previous steps, one has $\mathbf{X}^{(N)} = (x_1^{(N)}, \ldots, x_n^{(N)})$ redundant units in the system. For the next step one must calculate the relative increments to decide in which direction to move. For this purpose one must compute the following values for all i (including the one for which a redundant unit had been added at the previous step):

$$\gamma_i^{(N)}(x_i) = \frac{L(\mathbf{X}_i^{(N)}) - L(\mathbf{X}_i^{(N)})}{C(\mathbf{X}_i^{(N)}) - C(\mathbf{X}^{(N)})} = \frac{1}{c_i} [L_i(x_i^{(N)} + 1) - L_i(x_i^{(N)})] \tag{10.19}$$

Now let us describe the optimization algorithm step by step from the beginning.

1. For each i calculate

$$\gamma_i^{(0)} = \gamma_i^{(0)}(x_i^{(0)})$$

2. Find $\gamma^{(1)}$, the maximal of all $\gamma_i^{(0)}$. The corresponding subscript of the subsystem, say j, indicates that one has to add a redundant unit to the jth subsystem.

3. Find the new vector

$$\mathbf{X}^{(1)} = \mathbf{X}^{(0)} + e_j = \mathbf{X}_j^{(0)}$$

4. Repeat steps 1 to 3, changing N for $N + 1$

The stopping rule of the computing procedure is defined in the following way:

- For the direct problem, we stop at step N when the inequality

$$R(\mathbf{X}^{(N-1)}) < R_0 \leq R(\mathbf{X}^{(N)}) \tag{10.20}$$

is satisfied. The vector $\mathbf{X}^{(N)}$ is the solution.
- For the inverse problem, we stop at step N when the inequality

$$C(\mathbf{X}^{(N)}) \leq C_0 < C(\mathbf{X}^{(N+1)}) \tag{10.21}$$

is satisfied.

REMARK. Sometimes the SDM procedure requires the use of a very expensive unit at the current step because it produces the largest increment γ. Using the stopping rule (10.21), one must stop even though there are unspent resources. In this case it is still possible to improve the solution by using the best unit(s) among those whose cost is smaller than $C_0 - C(\mathbf{X}^{(N)})$.

Example 10.1   Consider a series system of three units. The units have the following parameters: $p_1 = 0.7$, $c_1 = 3$; $p_2 = 0.8$, $c_2 = 5$; $p_3 = 0.9$, $c_3 = 2$. It is possible to use an active redundancy to improve the system reliability. Find the optimal solutions for the two problems:

- Minimize the total system cost under the restriction on the system reliability of $Q_o = 0.01$.
- Maximize the system reliability index under the restriction on the total system cost of $C_0 - 25$.

Solution.   First of all, using the formulas for active redundancy given in Chapter 3, construct an auxiliary table (Table 10.1) of the values of $q_i^{x_i}$. On the basis of Table 10.1, one can construct Table 10.2 where values of $\gamma_i(x)$

TABLE 10.1   Values of q^* for Example 10.1

x	q_1^{x+1}	q_2^{x+1}	q_3^{x+1}
0	0.3	0.2	0.1
1	0.09	0.04	0.01
2	0.027	0.008	0.001
3	0.0081	0.0016	0.0001
4	0.00243	0.00032	0.00001
5	0.000729	0.000064	...
6	0.000021	...	
...	...		

TABLE 10.2   Values of $\gamma(x)$ for Example 10.1

x	$\gamma_1(x)$	$\gamma_2(x)$	$\gamma_3(x)$
1	(1) 0.07	(3) 0.032	(2) 0.045
2	(4) 0.009	(5) 0.0064	(7) 0.0045
3	(6) 0.0063	(9) 0.00128	(11) 0.00045
4	(8) 0.00189	(12) 0.000256	0.0000045
5	(10) 0.000567	0.0000512	0.0000045
6	(13) 0.000236	0.0000102	. . .
	. . .	. . .	. . .

TABLE 10.3   Step-by-Step Solution for Example 10.1

N	$x_1^{(N)}$	$x_2^{(N)}$	$x_3^{(N)}$	$C(X^{(N)})$	$Q(X^{(N)})$
1	1	0	0	3	0.39
2	1	0	1	5	0.30
3	1	1	1	10	0.14
4	2	1	1	13	0.077
5	2	2	1	18	0.045
6	3	2	1	21	0.0261
7	3	2	2	23	0.0171
8	4	2	2	26	0.01143
9	4	3	2	31	0.00503
10	5	3	2	34	0.00333
11	5	3	3	36	0.00243
12	5	4	3	41	0.00114
. . .	. . .	. . .	. . .	. . .	. . .

are collected. The values of $\gamma_i(x)$ are numbered in decreasing order for convenience in the construction of the next table (Table 10.3) containing all of the information needed for the solution.

From Table 10.3 one can find that the solution of the first problem $X^* = (4, 3, 2)$ has been obtained at the ninth step. At this step the system reliability index $[Q(X^*) = 0.00503]$ satisfies the given requirement for the first time. The cost for this variant is $C(X^*) = 31$ units.

The solution of the second problem was reached at the seventh step. At this step $X^* = (3, 2, 2)$ for which the total system cost equals $C(X^*) = 23$ and the system reliability index equals $Q(X^*) = 0.0171$.

In the latter case one can use the technique of the remaining resources distribution: $C_o - C(X^*) = 25 - 23 = 2$. It allows one to use one more additional unit of the first type. Thus, the final solution is $X^* = (3, 2, 3)$.

If the goal function $L(\mathbf{X}^*)$ is separable and if all $L_i(x_i)$ are convex, the solution $\mathbf{X}^*$ obtained with the help of the SDM is optimal if:

- The solution $R(\mathbf{X}^*)$ exactly coincides with R_o [or $Q(\mathbf{X}^*) = Q_o$] for the direct optimization problem.
- The solution $C(\mathbf{X}^*)$ exactly coincides with C_o for the inverse optimization problem.

The proof of these facts one can find in Barlow and Proschan (1975).

10.2.3 Approximate Solution

For practical purposes, an engineer sometimes needs to know an approximate solution which would be close to an optimal one a priori. Such a solution can be used as the starting point for the SDM calculation procedure. (Moreover, sometimes the approximate solution is a good practical solution if input data are too imprecise. In such a case an attempt to find an exact solution might be understood by reasonable people as a total absence of common sense!)

The proposed approximate solution is satisfactory for highly reliable systems. This means that the value of Q_0 is very small in the direct optimal redundancy problem. We could find that this is not a serious practical restriction. Indeed, if the investigated system is too unreliable, one should consider whether it is reasonable to improve its reliability at all. Maybe it would be easier to find another solution, for instance, to use another system.

For a highly reliable system one can write

$$Q_o \approx Q(\mathbf{X}^{(N)}) \approx \sum_{1 \le i \le n} Q_i(x_i^{(N)}) \tag{10.22}$$

at the stopping moment (the Nth step) of the optimization process. Now recall that for any i the current value of $\gamma_i^{(K)}$ is close enough to some constant $\gamma^{(K)} = \gamma^*$ depending on the number of the current step. This allows us to write

$$(1/c_i)\left[Q_i(x_i^{(N)}) - Q_i(x_i^{(N)} + 1)\right] \approx \gamma^* \tag{10.23}$$

For most practical cases (at least for the highly reliable systems mentioned above), one can write

$$Q_i(x_i) \gg Q_i(x_i + 1) \tag{10.24}$$

Inequality (10.24) allows us to rewrite (10.23) as

$$(1/c_i)Q_i(x_i^{(N)}) \approx \gamma^* \tag{10.25}$$

The solution of (10.25) and (10.22) gives us

$$Q_0 \approx \gamma^* \sum_{1 \le i \le n} c_i \qquad (10.26)$$

and, as the result,

$$\gamma^* \approx \frac{Q_0}{\sum\limits_{1 \le i \le n} c_i} \qquad (10.27)$$

Now we can substitute (10.27) into (10.25) and obtain

$$Q_i\left(x_i^{(N)}\right) \approx \frac{c_i Q_0}{\sum\limits_{1 \le i \le n} c_i} \qquad (10.28)$$

If the functions $Q_i(x)$ are known, it is easy to find $x_i^{(N)}$. The meaning of this approximate solution is as follows: the "portion" of the failure probability for each subsystem (redundant group) should be proportional to the cost of its redundant unit.

Incidentally, this approximate solution, in some sense, explains a very close connection between the two methods: LMM and SDM. Indeed, the value γ^* is an analogue of the Lagrange multiplier λ introduced in Section 10.2.1.

Unfortunately, all attempts to find such a simple solution for the inverse optimal redundancy problem have been unsuccessful. But, of course, one can use a very simple iterative procedure to find the solution. Below we describe a possible algorithm.

1. Find approximate starting values of the x_i's:

$$x_1^{(1)} = x_2^{(1)} = \cdots = x_n^{(1)} = \frac{C_0}{\sum\limits_{1 \le i \le n} c_i}$$

2. Use these x_i's to calculate $Q^{(1)}$ as

$$Q^{(1)} = \sum_{1 \le i \le n} Q_i\left(x_i^{(1)}\right)$$

3. Calculate $\gamma^{(1)}$ as

$$\gamma^{(1)} = \frac{Q^{(1)}}{\sum\limits_{1 \le i \le n} c_i}$$

4. For the $\gamma^{(1)}$ determine $x_i^{(2)}$ for all i from the equation

$$Q_i\left(x_i^{(2)}\right) = c_i\gamma^{(1)}$$

5. For x_i's so obtained calculate the controlling total cost of the system's redundant units by

$$C^{(1)} = \sum_{1 \le i \le n} c_i x_i^{(2)}$$

6. If $C^{(1)} > C_0$ one sets a new $\gamma^{(2)} > \gamma^{(1)}$; if $C^{(1)} < C_0$ one sets a new $\gamma^{(2)} < \gamma^{(1)}$.

After this, the procedure continues from the third step.

10.2.4 Dynamic Programming Method

As mentioned above, the problem has an essentially discrete nature, so the SDM cannot guarantee the accuracy of the solution. Thus, if an exact solution of the optimal redundancy problem is needed, one generally needs to use the *dynamic programming method* (DPM).

The DPM provides an exact solution of discrete optimization problems. In fact, it is a well-organized method of direct enumeration. For exact solutions one must pay a high calculation cost in time and computer memory. For these reasons, the DPM is inadmissible for large-dimensional problems.

To solve the direct optimal redundancy problem, let us construct Bellman's function, $B_k(r)$. This function reflects the optimal value of the goal function for a system of k redundant groups and a specified restriction r:

$$B_k(r) = \min\left\{ \sum_{1 \le i \le k} c_i x_i \middle| \prod_{1 \le i \le k} R_i(x_i) \le r\right\} \qquad (10.29)$$

The function $B_k(r)$, $R_0 < r < 1$, is constructed by the recurrence equation

$$B_k(R) = \min_{x_k}\left\{B_{k-1}(r') + c_k x_k \middle| r' R_k(x_k)\right\} \qquad (10.30)$$

where for $B_1(r)$,

$$B_1(r) = \min_{x_1}\left\{c_1 x_1 \middle| R_1(x_1) \ge r\right\} \qquad (10.31)$$

The solution proceeds, in some sense, backwards. First, one solves (10.31) for different values of r, $R_0 < r < 1$, and finds a set x_1. Then one finds $B_2(r')$ for different values of r', $R_0 < r' < 1$, that is, the best pair of x_1 and x_2 for each r'. After this, one finds $B_3(r')$ for different values of r',

$R_0 < r' < 1$, that is, the best pair of x_3 and B_2 for each r'. (A knowledge of B_2 immediately gives us the corresponding values of x_1 and x_2.) The procedure continues until one finds x_n and B_{n-1}.

One must deal with a different Bellman function to solve the inverse optimal redundancy problem

$$\tilde{B}_k(c) = \max_{\mathbf{X}} \left\{ \prod_{1 \le i \le k} R_i(x_i) \,\middle|\, \sum_{1 \le i \le k} c_i x_i \le c \right\}$$

which can be recurrently constructed with the use of the expressions

$$\tilde{B}_1(c) = \max_{x_1} \{R_1(x_1)|c_1 x_1 \le c, 0 \le c \le C_0\}$$

$$\tilde{B}_k(c) = \max_{x_k} \{\tilde{B}_{k-1}(c) R_k(x_k)|c + c_k x_k \le c', 0 \le c' \le C_0\}$$

$$(10.32)$$

The best explanation of the dynamic programming method can be found by considering a simple illustrative example.

Example 10.2   Let us consider a very simple series system consisting of three units with the characteristics: $p_1 = 0.7$, $p_2 = 0.8$, $p_3 = 0.9$, and $c_1 = c_2 = c_3 = 1$. Active redundancy is permitted to improve the system reliability. There are six units of cost that should be used in an optimal way.

Solution.   During the solution we will use (for simplicity of the numerical calculations) the approximation

$$Q_{\text{syst}}(\mathbf{X}) = \sum_{1 \le i \le n} q_i^{x_i}$$

although all q^x's are not always small.

With the help of Table 10.1 we construct Table 10.4 where all possible allocation variants of the redundant units for x_1 and x_2 are analyzed for all possible costs. Conditionally optimal solutions, that is, the best pairs (x_1, x_2) for different values of c, $1 \le c \le 6$, have been found. These solutions are marked with an asterisk and denoted by y_k. For example, y_3 denotes the pair $(x_1 = 2, x_2 = 1)$ and signifies that this is the best allocation of 3 cost units with resulting $Q = 0.067$.

TABLE 10.4 Stages of the Solution by the Dynamic Programming Method for Example 10.2

C	x_1	x_2	Q	Chosen Variant
1	1	0	0.290	*
	0	1	0.340	
2	2	0	0.227	
	1	1	0.130	*
	0	2	0.308	
3	3	0	> 0.2	
	2	1	0.067	*
	1	2	0.098	
	0	3	> 0.3	
4	4	0	> 0.2	
	3	1	> 0.04	
	2	2	0.035	*
	1	3	> 0.09	
	0	4	> 0.3	
5	5	0	> 0.2	
	4	1	> 0.04	
	3	2	0.0161	*
	2	3	0.0286	
	1	0	> 0.09	
	0	5	> 0.3	
6	6	0	> 0.2	
	5	1	> 0.04	
	4	2	0.1043	
	3	3	0.0097	*
	2	4	0.02769	
	1	5	> 0.09	
	0	6	> 0.3	

The next step of the solution is the construction of Table 10.5 where the above-mentioned conditionally optimal solutions convolute with x_3 to obtain the final result. The best solution, marked with an asterisk, is ($y = 5$, $x_3 = 1$). From Table 10.4 one can find that $y = 5$ corresponds to ($x_1 = 3$, $x_2 = 2$), and so the solution of the entire problem is ($x_1 = 3$, $x_2 = 2$, $x_3 = 1$).

This simple example is given only as an illustration of the solution procedure. The direct problem solution is more complicated in terms of the calculations. It even may lead to confusion among those who do not have experience in solving analogous numerical problems.

TABLE 10.5   Final Stage of the Solution for Example 10.2

C	y	x_3	Q	Chosen Variant
6	0	6	> 0.5	
	1	5	> 0.290	
	2	4	> 0.130	
	3	3	0.0671	
	4	2	0.036	
	5	1	0.0261	*
	6	0	0.1097	

10.2.5   Kettelle's Algorithm

If a researcher is not satisfied by a particular DPM solution for some specified restrictions and decides to change some of the restrictions, it may lead to a complete re-solving of the problem, or at least, to some rearranging of data files in a computer.

In our opinion, *Kettelle's algorithm*, being, in fact, a modification of the DPM, is very effective for the exact solution of engineering problems. One of its obvious advantages is its clarity and calculational flexibility.

As with the SDM, this method also allows one to construct an *undominated sequence*. But, in contrast with the SDM, Kettelle's algorithm allows one to build a complete undominated sequence. We emphasize that such a solution is often very convenient for practical engineering purposes. One does not need to solve the entire problem every time for each set of restrictions. Moreover, one can easily switch from the direct optimal redundancy problem to the inverse one. Of course, DPM and any of its modifications require more computer time and memory than the SDM but they give accurate solutions. (Recall that one should always decide in advance on an appropriate accuracy level.)

We shall describe Kettelle's algorithm step by step.

1. For each ith redundant group, one constructs a table of pairs, each consisting of the reliability index and its corresponding cost: $(R_i(1), C_i(1)), (R_i(2), C_i(2)), \ldots, (R_i(x_i), C_i(x_i))$. All of these pairs constitute the ith undominated sequence, where i is the number of its representative in order of increasing cost. (Note that the number of undominated sequence pairs might not be restricted in advance and can be extended during the procedure.)

2. For any two redundant groups, say n and $n - 1$, one constructs Table 10.6 reflecting the composition of these groups. This composition includes all possible combinations of the pairs of x_n and x_{n-1}. Table 10.6 contains pairs of reliability indexes and costs. For example,

TABLE 10.6   Composition of Units n and $n-1$

x_{n-1}	x_n			
	0	1	2	$\cdots$
0	$R^*_{n-1}(0,0)$ $C^*_{n-1}(0,0)$	$R^*_{n-1}(0,1)$ $C^*_{n-1}(0,1)$	$R^*_{n-1}(0,2)$ $C^*_{n-1}(0,2)$	$\cdots$
1	$R^*_{n-1}(1,0)$ $C^*_{n-1}(1,0)$	$R^*_{n-1}(1,1)$ $C^*_{n-1}(1,1)$	$R^*_{n-1}(1,2)$ $C^*_{n-1}(1,2)$	$\cdots$
$\cdots$	$\cdots$	$\cdots$	$\cdots$	$\cdots$

for $x_n = i$ and $x_{n-1} = j$, such a pair is

$$\left(R^*_{n-1}(i,j), C^*_{n-1}(i,j) \right)$$

where

$$R^*_{n-1}(i,j) = R_n(i)R_{n-1}(j) \tag{10.33}$$

and

$$C^*_{n-1}(i,j) = C_n(i) + C_{n-1}(j) \tag{10.34}$$

The asterisk indicates that this parameter relates to a new equivalent system unit, in this particular case to the $(n-1)$th unit.

3. On the basis of Table 10.6, one constructs Table 10.7 containing only undominated reliability–cost pairs. Let us enumerate representatives of this undominated sequence

$$\left(R^*_{n-1}, C^*_{n-1} \right)$$

with the numbers y_{n-1}. Each number y_{n-1} contains a pair of corresponding numbers x_n and x_{n-1}. If $x_n = i$ and $x_{n-1} = j$ for some $y_{n-1} = k$, then

$$R^*_{n-1}(k) = R^*_{n-1}(i,j)$$

and

$$C^*_{n-1}(k) = C^*_{n-1}(i,j)$$

which are defined in (10.33) and (10.34), respectively.

TABLE 10.7   Undominated Reliability – Cost Pairs

y_{n-1}	1	2	$\cdots$
$C^*_{n-1}(y_{n-1})$ $R^*_{n-1}(y_{n-1})$	$C^*_{n-1}(x'_{n-1}, x'_n)$ $R^*_{n-1}(x'_{n-1}, x'_n)$	$C^*_{n-1}(x''_{n-1}, x''_n)$ $R^*_{n-1}(x''_{n-1}, x''_n)$	$\cdots$

4. One considers the reduced system consisting of only $n - 1$ units. The next iteration of the optimization process consists in combining the $(n - 1)*$th and $(n - 2)$th units into a new equivalent unit, namely, the $(n - 2)*$th (see Table 10.8).

   In the new undominated sequence, each number y_{n-2} contains pairs of corresponding numbers x_{n-2} and y_{n-1}. If $x_{n-2} = i$ and $y_{n-1} = j$ for some $y_{n-2} = k$, then

$$R^*_{n-2}(k) = R^*_{n-2}(i, j) = R_{n-2}(i) R^*_{n-1}(j)$$

and

$$C^*_{n-2}(k) = C^*_{n-2}(i, j) = C_{n-2}(i) + C^*_{n-1}(j)$$

5. The procedure continues until, finally, the reduced system can be represented as a single equivalent unit. This unit will be determined by its undominated sequence

$$(R^*_1(1), C^*_1(1)), (R^*_1(2), C^*_1(2)), \ldots, (R^*_1(k), C^*_1(k)), \ldots$$

Each number y_1, $y_1 = 1, 2, \ldots,$ contains corresponding values of x_1 and y_2. Thus, the solution of the given problem of optimal redundancy can be found at step N which, for the direct task, is the step satisfying the condition

$$C^*_1(N) \leq C_0 < C^*_1(N + 1)$$

or, for the inverse task,

$$R^*_1(N - 1) < R_0 \leq R^*_1(N)$$

From y_1, one can determine the corresponding numbers x_1 and y_2. The value of y_2 allows one to find x_2 and y_3, and so on. At last, one has determined all x_i's satisfying the optimal solution of the problem under consideration.

TABLE 10.8   Composition of Equivalent Group ($n - 1$)* and Unit $n - 2$

	y_{n-1}			
x_{n-2}	1	2	3	$\cdots$
0	$R^*_{n-2}(0, 1)$ $C^*_{n-2}(0, 1)$	$R^*_{n-2}(0, 2)$ $C^*_{n-2}(0, 2)$	$R^*_{n-2}(0, 3)$ $C^*_{n-2}(0, 3)$	$\cdots$
1	$R^*_{n-2}(1, 1)$ $C^*_{n-2}(1, 1)$	$R^*_{n-2}(1, 2)$ $C^*_{n-2}(1, 2)$	$R^*_{n-2}(1, 3)$ $C^*_{n-2}(1, 3)$	$\cdots$

REMARK. Suppose that there are two neighboring representatives in the current undominated sequence which have very close reliability index values, for example, $R(N + 1) - R(N) = \varepsilon$, where ε is small, but have an essential difference in cost. In this case the second representative might be excluded as unimportant. (Of course, the choice of ε depends on the required accuracy of the analysis.) The same situation is observed if $C(N + 1) - C(N) = \varepsilon$, where ε is small, but there is an essential difference in the reliability indexes. Such a reduction of undominated sequences during each step of the procedure is very reasonable, because it can essentially reduce the total volume of calculations.

While Kettelle's algorithm may be very clear, we will still demonstrate it with an example.

Example 10.3 Consider a system consisting of three units with parameters: $p_1 = 0.7$, $c_1 = 3$; $p_2 = 0.8$, $c_2 = 5$; $p_3 = 0.9$, $c_3 = 2$. The system's reliability can be improved by means of active redundancy. Find the optimal allocation of redundant units for two tasks: (a) $R_0 \geq 0.98$ and (b) $C_0 \leq 25$.

Solution. Using Table 10.1, we construct Table 10.9 containing the above-described composition of the second and third redundant groups. Note that in the direct task the required level of reliability is high enough (as well as the given restriction on the resource volume in the inverse task), thus one can again use an approximation for calculating Q. Then the problem of maximiz-

TABLE 10.9 Composition of Units 2 and 3 into Equivalent Unit 2*
for Example 10.3

x_2	0	1	2	3	4	$\cdots$
			x_3			
0	0	2	4	6	8	
	0.3 ①	0.21 ②	0.201 ③	0.200	0.200	$\cdots$
1	5	7	9	11	13	
	0.14 ④	0.05 ⑤	0.041 ⑥	0.040 ⑦	0.040	$\cdots$
2	10	12	14	16	18	
	0.108	0.018 ⑧	0.009 ⑨	0.0081 ⑩	0.0081	$\cdots$
3	15	17	19	21	23	
	0.1016	0.0116	0.0026 ⑪	0.0017 ⑫	0.0017	$\cdots$
	20	22	24	26	8	
	0.10003	0.0003	0.0013 ⑬	0.0004	0	$\cdots$
$\cdots$	$\cdots$	$\cdots$	$\cdots$	$\cdots$	$\cdots$	$\cdots$

TABLE 10.10 Final Result for Example 10.3

	y												
x_1	1	2	3	4	5	6	7	8	9	10	11	12	13
0	0 0.6	2 0.51	4 0.501	5 0.44	7 0.35	9 0.341	11 0.340	12 0.318	14 0.309	16 0.308	19 0.303	21 0.302	24 0.30
1	3 0.39 (3)	5 0.30 (4)	7 0.291 (5)	8 0.23 (6)	10 0.14 (7)	12 0.131 (8)	14 0.130	15 0.108	17 0.099	19 0.098	22 0.093	24 0.092	27 0.09
2	6 0.321	8 0.237	10 0.228	11 0.167	13 0.077 (9)	15 0.068 (10)	17 0.067	18 0.045 (12)	20 0.036 (13)	22 0.035	25 0.030	27 0.029	30 0.02
3	9 0.308	11 0.218	13 0.209	14 0.148	16 0.058 (11)	18 0.049	20 0.048	21 0.026 (14)	23 0.017 (15)	25 0.016 (16)	28 0.011	30 0.010	33 0.00
4	12 0.302	14 0.212	16 0.203	17 0.142	19 0.052	21 0.043	23 0.042	24 0.020	26 0.011 (17)	28 0.010 (18)	31 0.005 (19)	33 0.004 (20)	36 0.00
5	15 0.301	17 0.211	19 0.202	20 0.141	22 0.051	24 0.042	26 0.041	27 0.019	29 0.010	31 0.009	34 0.004	36 0.003	39 0.00

ing the probability of successful operation is transformed into the problem of minimizing the probability of failure Q. In Table 10.9 the numbers of y are circled in the corresponding cells. Then one constructs the final Table 10.10.

From Table 10.10 one can see that the probability of failure becomes smaller than 0.02 for $y_1 = 15$ in the final undominated sequence (the corresponding number is circled). It corresponds to $Q_1^* = 0.017$ and $C_1^* = 23$. The chosen cell is located at the intersection of the column with $y_2 = 0$ and the row with $x_1 = 3$. With y one can find in Table 10.9 that $x_2 = 2$ and $x_3 = 2$ (see again $y = 9$ which is circled).

For the second task with $C_0 \leq 25$, one finds $y = 16$ with $C_1^* = 25$ and $Q_1^* = 0.016$. Then $y = 16$ corresponds to $x_1 = 3$ and $x_2 = 10$. From Table 10.9 it follows that $x_2 = 2$ and $x_3 = 3$ for $y = 10$.

REMARK. There are two ways of ordering units (redundant groups) of the system for constructing an equivalent unit at each current stage of the procedure. It is not necessary to combine units only in the order: unit n with unit $(n - 1)$, then the equivalent unit $(n - 1)*$ with unit $(n - 2)$, and so on. This order is natural for calculations with a computer. Without a computer, it is better to combine units n and $(n - 1)$, then unit $(n - 2)$ and unit $(n - 3)$, and so on. After such a first "circle of combining," one has a system of $n/2$ units when n is even. The procedure then continues in the same dichotomous way. Two of the different possible ways of combining the units are shown in Figure 10.1.

It is interesting to emphasize the following point. If redundant groups use ordinary types of redundancy (active or standby), then the composition of two initially redundant groups will always generate an undominated sequence

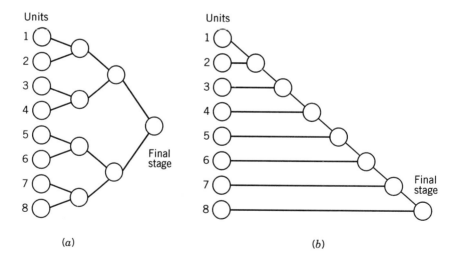

Figure 10.1. Two possible orders of a unit's composition.

which forms a uniquely connected subset of cells in the corresponding table. (As an example, see Table 10.9 where all cells have formed diagonal types of sets of neighbors; there are no cells with undominated terms "outside" this compact set and there are no cells with dominated terms "inside" this set. This property of combining two undominated sequences of initial redundant groups makes Figure 10.1a more preferable.

10.2.6 Method of Generalized Generating Sequences

The method of generalized generating sequences was described in Chapter 1. As we mentioned there, this method is very convenient for a computerized solution of different enumeration problems, in particular, for the solution of discrete optimization problems including the optimal redundancy problem. At first, to facilitate the description, we will use a polynomial representation of the *generalized generating sequence* (GGS).

Let us present the GGS of each redundant group of the system as a legion of the following type (the terminology that we use is explained in Chapter 1):

$$L_i = \left[(R_{1i}, c_{1i}, x_{1i}), \ldots, (R_{ki}, c_{ki}, x_{ki}), \ldots, (R_{n_i i}, c_{n_i i}, x_{n_i i}) \right]$$

where n_i is the number which, in fact, should be chosen in advance, depending on the dimension of the problem being solved.

The legion interaction can be arranged in two ways:

1. Only one pair of legions interacts at a time

$$L_{n-1}^* = \Omega^L(L_n, L_{n-1}) \tag{10.35}$$

and, after such an interaction, a system of order n reduces to a system of order $n - 1$ (in other words, a system of n units reduces to a system of $n - 1$ units). Notice that again the reduction procedure might be set up in correspondence with the two methods described in Figure 10.1.

2. All of the legions interact simultaneously

$$L^{\text{syst}} = \Omega^L_{1 \le i \le n} L_i \tag{10.36}$$

Consider now the interaction between the cohorts. Let us take the cohort (R_{ki}, c_{ki}, x_{ki}). Here we use the following new notation:

R_{ki} is the PFFO of the ith redundant group if it has $x_i = k$ redundant units, that is, in the previous notation $R_{ki} = R_i(k)$.

c_{ki} is the ith redundant group's cost when $x_i = k$, that is, in the previous notation, $c_{ki} = kc_i$.

x_{ki} is the number of redundant units of the ith redundant group.

Let us describe the interaction of the maniples. For the interaction of the two maniples of the first type we have

$$\Omega_R(R_{ki}, R_{mj}) = R_{ki} R_{mj}$$

and, for the interaction of n such maniples,

$$\Omega_R \ R_{ji_j} = \prod_{1 \le j \le n} R_{ji_j} \approx 1 - \sum_{1 \le j \le n} Q_{ji_j}$$

where, for a highly reliable system when

$$\max_{1 \le j \le n} Q_{ji_j} \ll \frac{1}{n}$$

we use the corresponding approximation. For two maniples of the second type, we have

$$\Omega_c(c_{ki}, c_{mj}) = c_{ki} + c_{mj}$$

and, for the interaction of n such maniples,

$$\Omega_c^M \ c_{ji_j} = \sum_{1 \le j \le n} c_{ji_j}$$

Maniples of the third type represent a special kind of interaction. For two of them

$$\Omega_x(x_{ki}, x_{mj}) = (x_{ki}, x_{mj})$$

and for n of them

$$\Omega_x^M \ x_{ji_j} = \mathbf{X}_i = (x_{1i_1}, x_{2i_2}, \ldots, x_{ni_n})$$

that is, the resulting maniple of this type is the file of all x_{ki}'s used in forming this particular maniple.

Thus, the interaction of cohorts Ω^M is that their maniples interact in the above-mentioned way.

The interaction between legions (Ω^L) is the above-mentioned kind of interaction between cohorts and the consequential exclusion of all such cohorts which have been dominated by others. The *domination property* states the following. Assume that two neighboring cohorts of the final legion L^{syst} are $C_k^{\text{syst}} = (R_k^{\text{syst}}, c_k^{\text{syst}}, \mathbf{X}_k^{\text{syst}})$ and $C_{k+1}^{\text{syst}} = (R_{k+1}^{\text{syst}}, c_{k+1}^{\text{syst}}, \mathbf{X}_{k+1}^{\text{syst}})$. Let the cohorts be ordered by the values c_k^{syst} and for these cohorts we have $R_k^{\text{syst}} > R_{k+1}^{\text{syst}}$. In this case the cohort C_{k+1}^{syst} must be excluded from the final legion. Such exclusion continues until all dominated cohorts are excluded. The legion L_*^{syst} obtained from the initial legion L^{syst} is the desired solution.

The above-described deletion of dominated cohorts corresponds to an obvious condition: more expensive but less reliable variants of redundancy must be excluded from consideration.

If one is interested in the solution of the direct optimal redundancy problem, then the operator Ω^L additionally includes an operation of finding such a cohort C_{k*} in the resulting legion L_*^{syst} where

$$R_{(k-1)*} < R^o \le R_{k*}$$

The solution of the inverse optimal redundancy problem is similar and we omit its description.

One sees that this procedure is a reformulation of Kettelle's algorithm. There are several computational advantages in the proposed modified algorithm:

1. Using (10.35) allows one to apply an effective method in the selection of undominated cohorts.
2. The procedure of eliminating dominated cohorts one by one during the computational procedure allows one to minimize the memory used for a computer program. This procedure can be arranged in the following way.

Assume that we are in the middle of the procedure of obtaining an optimal solution. At the time we have some current results in the computer's memory. These results are in the computer as a file of undominated cohorts ordered by the maniple c (cost). At the current step of solution, we obtain a new cohort, say C_k. This cohort is located in an appropriate place in the file of cohorts in accordance with the ordering rule by increasing value of c. Let this new cohort be in the kth place. (This means that on the left there is a cohort C_{k-1} with a smaller value of c and on the right there is a cohort C_{k+1} with a large value of c.) Then we have the following possibilities:

1. The cohort C_{k-1} has a maniple R_{k-1} larger than R_k of the cohort C_k. Then the new cohort is not included in the current set of cohorts.
2. The cohort C_{k+1} has a maniple R_{k+1} smaller than R_k of the cohort C_k. Then the new cohort C_k is included in the current set and the cohort C_{k+1} is eliminated from the current set. The comparison is repeated with the next cohort "on the right."
3. If $R_{k-1} < R_k < R_{k+1}$, the new cohort is inserted in the current subset and both old cohorts C_{k-1} and C_{k+1} remain in the set.

10.3 SEVERAL LIMITING FACTORS

10.3.1 Method of Weighing

A number of cases arise when one has to take into account several restrictions in solving the optimal redundancy problem. For example, various objects such as aircraft, satellites, submarines, and so on have restrictions on their cost and also on their weight, volume, required electric power, and so forth. (Apparently, the cost for most of these technical objects is an important factor, but, perhaps, less important than the others mentioned.)

In these cases one must use methods of optimization under several restrictions. In principle, this problem differs from the previous by dimension. But even a two-dimensional problem leads to enormous computation time and a huge computer memory requirement.

Now the problem is to optimize one of the factors, for example, maximize the system reliability index, under the conditions that the values of all of the other factors lie within specified limits. Of course, it is possible to minimize a system's weight under restrictions on cost and reliability or to formulate some other similar problem.

Consider a system consisting of n redundant groups connected in series. The optimization problem can be formulated as

$$\max_{\mathbf{X}} \left\{ R(\mathbf{X}) | C_{0j}(\mathbf{X}) \leq C_{0j}; 1 \leq j \leq M \right\} \tag{10.37}$$

where $C_j(\mathbf{X})$ is the expenditure of resource type j, $\mathbf{X}$ is the vector of redundant units: $\mathbf{X} = (x_1, x_2, \ldots, x_n)$, and M is the number of types of different resources. Let us assume that $C_j(\mathbf{X})$ is a linear function of the form

$$C_j(\mathbf{X}) = \sum_{1 \leq i \leq n} c_{ji} x_i \tag{10.38}$$

The most convenient way to solve this problem is to reduce it to a one-dimensional problem. To this end, we introduce "weight" coefficients d_j such that $0 \leq d_j \leq 1$ and

$$\sum_{1 \leq j \leq n} d_j = 1$$

Let $\mathbf{D} = (d_1, d_2, \ldots, d_M)$. In general, values of d_j may be arbitrary. We will consider an M-dimensional discrete lattice over which we define the vector $\mathbf{D}$. If we choose the number of gradations of the d_j's equal to some m, there are $K = m^M$ different vectors $\mathbf{D}^k = (d_1^k, d_2^k, \cdots, d_n^k)$, $1 \leq k \leq K$, which exhibit the distribution of weight coefficients between different factors subjected to restriction. Then for each unit we may compile K different one-dimensional "weighted" costs

$$c_i^k = \sum_{1 \le k \le K} c_{ji} d_j^k \qquad (10.39)$$

Of course, this "equivalent cost" demands a reasonable measure of each parameter subjected to restriction. Each parameter measure should be, for example, normalized by the maximal value. This gives us nondimensional values which might be weighed and added to obtain an equivalent cost. Sometimes, when the absolute size of the dimensional values of different parameters are of the same range, one can use their absolute value (using them as nondimensional).

For future use we introduce the notation

$$\mathbf{C}^k = \left(c_1^k, c_2^k, \ldots, c_n^k \right)$$

Now for each k we transform the original problem (10.37) into the one-dimensional problem:

$$\max_{\mathbf{X}^k} \left\{ R(\mathbf{X}^k) | C(\mathbf{X}^k) \in \left[C_j(\mathbf{X}^k) \le C_{0j}; 1 \le j \le M \right] \right\} \qquad (10.40)$$

where $\mathbf{X}^k = (x_1^k, x_2^k, \ldots, x_n^k)$ is the solution obtained for the "weighted" cost $\mathbf{C}^k$, and $[C_j(\mathbf{X}) \le C_{0j}; 1 \le j \le M]$ is the M-dimensional rectangular area in M-dimensional space of parameters subjected to restriction.

Thus, the solution of the original M-dimensional optimization problem is transformed into K one-dimensional problems. The new side of these problems in comparison with a "pure" one-dimensional problem is that one needs to control the violation of all M restrictions during the solving process. An illustrative example might be useful to demonstrate the method.

Example 10.4 Consider a series system consisting of three units with the characteristics given in Table 10.11. An active redundancy is permitted to improve the reliability of the system. The problem is to find the optimal

TABLE 10.11 Data for Example 10.4

Unit (i)	Reliability Index P_i	Cost C_{1i}	Weight C_{2i}
1	0.7	3	1
2	0.8	5	1
3	0.9	2	3

solution for each of two cases:

- Where there are restrictions on both the total system cost by $C_{01} = 25$ cost units and the total system weight by $C_{02} = 15$ weight units.
- Where there are restrictions on both the total system cost by $C_{01} = 25$ cost units and the total system weight by $C_{02} = 10$ weight units.

Solution. Choose the increment for each d_i equal to 0.25, that is, each of them might take one of the values 0, 1/4, 1/2, 3/4, 1. The equivalent costs can be calculated as: $c_i^1 = c_{1i}$; $c_i^2 = 0.75c_{1i} + 0.25c_{2i}$; $c_i^3 = 0.5c_{1i} + 0.5c_{2i}$; $c_i^4 = 0.25c_{1i} + 0.75c_{2i}$; and $c_i^5 = c_{2i}$. Then the following values for the equivalent costs:

$$c_1^2 = 2.5 \qquad c_2^2 = 4.0 \qquad c_3^2 = 2.25$$

$$c_1^3 = 2.0 \qquad c_2^3 = 3.0 \qquad c_3^3 = 2.5$$

$$c_1^4 = 1.5 \qquad c_2^4 = 2.0 \qquad c_3^4 = 2.75$$

$$c_1^5 = 1.0 \qquad c_2^5 = 1.0 \qquad c_3^5 = 3.0$$

are valid.

Now we separately solve all five problems for different equivalent costs. For simplicity, let us use the steepest descent method. For the first problem solution, one can use tables prepared for the one-dimensional case considered in Example 10.1. At each step of the procedure we must control the restriction C_{02}, and stop the procedure if this has been violated. Let us check the solution obtained with the help of Table 10.3: $x_1 = 3$, $x_2 = 2$, and $x_3 = 2$. For this case the second specified restriction of 10 costs units is not violated. In other words, this factor is *not critical* for the solution.

TABLE 10.12   Values of $\gamma_i^{(2)}(x_i)$ for Different Units for $\alpha_2^{(2)} = 0.25$

$x_i^{(2)}$	$\gamma_1^{(2)}(x_1)$	$\gamma_2^{(2)}(x_2)$	$\gamma_3^{(2)}(x_3)$	$\gamma_4^{(2)}(x_4)$	$\gamma_5^{(2)}(x_5)$
0	—	—	—	—	—
1	0.2351 (2)	0.4230 (1)	0.0474 (12)	0.500 (11)	0.2290 (3)
2	0.1288 (5)	0.1790 (4)	0.0045	0.0025	0.1175 (6)
3	0.0762 (7)	0.0727 (8)	0.0005	0.0001	0.0679 (9)
4	0.0527 (10)	0.0262 (16)	0	0	0.0412 (13)
5	0.0372 (14)	0.0081	—	—	0.0252 (17)
6	0.0275 (15)	0.0021	—	—	0.0155 (20)
7	0.0206 (18)	0.0005	—	—	0.0095 (22)
8	0.0158 (19)	0.0001	—	—	0.0058
9	0.0123 (21)	0.	—	—	0.0035
10	0.0095 (23)	—	—	—	0.0020

TABLE 10.13   Values of $\gamma_i^{(3)}(x_i)$ for Different Units for $\alpha_2^{(2)} = 0.50$

$x_i^{3)}$	$\gamma_1^{(3)}(x_1)$	$\gamma_2^{(3)}(x_2)$	$\gamma_3^{(3)}(x_3)$	$\gamma_4^{(3)}(x_4)$	$\gamma_5^{(3)}(x_5)$
0	—	—	—	—	—
1	0.2940 (4)	0.3818 (1)	0.0318 (16)	0.3333 (2)	0.3060 (3)
2	0.1520 (7)	0.1622 (5)	0.0032	0.0017	0.1568 (6)
3	0.0953 (8)	0.0655 (10)	0.0003	0.0001	0.0910 (9)
4	0.0650 (11)	0.0236 (18)	0	0	0.0548 (12)
5	0.0465 (13)	0.0073	—	—	0.0357 (14)
6	0.0344 (15)	0.0019	—	—	0.0207 (19)
7	0.0258 (17)	0.0005	—	—	0.0127 (22)
8	0.0198 (20)	0.0001	—	—	0.0077
9	0.0153 (21)	0	—	—	0.0047
10	0.0118 (23)	—	—	—	0.0027

TABLE 10.14   Values of $\gamma_i^{(4)}(x_i)$ for Different Units for $\alpha_2^{(2)} = 0.75$

$x_i^{(4)}$	$\gamma_1^{(4)}(x_1)$	$\gamma_2^{(4)}(x_2)$	$\gamma_3^{(4)}(x_3)$	$\gamma_4^{(4)}(x_4)$	$\gamma_5^{(4)}(x_5)$
0	—	—	—	—	—
1	0.3918 (2)	0.3490 (3)	0.0240 (19)	0.0250 (18)	0.4580 (1)
2	0.2028 (5)	0.1475 (6)	0.0022	0.0013	0.2350 (4)
3	0.1270 (8)	0.0595 (12)	0.0002	0.0001	0.1358 (7)
4	0.0866 (9)	0.0214 (20)	0	0	0.0820 (10)
5	0.0620 (11)	0.0066	—	—	0.0508 (13)
6	0.0458 (14)	0.0017	—	—	0.0310 (16)
7	0.0344 (15)	0.0004	—	—	0.0190 (22)
8	0.0264 (17)	0.0001	—	—	0.0115
9	0.0204 (21)	0.	—	—	0.0070
10	0.0158 (23)	—	—	—	0.0040

TABLE 10.15   Values of $\gamma_i^{(5)}(x_i)$ for Different Units for $\alpha_2^{(2)} = 1.00$

$x_i^{(5)}$	$\gamma_1^{(5)}(x_1)$	$\gamma_2^{(5)}(x_2)$	$\gamma_3^{(5)}(x_3)$	$\gamma_4^{(5)}(x_4)$	$\gamma_5^{(5)}(x_5)$
0	—	—	—	—	—
1	0.5870 (2)	0.3200 (4)	0.0191	0.0200	0.9150 (1)
2	0.3042 (5)	0.1350 (9)	0.0018	0.0010	0.4700 (3)
3	0.1900 (7)	0.0547 (15)	0.0002	0	0.2720 (6)
4	0.1300 (10)	0.0197	0	—	0.1645 (8)
5	0.0930 (12)	0.0061	—	—	0.1010 (11)
6	0.0687 (13)	0.0016	—	—	0.0620 (14)
7	0.0516 (16)	0.0004	—	—	0.0380 (18)
8	0.0396 (17)	0	—	—	0.0230 (21)
9	0.0306 (19)	—	—	—	0.0140
10	0.0237 (20)	—	—	—	0.0080

For the remaining four cases we construct separate tables of γ^k and tables for controlling the solution (see Tables 10.12 to 10.17). If the above-mentioned increment of d_i's is too large, the solution may be far from optimal. If the increment is too small, a better solution may (and usually will) be obtained but it may take an excessive amount of time and may require a large amount of computer memory. As in most practical cases, to find the appropriate choice of the increment is more of an art than a science.

TABLE 10.16   Step-by-Step Solutions for Different ha

Step Number	$\alpha_2^{(1)} = 0$		$\alpha_2^{(2)} = 0.25$		$\alpha_2^{(3)} = 0.5$		$\alpha_2^{(4)} = 0.75$		$\alpha_2^{(5)} = 1$	
	C_2	C_2	C_1	C_2	C_1	C_2	C_1	C_2	C_1	C_2
0	22	17	22	17	22	17	22	17	22	17
1	24	20	24	20	24	20	27	18	27	18
2	26	23	27	21	25	25	30	19	30	19
3	29	24	32	22	30	26	32	22	35	20
4	34	25	34	25	33	27	37	23	37	23
5	37	26	37	26	35	30	40	24	40	24
6	38	31	42	27	40	31	42	27	45	25
7	39	36	45	28	43	32	47	28	48	26
8	44	37	47	31	46	33	50	29	53	27
9	46	40	52	32	41	34	53	30	55	30
10	49	41	55	33	53	37	58	31	58	31
11	54	42	56	38	56	38	61	32	63	32
12	57	43	57	43	61	39	63	35	66	33
13	62	44	62	44	64	40	68	36	69	34
14	65	45	65	45	69	41	71	37	74	35
15	67	48	68	46	72	42	74	38	76	38
16	70	49	70	49	73	47	79	39	79	39
17	75	50	75	50	76	48	82	40	82	40
18	78	51	78	51	78	51	83	45	87	41
19	81	52	81	52	83	52	84	50	90	42
20	86	53	86	53	86	53	86	53	93	43

TABLE 10.17   Best Results Obtained for Different Values of α

$\alpha_2^{(j)}$	Stopping Step	x_1	x_2	x_3	x_4	x_5	L
0	9	2	3	1	1	2	1.509
0.25	10	4	3	0	0	3	2.112
0.50	10	3	3	0	1	3	1.142
0.75	9	4	2	0	0	3	1.275
1.00	9	3	2	0	0	4	1.241

Of course, one can use the SDM for searching for the best D^k itself. We explain this on a simple concrete example where the number of restrictions equals 4. Use the vectors $(1, 0, 0, 0)$, $(0, 1, 0, 0)$, $(0, 0, 1, 0)$, and $(0, 0, 0, 1)$ in that order. Let the best solution be given by the vector $(1, 0, 0, 0)$. Then at the next step one chooses vectors $(1 - 1/m, 1/m, 0, 0)$, $(1 - 1/m, 0, 1/m, 0)$, and $(1 - 1/m, 0, 0, 1/m)$. Let the best solution (including the one obtained at the first step) be given by the vector $(1 - 1/m, 0, 0, 1/m)$. At the next step of the procedure one should then try the vectors $(1 - 2/m, 0, 0, 2/m)$, $(1 - 3/m, 0, 0, 3/m)$, and so on. This procedure stops when the best d_1 and d_4 are found, for example, $(1 - k/m, 0, 0, k/m)$. Then one would try the two other vector components in a similar way. It is possible that one could return to the first parameters during these sequential attempts (as one does in any SDM solution). This procedure continues until the solution that is obtained at the current step is better than the one obtained at the previous step. This is the stopping rule. The solution finally obtained is taken to be "quasi-optimal." (It may even be optimal because of the discrete nature of the problem.) Such a method could decrease the number of iterations from m^M to approximately mM.

Note that this solution possesses all of the advantages and disadvantages of the common SDM which we used for the one-dimensional case.

REMARK. If for each ith-type unit there is a supplementary condition $c_{ij} = k_{j1}c_{i1}$ for all $1 \le i \le n$, that is, all costs are in the same proportion for all units, the multidimensional optimization problem can be reduced to a one-dimensional case with the strictest restriction.

10.3.2 Method of Generating Sequences

The problem treated above can be solved exactly with the GGS method (see Chapter 1 and Section 10.2.6). The legion for each ith redundant group is represented as the set of cohorts

$$L_i = \{C_{i1}, C_{i2}, \ldots, C_{iN_i}\}$$

where N_i is the number of cohorts in this legion. (In principle, the number of cohorts is unrestricted in this investigation.) Each cohort consists of $M + 2$ maniples:

$$C_{ik} = \left(R_{ik}, c_{ik}^1, \ldots, c_{ik}^M, x_{ik} \right)$$

where M is the number of restrictions. All maniples are defined as in the one-dimensional case that we considered above. A similar interaction is performed with the maniples:

$$\Omega_R^M(R_{ij}, R_{kl}) = R_{ij}R_{kl} \qquad \text{and} \qquad \Omega_R^M \underset{1 \le i \le n}{R_{ij_i}} = \prod_{1 \le i \le n} R_{ij_i}$$

$$\Omega_{c^k}^M(c_{ij_i}^k, c_{lj_1}^k) = c_{ij_i}^k + c_{lj_1}^k \qquad \text{and} \qquad \Omega_{c_k}^M \underset{1 \le i \le n}{c_i^k} = \sum_{1 \le i \le n} c_{ij_i}^k$$

$$\Omega_x^M(x_{ij_i}, x_{lj_1}) = (x_{ij_i}, x_{lj_1}) \qquad \text{and} \qquad \Omega_x^M \underset{1 \le i \le n}{x_{ij_i}} = (x_{lj_1}, x_{2j_2}, \dots, x_{nj_n}) = \mathbf{X}$$

where $\mathbf{J} = (j_1, \dots, j_n)$ is the set of subscripts.

The remaining formal procedures totally coincide with the one-dimensional case with one very important exception: instead of a scalar ordering, one must use the special ordering of the cohorts of the final legion.

It is difficult to demonstrate the procedure on a numerical example, so we give only a detailed verbal explanation. Suppose we have the file of current cohorts ordered according to increasing R. If we have a specified set of restrictions $[C_j(\mathbf{X}) \le C_{0j}$ for all j, $1 \le j \le M]$, then there is no cohort in this file which violates at least one of these restrictions. When a new cohort, say C_k, appears during the interaction procedure, it is put in the appropriate place in accordance with the value of its R-maniple. The computational problem is as follows.

1. Consider a part of the current file of cohorts for which the values of their R-maniples are less than the analogous value for C_k. If, among the existent cohorts, there is a cohort, say C^*, for which all costs are larger than those of C_k, this cohort C^* is excluded from the file.
2. Consider a part of the current file of cohorts for which the values of the R-maniples are larger than the analogous value for C_k. If between the existent cohorts there is a cohort, say C^{**}, for which all costs are smaller than those of C_k, the new cohort is not included in the file.
3. If neither 1 nor 2 takes place, the new cohort is simply added to the file in the appropriate place.

After a multidimensional undominated sequence is constructed, one easily finds the solution for the multiple restrictions: it is the cohort with the largest R-maniple value (in other words, a cohort on the right if the set is ordered by the values of R).

The stopping rule for this procedure is to find the size of each cohort which will produce a large enough number of cohorts in the resulting legion so as to contain the optimal solution. At the same time, if the numbers of cohorts in the initial legions are too large, the computational procedure will take too much time and will demand too large a memory space.

Of course, the simplicity of this description should not be deceptive. The problem is very bulky in the sense that the multidimensional restrictions and the large numbers of units in typical practical problems could require a huge memory and computational time. (But who can find a nontrivial multidimensional problem which has a simple solution?)

10.4 MULTICRITERIA OPTIMIZATION

10.4.1 Steepest Descent Method

The problem of multicriteria optimization is always concerned with a choice: only a user can choose a preferable variant of the system among "incomparable" ones. Each solution of a multioptimization problem gives us a set of different vectors, each component of which is an attained value of one of the parameters. One should compare several "acceptable" variants and choose the "best" one among them. The task of defining "the best variant" is that of the decision maker. Sometimes unformalized factors may prevail: a preference of some particular vendor, a subjective preference of a particular system configuration, some additional restrictions which never were mentioned before, and so on. Of course, the final decision lies outside of mathematics, but analytical methods can help to understand the problem: it is more difficult to make an incorrect decision if you know why and in what sense it is wrong!

We will demonstrate the situation on a simple (but frequently encountered) example. Consider proposals prepared for a project by several different vendors. All proposals are satisfactory according to the principal parameter, say reliability, but they differ in terms of their cost, power supply, size of equipment, and other restricting factors. None of these proposals is absolutely better than the others: one is cheaper, another needs less of a power supply, and so forth. By a lucky coincidence, your best friend is among the contractors! You may choose your friend's proposal without hesitation: as you can claim that his proposal perfectly satisfies your criteria. Nobody can object to this if there is no other proposal which is better in all parameters.

Consider a system that is simultaneously characterized by several restricting factors. The main goal is to design a system which satisfies the most important criterion, for example, the reliability index, $R(\mathbf{X})$. But, at the same time, the system should satisfy the remaining restrictions "in a best way." In what sense? In the sense of some compromise among the restrictions. As with all compromises, this one is full of subjective opinions (as we have picturesquely shown above).

Each possible solution for $R(\mathbf{X}) \geq R_0$ is characterized by a specific set of restricting parameters $c_1(\mathbf{X}), c_2(\mathbf{X}), \ldots, c_M(\mathbf{X})$. The designer is faced with the problem of choosing the most preferable solution. But first the designer must obtain a set of all principally acceptable solutions, that is, a set of solutions with an acceptable value of the reliability index and a different undominated set of other parameters.

The problem of constructing of a multidimensional Pareto set is always computationally difficult. As a matter of fact, the Pareto set of solutions serves as a working tool for decision makers, so it is important not only to obtain this set, but also to display it in a visually convenient form. Here we will discuss only the method of construction of the Pareto set, and not a computer visualization of its presentation.

In mathematical terms, the problem is formulated as

$$\underset{\mathbf{X}}{\text{MIN}}\{C_1(\mathbf{X}),\dots,C_M(\mathbf{X})|R(\mathbf{X}) \geq R_0\} \tag{10.41}$$

where the symbol MIN denotes an optimization in the Pareto sense.

One of the simplest ways to construct a subset of a Pareto set consists of solving one-dimensional problems with the use of "weighted" costs. (Note that this may not give an entire Pareto set.) Of course, the increments of each component of D_k must be chosen small enough. For each D_k one finds the solution

$$\{R(\mathbf{X}^k), c_1(\mathbf{X}^k),\dots,c_M(\mathbf{X}^k)\}$$

Then, for all solutions so obtained, one compares the vectors

$$\mathbf{c}(\mathbf{X}^k) = (c_1(\mathbf{X}^k),\dots,c_M(\mathbf{X}^k))$$

If there are identical solutions, one of them leaves and the remaining are omitted. If there are two vectors, say $\mathbf{c}(\mathbf{X}_k^i)$ and $\mathbf{c}(\mathbf{X}_k^j)$, such that $\mathbf{c}(\mathbf{X}_k^i) < \mathbf{c}(\mathbf{X}_k^j)$, then the second one is omitted. Assume all solutions for different $\mathbf{X}^k$ can be found. Among them we select solutions with incomparable vectors of different resources spent and satisfying, at the same time, the requirement $R(\mathbf{X}) \geq R_0$.

It is clear that the Pareto set obtained by the procedure just described will not be complete because the SDM, in general, does not allow one to obtain a completely undominated sequence. Additionally, the discrete nature of the vector D_k makes the solution even less accurate. Therefore, we have only the subset of the Pareto set. Fortunately, in engineering practice, this "mathematical tragedy" can usually go unnoticed.

10.4.2 Method of Generalized Generating Function

Now we describe how to construct the exact Pareto set using the generalized generating function method (GGFM). Let us return to the procedure described at the very end of Section 10.3.2. As a matter of fact, the procedure of building the Pareto set is based on the procedure of building a complete multidimensional undominated sequence.

Let us order the cohorts of the undominated sequence obtained so far according to increasing values of the R-maniples. Consider only the subset of cohorts with R-maniples larger than R^O. Take the first one, say C_k. For this cohort, all maniples characterizing costs of different types can be represented by the vectors $\mathbf{c}_k = (c_{1k}, c_{2k}, \ldots, c_{Mk})$. All cohorts satisfying the vector inequality $\mathbf{c}^* > \mathbf{c}_k$ are excluded from further consideration because they provide the same condition $R^* > R^O$ with a larger expense for all types of costs. Now take the second cohort from the remaining set and check the same condition. The procedure terminates after a finite number of such steps. All of the remaining cohorts constitute the Pareto set of solutions.

In case there are more than two restricting factors, for the convenience of subsequent comparisons of undominated cohorts, one can use the lexico-graphical ordering of cohorts by vectors $\mathbf{c}_k$.

To make the process of determining the Pareto set more productive, a decision maker should a priori determine the rank of acceptability of each restricted factor.

10.5 COMMENTS ON CALCULATION METHODS

Optimal redundancy is a very important practical problem. The solution of the problem allows one to improve the reliability at a minimal expense. But here, as in many other practical problems, questions arise. What is the confidence of the obtained results? What is the real effect of the use of sophisticated mathematics? These are not unreasonable questions!

We have already discussed what it means to design an "accurate" mathe-matical model. It is always better to speak about a mathematical model which more or less correctly reflects a real object. But let us suppose that we think that the model is perfect. What price are we willing to pay for obtaining numerical results? What method is best, and best in what sense?

The use of excessively accurate methods is, for practical purposes, usually not necessary because of the uncertainty of the statistical data. On the other hand, it is inexcusable to use approximate methods without reason. We compare the different methods in terms of their accuracy and ease of computation.

The Lagrange multiplier method (LMM) demands the availability of con-tinuous, differentiable functions. This requirement is met very rarely even if we relax the essentially discrete nature of the optimal redundancy problem. But this method can sometimes be used for a rough estimation of the desired solution.

The steepest descent method (SDM) is very convenient from a computa-tional viewpoint. It is reasonable to use this method if the resources that one might spend on redundancy are large. Of course, this generally coincides with the requirement of a high system reliability because this usually involves large expenditures of resources. As we mentioned above, the SDM may provide an

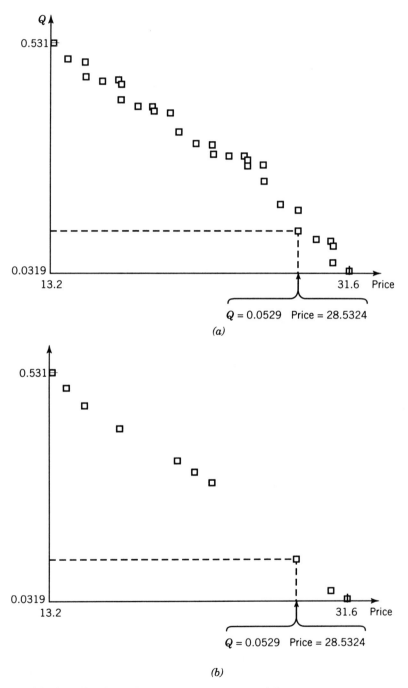

Figure 10.2. Sample of solutions by steepest descent (*a*) and dynamic programming (*b*) methods.

exact solution. This happens if the obtained solution coincides with the given restriction. But, unfortunately, it happens very rarely in practice. At any rate, one can use this approach to solve most practical problems without hesitation.

In Figure 10.2 one finds two solutions obtained by the steepest descent (*b*) and dynamic programming (*a*) methods. All dots of the upper plot are represented in the lower plot; that is, the upper set of dots is a subset of the lower set of dots.

The absolute difference between the costs of two neighboring SDM solutions cannot exceed the cost of the most expensive unit c^* value:

$$c^* = \max_{1 \le i \le n} c_i$$

It is clear that the larger the total cost of the system, the smaller is the relative error of the solution.

The dynamic programming method and its modifications (Kettelle's algorithm and the method of generalized generating sequences) are exact, but they demand more calculation time and a larger computer memory. As with most discrete problems requiring an enumerating algorithm, these optimal redundancy problems are extremely hard.

Of course, one of the problems of interest is the stability of the solutions. How does the solution depend on the accuracy of the input data? How are the solutions obtained by the use of different methods distinguished? How much do the numerical results of the solutions obtained by different methods vary? An illustration of the problem is given by numerical experiments. We present a simple example performed with the use of the software package.

Example 10.5 Consider the following illustrative numerical example. A series system consisting of three units is investigated. The input data are assumed to be uncertain: unit PFFOs and costs are known with an accuracy of 10%. The problem is to check the stability of the optimal solutions over the range of variability of the parameters.

Solution.   Consider the five systems in Table 10.18. We first compare the solutions for all five cases if the specified total system cost is to be at most 30 units. For each case we give two results: one obtained by the SDM and the second (marked with asterisk) obtained by the DPM. The results are presented in Table 10.19.

Table 10.19 shows that the only differences between the approximate and exact solutions are observed for the cases c and c*. But, even in this case, if we use an allocation rule based on the remaining resources, the solution will be the same. With an increase in spent resources, the relative difference between the solutions obtained by the SDM and the DPM will be increasingly smaller.

TABLE 10.18   Input Data for Systems Considered in Example 10.5

		1	2	3
(a)		1	2	3
	q	0.2	0.2	0.2
	c	1	1	1
(b)		1	2	3
	q	0.2	0.2	0.2
	c	0.9	1	1.1
(c)		1	2	3
	q	0.18	0.2	0.22
	c	0.9	1	1.1
(d)		1	2	3
	q	0.18	0.2	0.22
	c	1.1	1	0.9
(e)		1	2	3
	q	0.18	0.2	0.22
	c	1	1	1

TABLE 10.19   Results of Solutions for Five Systems of Example 10.5 for Cost Restrictions Equal to 30 Cost Units

Variant	Number of Redundant Units			Probability of System Failure	Factual Total Cost
	x_1	x_2	x_3		
a	10	10	10	$3.07 \cdot 10^{-7}$	30
a*		the same			
b	10	10	10	$3.07 \cdot 10^{-7}$	30
b*		the same			
c	9	10	10	$5.66 \cdot 10^{-7}$	29.1
c*	10	10	10	$4.04 \cdot 10^{-7}$	30
d	9	10	11	$3.59 \cdot 10^{-7}$	29.8
d*		the same			
e	9	10	11	$3.59 \cdot 10^{-7}$	29.8
e*		the same			

Now compare all of the solutions. They all are of the form $(10, 10, 10)$ or $(9, 10, 11)$. The latter is associated with the case where the difference in the unit reliability indexes is large.

We now analyze the solutions corresponding to a specified level of reliability. In Table 10.20, the corresponding results for $Q_0 = 1 \cdot 10^{-6}$ are collected.

Computational experiments with the same system for smaller and larger restrictions on the total system cost ($C_0 = 15$ and $C_0 = 50$) yield similar pictures emphasizing the closeness of the results.

TABLE 10.20   Results of Solutions for Five Systems of Example 10.5 for Restriction on Unreliability Equal to $1 \cdot 10^{-6}$

Variant	Number of Redundant Units			Probability of System Failure	Factual Total Cost
	x_1	x_2	x_3		
a	9	10	10	$7.17 \cdot 10^{-7}$	29
(the equivalent solutions are: 10, 9, 10 and 10, 10, 9)					
a*		the same			
b	10	10	9	$7.17 \cdot 10^{-7}$	28.9
b*		the same			
c	9	9	10	$9.76 \cdot 10^{-7}$	28.1
c*		the same			
d	9	9	10	$9.76 \cdot 10^{-7}$	27.9
d*		the same			
e	9	9	10	$9.76 \cdot 10^{-7}$	28.0
e*		the same			

Such numerical examples often help one to sharpen one's computational intuition. For example, one could decide that, for the approximate estimates, it is possible to concentrate different units in one of just a few groups so as to make the problem simpler in a computational sense. Therefore, suppose the system consists of six units with parameters

$$q_1 = 0.11, \quad c_1 = 0.9 \qquad q_2 = 0.11, \quad c_2 = 1.0 \qquad q_3 = 0.11, \quad c_3 = 1.1$$
$$q_4 = 0.10, \quad c_4 = 0.9 \qquad q_5 = 0.10, \quad c_5 = 1.0 \qquad q_6 = 0.10, \quad c_6 = 1.1$$
$$q_7 = 0.9 \,, \quad c_7 = 0.9 \qquad q_8 = 0.9 \,, \quad c_8 = 1.0 \qquad q_9 = 0.9 \,, \quad c_9 = 1.1$$

Suppose one is interested in finding the optimal allocation of resources under the restriction that the total system cost C_0 be at most 90. Since the parameters are close, one can simply guess that the solution is $x_1 = x_2 = \cdots = x_9 = 90/9 = 10$.

If, for the same system, $Q_0 \le 0.001$ is desired, the solution can be found in the following way. Assume that we use a loaded redundancy. For each redundant group, $Q_{0i} \le 0.00011$. Thus, $q^* \approx 0.0001$ leads to $x = 4$ for all types of units.

Numerical computer experiments and practical experience in finally solution of the optimal redundancy problem could develop a keen engineering intuition in the approximate solving of such problems and their sensitivity analysis.

CONCLUSION

The first papers dedicated to the optimal redundancy problem appeared in the late 1950s [see, e.g., Moskowitz and McLean (1956)]. All early papers in the field were methodological in character. They formulated the problem but did not give a constructive practical solution. At any rate, they were very important first steps.

The SDM was one of the first methods introduced to solve the optimal redundancy problem. Later in Barlow and Proschan (1975) it was shown that the SDM allows one to obtain a subset of undominated sequences for a wide class of practical methods of redundancy with a logarithmic convexity of a goal function.

The earliest publication concerning a DPM application to optimal redundancy problems can be found in Bellman and Dreyfus (1958). Kettelle's algorithm was published in connection with the spare-parts supply problem [Kettelle (1962)]. It seems that this is one of the most convenient algorithms of discrete optimization for an optimal redundancy problem. Probably, this algorithm is not widely used only because it was directed to the solution of a particular applied problem. (Alas, extremely "pure" mathematicians sometimes pay no attention to publications in technical and applied journals . . .). The further development of Kettelle's algorithm with the use of the generalized generating function can be found in Ushakov (1986, 1987a, 1987b, 1988).

The SDM and Kettelle's algorithm are systematically and comprehensively presented in Ushakov (1969), Kozlov and Ushakov (1970), Barlow and Proschan (1975), and Ushakov (1985, 1994).

In 1963 Everett introduced a generalization of the Lagrange multiplier method for the solution of the optimal redundancy problem. This paper spent a long time in obscurity, possibly because the author also treated a specially structured problem of reliability. Later this idea was effectively explored for the estimation of bounds on possibly optimal solutions for different discrete optimization problems.

Systematic consideration of the problem can be found in Barlow and Proschan (1975), Ushakov (1970), and Volkovich et al. (1992).

New directions in optimal redundancy include the analysis of multipurpose systems [Ushakov (1971); Karshtedt and Kogan (1971); Genis and Ushakov (1984)].

Using Monte Carlo simulation in a nonstandard way for the optimal redundancy problem was introduced in Ushakov and Yasenovets (1977) and Ushakov and Gordienko (1978). This opens new possibilities for solving different tasks in this field. By its nature, it reminds one of adaptive procedures which are used in everyday practice: on the basis of observed data, one generates the decision which fits these particular data best of all. Then the solution may be corrected by further observations. This method will be considered in the second volume.

Some important practical problems of spare-parts supply will be considered in Chapter 13. The complexity of the structure of real inventory systems and their inventory control processes leads, by necessity, to more sophisticated methods combined with a heuristic.

REFERENCES

Aggarval, K. V. (1976). Redundancy optimization in general systems. *IEEE Trans. Reliability*, vol. R-25, no. 5.

Barlow, R. E., and F. Proschan (1975). *Statistical Theory of Reliability and Life Testing Probabilistic Models*. New York: Holt, Reinhart, and Winston.

Bellman, R., and S. Dreyfus (1958). Dynamic programming and reliability of multicomponent devices. *Oper. Res.*, vol. 6, no. 2.

Everett, H. (1963). Generalized Lagrange multiplier method for solving problems of optimum allocation of resources. *Oper. Res.*, vol. 11, no. 3.

Genis, Ya. G., and I. A. Ushakov (1984). Optimization of multipurpose systems. *Engrg. Cybernet.* (USA), vol. 21, no. 3.

Gertsbakh, I. B. (1966). Optimum use of reserve elements. *Engrg. Cybernet.*, vol. 4, no. 5.

Gertsbakh, I. B. (1970). Dynamics reserves: optimal control of spare items (in Russian). *Avtomat. i Vychisl. Tekhn.*, no. 1.

Gnedenko, B. V., ed. (1983). *Aspects of Mathematical Theory of Reliability* (in Russian). Moscow: Radio i Sviaz.

Karshtedt, I. M., and L. M. Kogan (1971). Optimal standby with a choice of setups. *Engrg. Cybernet.* (USA), vol. 9, no. 2.

Kettelle, J. D. (1962). Least-cost allocation of reliability investment. *Oper. Res.*, vol. 10, no. 2.

Konev, U. V. (1974). Optimal inclusion of standby elements. *Engrg. Cybernet.* (USA), vol. 12, no. 4.

Kozlov, B. A., and I. A. Ushakov (1970). *Reliability Handbook*. New York: Holt, Reinhart, and Winston.

Mandel, A. S., and A. L. Raykin (1967). The optimal plan of switching on spare units. *Automat. Remote Control*, no. 5.

Moskowitz, F., and J. McLean (1956). Some reliability aspects of system design. *IRE Trans.*, vol. PGRQC-8.

Pestov, G. G., and L. V. Ushakova (1971). Optimum strategies in dynamic replacement. *Engrg. Cybernet.* (USA), no. 5.

Tillman, F. A., H. Ching-Lai, and W. Kuo (1980). *Optimization of System Reliability*. New York: Marcel Dekker.

Ushakov, I. A. (1965). Approximate solution of the optimal redundancy problem (in Russian). *Radiotechnika*, no. 12.

Ushakov, I. A. (1969). *Solution of Optimal Redundancy Problems Under Constraints* (in Russian). Moscow: Sovietskoe Radio.

Ushakov, I. A. (1971). Approximate solution of optimal redundancy problem for multipurpose systems. *Engrg. Cybernet.* (USA), vol. 9, no. 2.

Ushakov, I. A. (1972). A heuristic method of optimization of the redundancy of multifunction system. *Engrg. Cybernet.* (USA), vol. 10, no. 4.

Ushakov, I. A. (1981). Method of approximate solution of problem of dynamic standby. *Engrg. Cybernet.* (USA), vol. 19, no. 2.

Ushakov, I. A., ed. (1985). *Reliability of Technical Systems*: *Handbook*. Moscow: Radio i Sviaz.

Ushakov, I. A. (1986). A universal generating function. *Soviet J. Comput*. (in Russian) *Systems Sci*., vol. 24, no. 5.

Ushakov, I. A. (1987a). Solution of multicriteria discrete optimization problem using a universal generating function. *Soviety J. Comput. Systems Sci*., vol. 25, no. 5.

Ushakov, I. A. (1987b). Optimal standby problems and a universal generating function. *Soviet. J. Comput. Systems Sci*., vol. 25, no. 4.

Ushakov, I. A. (1988). Solving an optimal redundancy problem by means of generalized generating function. *J. Inform. Process. Cybernet*., vol. 24, no. 4–5.

Ushakov, I. A., ed. (1994). *Handbook of Reliability Engineering*. New York: Wiley.

Ushakov, I. A., and E. I. Godienko (1978). On statistical simulation approach to solution of some optimization problems. (in Russian). *Elektron. Informationsverarb. Kybernet*., vol. 14.

Ushakov, I. A., and A. V. Yasenovets (1977). Statistical methods of solving problems of optimal standby. *Engrg. Cybernet*. (USA), vol. 15, no. 6.

Volkovich, V. L., A. F. Voloshin, V. A. Zaslavsky, and I. A. Ushakov (1992). Models and Methods of Optimization of Complex Systems Reliability (in Russian). Kiev: Naukova Dumka.

EXERCISES

10.1   Consider a series system consisting of three units. Each unit has an exponentially distributed TTF with parameters (in $1/\text{hour}$) $\lambda_1 = 0.0001$, $\lambda_2 = 0.0002$, and $\lambda_3 = 0.0003$, respectively. The costs of the units are $c_1 = 1$, $c_2 = 3$, and $c_3 = 2$, respectively. Using the SDM find an optimal allocation of spare units with a minimal cost for the following cases:

(a) The system is supposed to be supplied with spare units for $t_0 = 200$ hours. The probability of successful operation must not be less than 0.95. (*Hint*: use the Poisson distribution.)

(b) The system is supposed to be supplied with spare units for $t_0 = 20{,}000$ hours. The probability of successful operation must not be less than 0.95. (*Hint*: use the normal distribution approximation.)

Compute the minimal cost of the spare units in these cases.

10.2   For the conditions of the previous exercise, find an optimal allocation of spare units with the maximal probability of successful system operation for the following cases:

(a) The system is supposed to be supplied with spare units for $t_0 = 200$ hours. The total system cost (main and spare units) must not exceed 12 cost units.

(b) The system is supposed to be supplied with spare units for $t_0 =$ 20,000 hours. The total system cost (main and spare units) must not exceed 30 cost units.

10.3 For the conditions of Exercise 10.1, find a solution using the approximate method described in Section 10.2.3.

10.4 For the conditions of Exercise 10.2, find a solution using the approximate method described in Section 10.2.3.

CHAPTER 11

OPTIMAL TECHNICAL DIAGNOSIS

11.1 GENERAL DESCRIPTION

Modern technical systems have a complex structure and consist of a huge number of components, modules, and blocks. In order to repair such systems in a timely manner, it is of the utmost importance to be able to perform a swift and competent diagnosis. A technical diagnosis of a system consists of recognizing that the system has failed and then determining the location of the failed part of the system (unit, module, block, etc.) with a reasonable accuracy. Most modern systems are designed with a modular construction which allows an operationally simple replacement of the failed entity. Thus, for a user, the accuracy of a technical diagnosis should correspond to the "level" of the part (chip, block, subsystem) which can be replaced and/or repaired. Most modern systems are designed with replaceable modules. After a failure, the failed module is sent to a repair shop. Here the diagnosis might continue: the failed module itself could be considered an independent object with the concomitant problem of determining the source of the local failure (if repair is possible). Some systems are monitored continuously; others need a periodic checking for a diagnosis of the system's current state.

For a diagnostic analysis, a system may be represented by n independent units connected to each other by functional ties. Each unit may be in one of two possible states: an operational state and a failed state. Let p_i denote the probability that unit i is in an operational state and let $q_i = 1 - p_i$.

A technical diagnosis of such a system is based on functional tests. Each test serves to determine the state of a subset of units which are usually performing some particular functional task. The technical diagnosis has two main goals: (1) to determine if the system is up or down and (2) to localize all failed units if the system is down.

The overall failure or operability of a system is ordinarily determined by a "global" test. However, even when a global test exists, it might require an excessive amount of time or other resources, and one might prefer a sequence of ordinary tests.

For diagnosis, a set of tests T_i, $i = 1, \ldots, m$, each of which checks a subset of units Ω_i, might be used. Let the entire set of system units be denoted by Ω. The complementary subset to Ω_i is denoted by $\overline{\Omega}_i$, $\overline{\Omega}_i = \Omega \setminus \Omega_i$. The results of the tests are classified as "successful" if all units belonging to a subset Ω_i are operational, or "unsuccessful" if at least one unit of the set has failed. Each test is characterized by a cost c_i. The "cost" is understood in general terms; that is, it may be measured by the duration of its diagnosis, by the salaries of the personnel required for its application, by the cost of the test equipment needed, or by a combination of these resources. The set of tests can be described by the $m \times n$ "incidence" matrix $T = \|T_{ij}\|$, where

$$T_{ij} = \begin{cases} 1 & \text{if} \quad j \in \Omega_i \\ 0 & \text{if} \quad j \in \overline{\Omega}_i \end{cases} \qquad i = \overline{1, m}; \, j = \overline{1, n}$$

The row vector $\mathbf{c} = \{c_1, \ldots, c_m\}$ describes of the costs of the tests. The available set of tests must be sufficient for successful performance of the diagnostic procedure.

The optimization of a diagnostic procedure is a special engineering problem. We shall present some simple mathematical models and algorithms that can be used to design equipment and diagnostic procedures. We would like to emphasize that an optimal technical diagnostic problem is, primarily, an engineering problem, not an applied mathematical problem.

11.2 ONE FAILED UNIT

Sometimes we know in advance that a system under diagnosis has a single failed unit. Such a situation exists, for instance, if a system stops immediately after the failure of one of its units. As an example, consider a series system under a continuous and absolutely reliable monitoring. Let the test matrix $\mathbf{T}$ and the probabilities q_i be given. Then the optimization problem is to select the set of tests and their order which minimizes the total expense of the diagnosis.

The application of a test T_i divides the set of system units Ω into two subsets: Ω_i and $\tilde{\Omega}_i$. If the test is "unsuccessful," the failed unit is in the subset Ω_i, and if the test is "successful," the failed unit is in the subset $\tilde{\Omega}_i$. To isolate the failed unit, one can use tests $T_e \in \mathbf{T}$, which divide the subset Ω_i or $\tilde{\Omega}_i$. We call a test T_e *essential* for a subset Ω_i if and only if

$$\Omega_e \cap \Omega_i \neq \Omega_i$$

and

$$\Omega_e \cap \Omega_i \neq \varnothing$$

where $\cap$ is an intersection sign. The set of essential tests for a subset Ω_i is denoted by $\mathbf{T}_i$. If, for two tests T_u and T_v from $\mathbf{T}_i$,

$$\Omega_u \cap \Omega_i = \Omega_v \cap \Omega_i$$

these tests are equivalent in terms of their "resolution ability." Then one of then, which is characterized by a larger cost, is excluded from the set $\mathbf{T}_i$. The set $\mathbf{T}_i$ can be depicted by a matrix whose rows correspond to the unit sets Ω_i which are covered by the related test. The matrix columns correspond to the system units. Figure 11.1 explains the details. Each kth row of this matrix corresponds to the test $T_{u(i)}^k$, $k = 1, \ldots, m_i$, which checks the operational state of the subset

$$\Omega_{u(i)}^k = \Omega_u \cap \Omega_i$$

with cost $c_{u(i)}^k = c_u$. The subscript without parentheses identifies each test of the initial matrix $\mathbf{T}$. For simplicity, we will sometimes omit this subscript.

Test #	Unit # 1	2	3	4	5	6	7	8
1	1	1			1	1		
2		1		1			1	
3			1		1	1		
4	1					1		1
5	1		1	1	1			
6					1		1	1

Figure 11.1.

If a failed unit is in the subset $\tilde{\Omega}_i$, the test matrix for this subset can be formed analogously. If there is only one failed unit, the necessary condition of the sufficiency of the matrix $\mathbf{T}$ is the difference between all of its pairs of columns.

11.2.1 Dynamic Programming Method

First, we introduce some additional notation. The probability that the test for the subset Ω_i is unsuccessful is

$$Q_i = \sum_{j \in \Omega_i} \tilde{q}_j$$

If the previous steps have confirmed that the failed unit is in the subset Ω_i, the probability that a unit is in a subset Ω_u, $\Omega_u \subset \Omega_i \neq \varnothing$, is

$$\Omega_{u(i)} = \frac{Q_u}{Q_i}$$

Assume that the procedure (or program) for determining the locality of a failed unit in the subset Ω_i is known. Denote this program by $\sigma(\Omega_i)$. At the first step of the procedure, suppose we use a test $T_{(i)}^k \in \mathbf{T}_i$. Then the program $\sigma(\Omega_i)$ can be represented as

$$\sigma(\Omega_i) = \left[T^k(i), \sigma\left(\Omega_{(i)}^k\right), \sigma\left(\overline{\Omega}_{(i)}^k\right)\right]$$

Thus, if one begins to analyze the subset Ω_i and applies the test $T_{(i)}^k$, there are two possible ways to continue the diagnostic procedure: $\sigma(\Omega_{(i)}^k)$ and $\sigma(\overline{\Omega}_{(i)}^k)$. Which one should be chosen depends on the locality of the failed unit, that is, on the result of the previous test.

The expected value of the cost of the program $\sigma(\Omega_i)$ can be found from the recurrent expression:

$$C[\sigma(\Omega_i)] = c_{(i)}^k + Q_{(i)}^k C\left[\sigma\left(\Omega_{(i)}^k\right)\right] + \left(1 - Q_{(i)}^k\right)C\left[\sigma\left(\overline{\Omega}_{(i)}^k\right)\right] \quad (11.1)$$

The calculations start with two-unit sets. For each of the units, the cost of locating a failed unit equals the minimal cost of the test which allows us to divide this subset into two separate one-unit subsets.

The dynamic programming method is based on a systematic consideration of the subsets $\Omega_i \in \Omega$ and on the determination of a test $T_{(i)}^k \in \mathbf{T}_i$ which divides this subset in an optimal way.

If $\sigma^*(\Omega_i)$ is the optimal program for a subset Ω_i containing a failed unit, and if $C[\sigma^*(\Omega_i)]$ is the cost of this program, the procedure for finding $\sigma^*(\Omega)$ can be represented as a recurrent procedure for finding $\sigma^*(\Omega_i)$ for all subsets of x units ($x = 2$ for the first iteration). We increase x by 1 with each iteration until an optimal diagnostic procedure has been constructed. At each step of the procedure, one finds a test $T_{(i)}^k \in \mathbf{T}_i$ such that

$$C[\sigma^*(\Omega_i)] = \min_k \left\{ c_{(i)}^k + Q_{(i)}^k C[\sigma^*(\Omega_{(i)}^k)] + \left(1 - Q_{(i)}^k\right) C[\sigma^*(\overline{\Omega}_{(i)}^k)] \right\}$$

To construct an optimal diagnostic program by dynamic programming, all tests that divide all current subsets of the set Ω in an optimal fashion must be found and recorded. While this method does provide an exact optimal solution, its application for problems of any practical dimension is generally unrealistic. Thus, we must design a heuristic algorithm (see Chapter 13) to obtain an approximate solution. Below we consider the exact method for one very particular case.

11.2.2 Perturbation Method for One-Unit Tests

If one-unit tests are available, a simple solution by dynamic programming can be proposed. This method is based on the fact that by a finite number of transpositions of neighboring tests one can obtain any specified in advance sequence of tests (including the optimal sequence). The problem is to find a monotone procedure of such transposition (reordering) which will produce an optimal procedure.

Of course, one easily suspects that if all tests are equal in cost, the tests should be ordered according to increasing unit reliability. Indeed, the least reliable unit is most likely to fail, so it must be checked first. On the other hand, if all units are equally reliable, then the tests should be ordered according to increasing test costs as it is natural to try the test that costs the least before trying other more expensive tests. Nontrivial problems demand more interesting mathematics.

For an arbitrary initial ordering of the tests, the cost of the test procedure is expressed by

$$C[\sigma(\Omega)] = c_1 + \tilde{q}_1 C[\sigma(e_1)] + \tilde{p}_1 C[\sigma(\Omega \setminus e_1)] = c_1 + \tilde{p}_1 C[\sigma(\Omega \setminus e_1)]$$

where e_1 is the first one-unit subset (unit 1) that does not affect the costs of future checking, $C[\sigma(e_1)] = 0$. The probability $\tilde{q}_i = 1 - \tilde{p}_i$ is the probability that the ith unit has failed if it is known that exactly one unit of the system has failed:

$$\tilde{q}_i = \frac{q_i \displaystyle\prod_{\substack{1 \le j \le n \\ j \ne i}} p_j}{\displaystyle\sum_{1 \le j \le n} q_j \displaystyle\prod_{\substack{1 \le k \le n \\ k \ne j}} p_k} = \frac{\dfrac{q_i}{p_i}}{\displaystyle\sum_{1 \le j \le n} \dfrac{q_j}{p_j}}$$

It is clear that if unit 1 is tested, the cost c_1 is paid; then, if this unit fails, the procedure is stopped; otherwise, with probability $\tilde{p}_1$, the diagnostic procedure for the remaining subset of units must be continued. In the latter case,

$$C[\sigma(\Omega \setminus e_1)] = c_2 + \tilde{p}_2 C[\sigma(\Omega \setminus (e_1 \cup e_2))]$$

Finally, the recurrent expression of the total diagnostic procedure cost is

$$C = c_1 + \tilde{p}_1(c_2 + \tilde{p}_2(c_3 + \cdots + \tilde{p}_k(c_k + \tilde{p}_{k+1}(c_{k+1} + \cdots)) \cdots)) \quad (11.2)$$

or, in a more convenient form,

$$C(k, k+1) = c_1 + \sum_{1 \le i \le k-1} c_i \prod_{1 \le j \le i-1} \tilde{p}_j + c_k \prod_{1 \le j \le k-1} \tilde{p}_j$$

$$+ c_{k+1} \prod_{1 \le j \le k} \tilde{p}_j + \sum_{k+2 \le i \le n} c_i \prod_{1 \le j \le i-1} \tilde{p}_j$$

$$= A_* + c_k A_{k-1} + c_{k+1} \tilde{p}_k A_{k-1} + A^* \quad (11.3)$$

where

$$A_* = c_1 + \sum_{1 \le i \le k-1} c_i \prod_{1 \le j \le i-1} \tilde{p}_j$$

$$A_{k-1} = \prod_{1 \le j \le k-1} \tilde{p}_j$$

$$A^* = \sum_{k+2 \le i \le n} c_i \prod_{1 \le j \le i-1} \tilde{p}_j$$

Thus, we have grouped all terms of the sum except the two neighboring terms k and $k+1$. These two are especially distinguished for further discussion. The notation we use is obvious.

Now change the order of the units k and $k+1$ for the test procedure. Using the last part of (11.3), we have

$$C(k+1, k) = A_* + c_{k+1} A_{k-1} + c_k \tilde{p}_{k+1} A_{k-1} + A^* \quad (11.4)$$

Determine when $C(k, k+1) \ge C(k+1, k)$. From (11.3) and (11.4) it follows that this condition corresponds to

$$c_k + c_{k+1} \tilde{p}_k \le c_{k+1} + c_k \tilde{p}_{k+1}$$

that is, the final form of the condition is found in the simple form

$$c_k \tilde{q}_{k+1} \le c_{k+1} \tilde{q}_k$$

or

$$\frac{c_k}{\tilde{q}_k} \le \frac{c_{k+1}}{\tilde{q}_{k+1}} \tag{11.5}$$

From (11.5) one can see that all units must be ordered in correspondence with nondecreasing values of $c_k/\tilde{q}_k$. This order corresponds to the optimal order of the tests in the diagnostic procedure.

This solution is extremely simple. But, unfortunately, this degenerate case is seldom encountered in practice. Again, we have an unfortunate situation: this case is practically impossible but the solution is elegant!

11.2.3 Recursive Method

For the exact solution in the general case, one can use a recursive method, which, in practice, can be more effective than direct dynamic programming.

A search for a failed unit in the set Ω can be regarded as a multistep process of a sequential division of its subsets until subsets consisting of a single unit are obtained. The essence of the method consists in the construction of a "decision tree" by a sequential division of Ω. Each branch of this tree is dichotomously divided until the best solution appears at the last layer where only one-unit subsets are located. The optimal solution can be found by a backward movement from this last layer to the root of the decision tree.

At the first step of the procedure, we can use any test from the matrix **T** for dividing the set Ω into two subsets. Consider test T_1 which divides the set Ω into the two subsets Ω_1 and $\tilde{\Omega}_1$: $\sigma_1^*(\Omega_1)$ if the test T_1 is unsuccessful and $\sigma_1^*(\tilde{\Omega}_1)$ if the test T_1 is successful are known. Then one can construct the conditionally optimal program

$$\sigma_1^*(\Omega) = \left[T_1, \sigma^*(\Omega_1), \sigma^*(\overline{\Omega}_1) \right]$$

and calculate the expected cost of this procedure. Then $\sigma_1^*(\Omega_1)$ and $\sigma_1^*(\overline{\Omega}_1)$ are analyzed in the same manner. As a result, a sequence of the applied tests has been found with the condition that the procedure start with test T_1. In other words, we have obtained the best conditional (respective to the specified starting test T_1) test sequence.

After this, in a similar manner, we construct the best test sequence for the conditionally optimal program $\sigma_2^*(\Omega)$. This search stops when each test is tried as a starting one. The test sequence which is characterized by the lowest cost is chosen as the solution of the problem.

Assume that, for some subset Ω_i, the program $\sigma^*(\Omega_i)$ needed to construct $\sigma_i^*(\Omega_i)$ is unknown. Select the best of the previously discovered $i - 1$ conditionally optimal programs and consider all possible divisions of the subset Ω_i. Construct the matrix of the essential tests, $\mathbf{T}_i$, and, for each test $T_{(i)}^k$ from this matrix, find the conditionally optimal program:

$$\sigma^{k*}(\Omega_i) = \left\{ T_{(i)}^{(k)}, \sigma^*\left(\Omega_{(i)}^k\right), \sigma^*\left(\overline{\Omega}_{(i)}^k\right) \right\} \qquad k = 1, \ldots, m$$

Then find the optimal $\sigma^*(\Omega)$ as

$$C\left[\sigma^*(\Omega_i)\right] = \min_{1 \le k \le m_i} C\left[\sigma^{k*}(\Omega_i)\right]$$

The calculations can be continued in the same way.

The procedure is systematically applied to each subset, obtained by division of the set Ω, and leads to the construction of an optimal program for the entire system.

The efficiency of the method can be increased by eliminating those candidates from the program which, by a current evaluation, cannot lead to an optimal solution. This can be done, as usual, by comparing the candidate obtained at the current step with the best program found during previous steps.

The conditionally optimal programs $\sigma^{k*}(\Omega_i)$, $k = 1, \ldots, m_i$, which are the basis for finding the optimal program $\sigma^*(\Omega_i)$, can be constructed by properly ordering the tests $T_{(i)}^k$ o f the matrix $\mathbf{T}_i$. To improve the procedure of eliminating ineffective branches, we should try to find an approximate ordering of tests with increasing costs. Then one should use any method for the estimation of the lower bound of the cost of the remaining part of the diagnostic procedure. These devaluations might help to select and exclude from further consideration nonprospective variants. At some step $k - 1$ conditionally optimal programs might be found and the best one $\sigma^*(\Omega_i)$ which is characterized by the cost $C[\sigma^*(\Omega_i)]$. Let $\Gamma_{(i)}^k$ be a lower bound on the cost of a conditionally optimal program $\sigma^{k*}(\Omega_i)$. Obviously, the program $\sigma^{k*}(\Omega_i)$ cannot be the best one if

$$C\left[\sigma^{k*}(\Omega_i)\right] \ge \Gamma_{(i)}^k$$

Specific procedures that are best for the selection of the test sequence depend on the equipment in question. But some simple general recommendations can be formulated. The tests should be ranked in order of nondecreasing cost. To determine one possible bound $\Gamma_{(i)}^k$, one can use a simple and obvious rule: the cost of the conditionally optimal program $\sigma^{k*}(\Omega_i)$ cannot be less than $c_{(i)}^k$ (the cost of the test $T_{(i)}^k$ which is used at the first step of this procedure. In other words, one may take $\Gamma_{(i)}^k = c_{(i)}^k$.

The logic behind the construction of a diagnostic program is the same for all layers: this follows from the recurrent structure of the problem.

11.3 SEQUENTIAL SEARCHES FOR MULTIPLE FAILED UNITS

11.3.1 Description of the Procedure

There is often the possibility of more than one failure in a system. Such a situation occurs if there is no continuous monitoring. In this case, after each failed unit is found, it is replaced, and then the diagnostic procedure continues. The subset Ω_*, which includes the replaced unit, is tested. If the subset Ω_* still contains at least one failed unit, the search continues within this subset in accordance with the same rules. If the test for the subset Ω_* is successful, the search continues throughout the remaining part of the system. The procedure continues until all failed components have been replaced and, consequently, the system will be totally repaired.

We emphasize that the replacement cost does not affect the diagnostic procedure. Indeed, any failed component must be replaced or repaired regardless of the diagnostic method.

In many practical cases we can use a global test to determine if a system has any failed units. Then, after each failed component is replaced, we can use such a global test to make sure that there are no more failures in the system. Sometimes, however, a global test might cost too much relative to the complete sequence of subset tests. In particular, such a situation may appear at the final steps of the diagnostic procedure: one needs to test only a very small part of the system, and it is easier to use several local tests rather than a global test.

11.3.2 Perturbation Method

Consider the case where each of a system's units can be checked separately. The cost of checking unit i equals c_i. In addition, one can apply a global test to check the state of the entire system, and this test costs c_0.

Usually, one applies the global test first. If the system is found to be in a failed state, the search begins. After the failed unit has been found and replaced, one may choose to use the global test or to continue sequentially testing the individual units. If at least one failed unit remains elusive, the procedure continues. The problem is to find a simple rule of ordering tests.

As above, suppose that all test are numbered arbitrarily. First, we consider the case where the global test is used after each failed unit's replacement. Consider a solution of the problem which differs from the above-used solution for the case of one failed unit. This solution differs from the previous one in a methodological way. The cost of the unit k test depends on the previous steps as follows.

1. $C^{(0)} = 0$. This may occur if, during the diagnostic procedure, some unit i, $i < k$, has failed. This unit is replaced, and afterwards a global test shows that the system is operational. This means that the procedure stopped before it reached unit k.

2. $C^{(i)} = c_k$. This occurs if unit k is operational but must be checked because the procedure has not finished. In other words, at some previous step of the procedure, a global test shows that there are failed units among remaining units, and after that moment no failed units have been found, that is, at least one unit, say i, is failed among remaining ones, $i \geq k$. (This case includes a situation at the beginning of the procedure: one knows a priori that there are failed units in the system.)

3. $C^{(2)} = c_k + c_0$. Unit k has failed, thus it must be first checked and then replaced. (In this case, nothing depends on the states of all of the other units.) After this a global test is applied and if there are failed units, the procedure continues.

We are interested in the average cost of the test of the kth unit. This can be calculated as

$$c_k^* = P^{(0)}C^{(0)} + P^{(1)}C^{(1)} + P^{(2)}C^{(2)}$$

The problem is to find the probabilities $P^{(0)}$, $P^{(1)}$, and $P^{(2)}$. Notice that we do not need to find $P^{(0)}$ because $C^{(0)} = 0$. Now

$$P^{(1)} = p_k\left(1 - \prod_{k+1 \leq i \leq n} p_i\right)$$

and

$$P^{(2)} = q_k$$

Thus, the mean cost related to the kth unit test is

$$c^{(k)} = c_k\left(1 - \prod_{k+1 \leq i \leq n} p_i\right) + (c_k + c_0)q_k$$

The mean expenditure for the test of the entire system with an arbitrary (but fixed) order of units can be calculated as the sum of all individual average values of $c^{(k)}$:

$$C(k, k+1) = A - \sum_{1 \leq i \leq n} c_i \prod_{i+1 \leq j \leq n} p_j \tag{11.6}$$

where A denotes the term that does not depend on the order of the units. We present (11.6) in the following form:

$$C(k, k + 1) = A - \sum_{1 \le i \le k-1} c_i \prod_{i+1 \le j \le n} p_j - c_k \prod_{k+1 \le j \le n} p_j$$
$$- c_{k+1} \prod_{k+2 \le j \le n} p_j - \sum_{k+2 \le i \le n} c_i \prod_{k+3 \le j \le n} P_j \quad (11.7)$$

Extract terms depending on the order of units k and $k + 1$:

$$C(k, k + 1) = A^* - c_k \prod_{k+1 \le j \le n} p_j - c_{k+1} \prod_{k+2 \le j \le n} p_j \quad (11.8)$$

Here A^* is the sum of all terms that do not depend on the order of units k and $k + 1$.

Now let us change the checking order of units k and $k + 1$. The expression similar to (11.8) is

$$C(k + 1, k) = A^* - c_{k+1} p_k \prod_{k+2 \le j \le n} p_j - c_k \prod_{k+2 \le j \le n} p_j \quad (11.9)$$

Now compare (11.8) and (11.9). The equation having the minimal cost value corresponds to a better ordering. If $\Delta = C(k, k + 1) - C(k + 1, k) \le 0$, then the numbering $\ldots k, k + 1, \ldots$ is more preferable than the numbering $\ldots k + 1, k \ldots$. Compute the value of Δ:

$$\Delta = A^* - c_k \prod_{k+1 \le j \le n} p_j - c_{k+1} \prod_{k+2 \le j \le n} p_j$$
$$- \left(A^* - c_{k+1} p_k \prod_{k+2 \le j \le n} p_j - c_k \prod_{k+2 \le j \le n} p_j \right)$$
$$= -c_k p_k p_{k+1} B - c_{k+1} p_{k+1} B - (-c_{k+1} p_k p_{k+1} B - c_k p_{kB}) \le 0$$

where B is

$$B = \prod_{k+2 \le j \le n} p_j$$

The latter expression can be rewritten as

$$-c_k p_k p_{k+1} - c_{k+1} p_{k+1} - (-c_{k+1} p_k p_{k+1} - c_k p_k)$$
$$= c_k p_k q_{k+1} - c_{k+1} p_{k+1} q_k \le 0$$

This inequality is equivalent to

$$\frac{c_k p_k}{q_k} \le \frac{c_{k+1} p_{k+1}}{q_{k+1}}$$

Thus, the rule has been found: all unit tests must be ordered corresponding to nondecreasing values of $\tilde{p}_k c_k / \tilde{q}_k$. This order corresponds to the optimal order of the diagnostic tests.

The following problem may arise. If during the test procedure, a unit k has been found to have failed, is it reasonable to perform a global test? Is it better to continue separate unit tests? The answer can be easily found. The number k, after which there is no reason to perform a global test, can be determined from the condition:

$$\sum_{k \le j \le n} c_k \le c_0 \le \sum_{k-1 \le i \le n} c_i$$

11.3.3 Recursive Method

Consider the case where no global test is available. Suppose a test T_i divides the set Ω into Ω_i and $\tilde{\Omega}_i$. If the test is unsuccessful, that is, there are failures in the subset Ω_i, this subset must be divided again. For this purpose, one must use only the essential tests T_* from $\mathbf{T}$ for which $\Omega_* \subset \Omega_i$. (If the tests are chosen differently, the indication "test is unsuccessful" might be caused by failed units outside of Ω_i.) We denote the matrix of these tests by $\mathbf{T}_i$. One can apply similar arguments to the set $\tilde{\Omega}_i$ if the first test is successful.

The subset Ω_i is called *localizable* if there exists $\mathbf{T}_i$ which allows one to identify all failed units under condition that during the diagnostic procedure failed units are replaced. The matrix $\mathbf{T}$ is "sufficient" when it is possible to cover the set Ω with the subsets Ω_i so that each subset is localizable. In a subset of two units a test might check a state of any of these units, and the state of another can be deductively found. A set of three units requires the capability to localize any of its two-unit subsets. Requirements for larger subsets may be found by the deduction.

The sufficiency of the matrix $\mathbf{T}$ can be checked systematically. Consider a test T_i covering more than one unit. The new matrix $\mathbf{T}_i$ of the essential tests for the subset Ω_i is then developed in accordance with the above-described procedure. If there are no rows in a new matrix (no tests for dividing the subset of units), the subset Ω_i is not localizable. In the situation one must add new tests to the initial $\mathbf{T}$ or stop the procedure of failure localization in the subset Ω_i. If $\mathbf{T}_i$ contains several rows, we choose one which divides the subset into two parts as equally as possible.

If the localization property of this subset is not obvious, we need to construct the essential test matrix and check again. This process continues until we find that the subset Ω_i either can be localized or cannot be localized (or can be localized only in part).

Let $\Psi(\Omega_i)$ denote the sequential diagnostic program if failed units are in a subset Ω_i. Assume that, at the first step of this program, the test $T_i^k \in \mathbf{T}_i$ is used. This test divides the entire set Ω into two subsets: $\Omega_{(i)}^k$ and $\tilde{\Omega}_{(i)}^k$. We know that in this case $\Psi(\Omega_{(i)}^k)$ and $\Psi(\tilde{\Omega}_{(i)}^k)$ are used for continuation of the optimal diagnostic procedure. Then for $\Psi(\Omega_i)$ we can write

$$\Psi(\Omega_i) = \left[T_{(i)}^k, \Psi\left(\Omega_{(i)}^k\right), \left(\tilde{\Psi}_{(i)}^k\right) \right] \tag{11.10}$$

The cost of this program is a random variable depending on the current state of the system's units. The expected value of the cost is

$$C[\Psi(\Omega_i)] = c_{(i)}^k + \tilde{Q}_{(i)}^k \left\{ C\left[\Psi\left(\Omega_{(i)}^k\right)\right] + c_{0(i)} \right\} + \left(1 - \tilde{Q}_{(i)}^k\right) C\left[\Psi\left(\tilde{\Omega}_{(i)}^k\right)\right]$$

where $C_{0(i)}$ is the cost of replacement of unit i, and $\tilde{Q}_{(i)}^k$ is the posterior probability of the failure of at least one unit in the subset $\Omega_{(i)}^k$ when a failure has been detected in the subset Ω_i:

$$\tilde{Q}_{(i)}^k = \frac{1 - \displaystyle\prod_{j \in \Omega_{(i)}^k} p_i}{1 - \displaystyle\prod_{j \in \Omega_i} p_i}$$

This process essentially coincides with the one which was described in the previous section. The only important difference is in checking the effectiveness of the conditionally optimal programs. In this algorithm each conditionally optimal continuation of the diagnostic procedure is checked for the possibility of a prospective future use. For excluding nonprospective conditionally optimal programs $\Psi^{k*}(\Omega_i)$, we use the lower bound of the program cost calculated by

$$\Gamma_{(i)}^k = c_{(i)}^k + c_{0(i)} \tilde{Q}_{(i)}^k$$

11.4   SYSTEM FAILURE DETERMINATION

Usually, to determine if a system has failed, a global test is used. But if such a test is not available, we need to determine a system's failure with the help of partial tests. Here we consider an exact solution for tests which cover disjoint subsets of units.

One can introduce conditional units, each of which is equivalent to the corresponding set. In this case the previous notation and terminology can be used. Consider a system of n conditional units, each of which has a failure probability equal to q_i, $1 \le i \le n$. We need to find an ordering of tests to obtain the minimal cost of determining a system's failure.

Notice that the procedure is stopped as soon as the first failed unit has been found. In other words, the procedure stops at step k with probability

$$\prod_{1 \le j \le k-1} p_i$$

and cost equal to

$$\sum_{1 \le j \le k} c_i$$

Obviously, the procedure may continue to the very end if there are no failed units in the system at all. Indeed, we do not know in advance the real state of the system and, in fact, the system's state might be totally operational.

After these remarks, let us express the test procedure's cost for some arbitrary, but fixed numbering of units:

$$C(k, k+1) = q_1 c_1 + p_1 q_2 (c_1 + c_2) + p_1 p_2 q_3 c_3 + \cdots$$
$$+ q_k \prod_{1 \le j \le k-1} p_j \sum_{1 \le i \le k} c_i + q_{k+1} \prod_{1 \le j \le k} p_j \sum_{1 \le i \le k+1} c_i + \cdots$$
$$+ \prod_{1 \le j \le n-1} p_j \sum_{1 \le i \le n} c_i \qquad (11.11)$$

Rewrite (11.11) in the more convenient form

$$C(k, k+1) = A_* + q_k B(C + c_k) + q_{k+1} p_k B(C + c_k + c_{k+1}) + A^* \qquad (11.12)$$

where

$$A_* = q_1 c_1 + p_1 q_2 (c_1 + c_2) + p_1 p_2 q_3 c_3 + \cdots + q_{k-1} \prod_{1 \le j \le k-2} p_j \sum_{1 \le i \le k-1} c_i$$

$$A^* = q_{k+2} \prod_{1 \le j \le k+1} p_j \sum_{1 \le i \le k+2} c_i + \cdots + \prod_{1 \le j \le n-1} p_j \sum_{1 \le i \le n} c_i$$

$$B = \prod_{1 \le j \le k-1} p_j$$

$$C = \sum_{1 \le i \le k-1} c_i$$

Now let us change the order of the units k and $k+1$ and calculate the new cost of the procedure:

$$C(k+1, k) = A_* + q_{k-1} B(C + c_{k+1}) + q_k p_{k+1} B(C + c_k + c_{k+1}) + A^* \qquad (11.13)$$

As mentioned above, the optimal ordering can be found from the condition $\Delta = C(k, k + 1) - C(k + 1, k) \leq 0$. For this particular case, we can write

$$\Delta = q_k B(C + c_k) + q_{k+1} p_k B(C + c_k + c_{k+1})$$
$$- [q_{k+1} B(C + c_{k+1}) + q_k p_{k+1} B(C + c_k + c_{k+1})] \leq 0$$

or

$$q_k c_k + q_{k+1} p_k (c_k + c_{k+1}) - q_{k+1} c_{k+1} - q_k p_{k+1}(c_k + c_{k+1})$$
$$\leq C[q_k + q_{k+1} p_k - q_{k+1} - q_k p_{k+1}] \tag{11.14}$$

It is easy to see that

$$q_k + q_{k+1} p_k - q_{k+1} - q_k p_{k+1} = 0$$

and so (11.1) transforms into

$$q_k c_k + q_{k+1} p_k (c_k + c_{k+1}) - q_{k+1} c_{k+1} - q_k p_{k+1}(c_k + c_{k+1}) \leq 0$$

After a simple transformation, the latter inequality can be written in the equivalent form

$$c_k q_{k+1} \leq c_{k+1} q_k$$

Thus, the optimal order of unit tests corresponds to nonincreasing values of c_k/q_k. Note that the result coincides with the result for finding a single failed unit in the system.

CONCLUSION

The study of technical diagnostic procedures represents a separate branch of reliability theory. Some specialists consider it as an independent direction of engineering design. It is difficult to disagree with such a viewpoint. At the same time, methodologically technical diagnosis might be considered within the scope of reliability theory. The best evidence of this can be found in early papers on the subject [Glass (1959); Winter (1960); and Belyaev and Ushakov (1964)].

We note that in this chapter the mean cost of a diagnostic procedure is considered as a "goal" function. Of course, one can consider a diagnostic procedure based on minimax principles: find the test procedure minimizing the maximal possible cost. For this case one can write

$$C[\sigma(\Omega_i)] = c_{(i)}^k + \max\left\{ C[\sigma(\Omega_i)], C(\sigma(\overline{\Omega}_i)] \right\}$$

where all values are determined and calculated as in the preceding discussion.

For more detailed discussions of technical diagnosis, one may refer to Ushakov (1985, 1994). A detailed description of the recursive method is given in Pashkovsky (1981).

We should emphasize that the main problem appears not with the problem of mathematical optimization but in the design of the tests. In our opinion, some of these works are very close to so-called "software reliability." In modern computerized equipment, the main problem is to construct a logically designed battery of tests. Usually, these approaches have no "time-probabilistic" basis, and, consequently, they lie outside the scope of reliability theory in terms of our current understanding. Even a more or less detailed review of these methods lies outside the scope of this book.

REFERENCES

Belyaev, Yu. K., and I. A. Ushakov (1964). Mathematical models in problems of failure searching and localization. In *Cybernetics in Service for Communism*, A. Berg, N. Bruevich, and B. Gnedenko, eds. (in Russian) Moscow: Energiya.

Glass, B. (1959). An optimum policy for detecting a fault in complex systems. *Oper. Res.*, vol. 7, no. 4.

Pashkovsky, G. S. (1971). Methods of optimization of sequential search of failures. *Engrg. Cybernet.* (USA), no. 2.

Pashkovsky, G. S. (1977). Optimization algorithm of discrete search of failures. *Engrg. Cybernet.* (USA), no. 3.

Pashkovsky, G. S. (1981). *Problems of Optimal Search and Determination of Failures in Radio and Electronics Equipment*. (in Russian) Moscow: Radio i Sviaz.

Ushakov, I. A., ed. (1985). *Reliability of Technical Systems: Handbook*. (in Russian) Moscow: Radio i Sviaz.

Ushakov, I. A., ed. (1994). *Handbook of Reliability Engineering*. New York: Wiley.

Winter, B. (1960). Optimal diagnostic procedure. *IRE Trans.*, vol. PRQC-9, no. 3.

CHAPTER 12

ADDITIONAL OPTIMIZATION PROBLEMS IN RELIABILITY THEORY

This chapter presents a sample of the many types of optimization problems that arise in reliability. In some sense, all mathematical models of reliability only serve as a first step in the improvement of system reliability and performance effectiveness. More picturesquely, probabilistic models of reliability theory serve, not as a post morten analysis, but rather as a cure. Probabilistic models allow a designer to understand the system which is being designed. Simultaneously, as a direct consequence, these models allow one to find ways of improving a system's performance. Of course, if possible, this improvement should be done under some specified restrictions on the available resources. Thus, an optimal (or at least a rational) solution in a cost-effectiveness sense is sought.

Sometimes an improvement in reliability might be attained without any special expenditure of resources. For example, one might simply change the order of the units' repair and so increase a system's availability, or one might choose appropriate switching or maintenance moments and so increase a system's dependability.

The sample of problems presented below is by no means complete. We only show a sample of several reliability problems and applications of mathematical tools for their possible solution.

12.1   OPTIMAL PREVENTIVE MAINTENANCE

As a rule, as time increases, the reliability parameters of an object will worsen. This phenomenon is very natural and is determined by different aging and wearing-out processes. Usually, the TTF distributions of such objects are IFR, or IFRA, or other "aging" distributions. If a failure rate in some way increases with time, it is reasonable to replace a unit in current use with a new one at an appropriate and convenient moment before the unit fails during its active operating time.

There are some instances when such a replacement is totally useless. For example, if a preventive replacement would lead to the same loss (expense) as a replacement during active operation, there is no reason for the preventive replacement. Also, there is the case where a unit's TTF has an exponential distribution. In this case the reliability parameters of the new unit are identical to those of the used one.

Denote the period of preventive replacement by θ, the penalty for a failure which occurs during a system's operation by c_1, and the cost of a preventive replacement by c_2. Consider the problem of finding the value of θ which minimizes the average total expense per unit of time for a system working in a stationary regime. For this purpose, find the mean length of the cycle between two neighboring replacements and the average expense for this cycle.

The cycle between replacements is constant and equals θ if there is no failure and the current replacement is preventive. If a failure has occurred, assume that a replacement is performed at a random moment $\xi < \theta$. The mean length of this interval is

$$E\{\xi\} = \theta P(\theta) + \int_0^\theta tf(t)\, dt = \int_0^\theta P(t)\, dt$$

Here θ is the period between checks and $f(t)$ is the density function of TTF. The mean expenses are

$$C(\theta) = c_1 F(\theta) + c_2 P(\theta)$$

Using the standard technique of setting the first derivative to 0, we can find the optimal value of θ from the equation

$$\frac{d}{d\theta}\left(\frac{c_1 F(\theta) + c_2 P(\theta)}{\int_0^\theta P(t)\, dt} \right) = 0$$

The same equation can be rewritten in the equivalent form

$$\left[c_1 f(\theta) - c_2 f(\theta)\right] \int_0^\theta P(t) \, dt - \left[c_1 F(\theta) + c_2 P(\theta)\right] P(\theta) = 0 \quad (12.1)$$

From (12.1) the equation for the failure rate can be written as

$$\lambda(\theta) = \frac{c_1 F(\theta) + c_2 P(\theta)}{(c_1 - c_2) \int_0^\theta P(t) \, dt}$$

This equation can be solved numerically. Incidentally, for this purpose it is better to present the equation in the form

$$\lambda(\theta) \int_0^\theta P(t) \, dt - F(\theta) = \frac{c_2}{c_1 - c_2}$$

12.2   OPTIMAL PERIODIC SYSTEM CHECKING

Many technical units and systems are assumed to have an exponential time to failure. In such cases preventive replacement is useless because of the memoryless property of the exponential distribution. But periodic checking of the unit or system state remains very important because it allows one to find hidden failures.

Let the distribution of the TTF, $F(t)$, be exponential with parameter λ. The test needed to find the state of the system (up or down) costs θ units of time. In other words, the system cannot be used for its main purpose during the test procedure. Checking tests can be performed regularly within a time period t_c.

The frequent use of checking tests leads to a quick detection of hidden failures, and the system's idle time decreases. But, at the same time, much useful time is spent for "self-service." Common sense will hint that there is an optimal period of checking a system which yields a maximal time of a system's useful operation.

If a failure appears inside an interval θ, the conditional distribution of the system's useful time is

$$f_0(t) = \frac{\lambda e^{-\lambda t}}{1 - e^{-\lambda \theta}} \qquad 0 \le t \le \theta$$

Of course, if there is no failure between two neighboring check points, the system's useful time is equal to θ.

The system's mean useful time between check points equals

$$T^* = \theta e^{-\lambda\theta} + \int_0^\theta t\lambda e^{-\lambda t}\, dt = \frac{1}{\lambda}\left(1 - e^{-\lambda\theta}\right)$$

The problem is to find a value of θ which yields the maximum of the function

$$K(\theta, t_c, \lambda) = \frac{T^*}{\theta + t_c} = \frac{1 - e^{-\lambda\theta}}{\lambda(\theta + t_c)}$$

One needs to solve the transcendental equation

$$e^{\lambda\theta} = 1 + \lambda^{(\theta + t_c)} \tag{12.2}$$

This equation can be solved in very simple terms if $\lambda\theta \ll 1$ (a condition which is very common for most practical cases). In this case the exponential term can be successfully approximated by three terms of a series representation. Thus, (12.2) transforms into

$$\frac{(\lambda\theta)^2}{2} \approx \lambda t_c \tag{12.3}$$

whose solution is

$$\theta = \sqrt{\frac{2t_c}{\lambda}}$$

12.3 OPTIMAL DISCIPLINE OF REPAIR

The "restricted repair" case appears very often in practice, where the number of failures might exceed the number of available service facilities. In this case there is a waiting line of failed units (blocks, modules, etc.), and the problem of choosing an optimal serving discipline arises. It is clear that the optimal discipline depends not only on the parameters of the system and the service facilities but also on the criteria of reliability which is chosen.

We do not consider complex models here. Such problems can be special topics for detailed investigations. Many mathematical approaches of queuing theory might be applied to related reliability tasks. Therefore, we should usually replace the term "a demand" with "a failure" and change "a service unit" to "a repair station."

Let us begin with a very simple example. Consider a system consisting of two different units. Both units have exponentially distributed TTFs and repair times. The parameters of these distributions are λ_1 and μ_1 and λ_2 and μ_2, respectively. Assume that the system does not switch off if only one of the units has failed. If there is a single repair facility and both units have failed, the problem of choosing an optimal order of the unit's repair arises. Should one continue to repair the first unit which fails? Should one begin to repair the most recently failed unit?

The system might be in four different states:

S_{11} is the system's state where both units are in operational states.

S_{01} is the system's state where unit 1 has failed and unit 2 is in an operational state.

S_{10} is the system's state where unit 2 has failed and unit 1 is in an operational state.

S_{00} is the system's state where both units have failed.

We will consider two types of systems: a redundant one with only one failure state, S_{00}; and a series one with only one operational state, S_{11}.

Maximization of the Availability Coefficient   Consider a redundant system. We wish to find the discipline of repair which delivers the maximal value to the system's availability coefficient. Let repair discipline 1 be such that when the system is in state S_{00}, unit 2 is always under repair. In other words, if first unit 1 has failed, then after unit 2's failure, the repair of unit 1 is stopped, and unit 2 begins to be repaired immediately. Discipline 2 is similar to discipline 1 but the unit's priority in state S_{00} is changed. The transition graphs for these two cases are presented in Figures 12.1a and b.

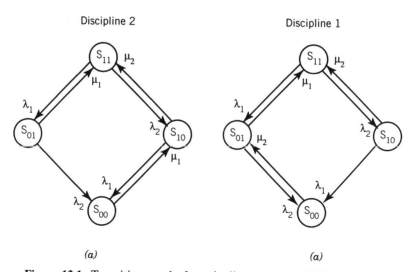

Discipline 2

Discipline 1

(a)   (a)

Figure 12.1. Transition graphs for a duplicate system of different units.

The system of algebraic equations for the stationary probabilities of discipline 1 has the form

$$
\begin{pmatrix}
-(\lambda_1 + \lambda_2) & \mu_1 & \mu_2 & 0 \\
\lambda_1 & -(\lambda_2 + \mu_1) & 0 & \mu_2 \\
\lambda_2 & 0 & -(\lambda_1 + \mu_2) & 0 \\
1 & 1 & 1 & 1
\end{pmatrix}
\begin{pmatrix}
p_{11}^1 \\
p_{01}^1 \\
p_{10}^1 \\
p_{00}^1
\end{pmatrix}
=
\begin{pmatrix}
0 \\
0 \\
0 \\
1
\end{pmatrix}
\qquad (12.4)
$$

From (12.4)

$$
P_{00}^1 = \frac{\Delta_{44}^1}{\Delta_{44}^1 + \mu_2 \Delta_{24}^1} \qquad (12.5)
$$

where Δ_{ij}^1 is the corresponding minor of the coefficient matrix in (12.4), namely,

$$
\Delta_{44} =
\begin{pmatrix}
-(\lambda_1 + \lambda_2) & \mu_1 & \mu_2 \\
\lambda_1 & -(\lambda_2 + \mu_1) & 0 \\
\lambda_2 & 0 & -(\lambda_1 + \mu_2)
\end{pmatrix}
$$

and

$$
\Delta_{24} =
\begin{pmatrix}
-(\lambda_1 + \lambda_2) & \mu_1 & \mu_2 \\
\lambda_2 & 0 & -(\lambda_1 + \mu_2) \\
1 & 1 & 1
\end{pmatrix}
$$

In an analogous way, the system for discipline 2 is

$$
\begin{pmatrix}
-(\lambda_1 + \lambda_2) & \mu_1 & \mu_2 & 0 \\
\lambda_1 & -(\lambda_2 + \mu_1) & 0 & 0 \\
\lambda_2 & 0 & -(\lambda_1 + \mu_2) & \mu_1 \\
1 & 1 & 1 & 1
\end{pmatrix}
\begin{pmatrix}
p_{11}^2 \\
p_{01}^2 \\
p_{10}^2 \\
p_{00}^2
\end{pmatrix}
=
\begin{pmatrix}
0 \\
0 \\
0 \\
1
\end{pmatrix}
\qquad (12.6)
$$

From (12.6) we obtain

$$
P_{00}^2 = \frac{\Delta_{44}^2}{\Delta_{44}^2 - \mu_1 \Delta_{34}^2} \qquad (12.7)
$$

where

$$\Delta_{34} = \begin{pmatrix} -(\lambda_1 + \lambda_2) & \mu_1 & \mu_2 \\ \lambda_1 & -(\lambda_2 + \mu_1) & 0 \\ 1 & 1 & 1 \end{pmatrix}$$

The availability coefficient for discipline i equals

$$K^{(i)} = 1 - p_{00}^{(i)}$$

Now we determine when discipline 1 is preferable to discipline 2. Obviously, the condition $K^{(1)} > K^{(2)}$ is equivalent to the condition $p_{00}^{(1)} < p_{00}^{(2)}$. The latter condition leads to the inequality

$$\mu_2 \Delta_{24} + \mu_1 \Delta_{34} > 0$$

which, in turn, leads to

$$\mu_2 - \frac{\lambda_2}{1 + \rho_1} > \mu_1 - \frac{\lambda_1}{1 + \rho_2} \tag{12.8}$$

where

$$\rho_i = \frac{\lambda_i}{\mu_i}$$

In practice, usually $\mu_i \gg \lambda_i$ and (12.8) can be written approximately as

$$\mu_2 - \lambda_2 > \mu_1 - \lambda_1 \tag{12.9}$$

An analogous result can be obtained for a system of n independent units: the first priority is the unit with the largest value of $\mu_i - \lambda_i$, the second priority is the unit with the next largest value of $\mu_i - \lambda_i$, and so on. Of course, in practice, when $\min \mu_i \gg \max \lambda_i$, (12.9) can be reduced to the condition

$$\mu_2 > \mu_1$$

For a series system, the system of equations is the same but the availability coefficient equals the probability of being in state S_{11}. From the same system of equations, we find

$$p_{11}^{(1)} = \frac{-\mu_1 \mu_2 (\lambda_1 + \mu_2)}{\Delta_{44} + \mu_2 \Delta_{24}}$$

From (12.4) and (12.6), we find

$$\Delta_{44} = \lambda_1\lambda_2(\lambda_1 + \lambda_2 + \mu_1 + \mu_2)$$

$$\Delta_{34} = \mu_1\mu_2 + \mu_2(\lambda_1 + \lambda_2) + \mu_1\lambda_2 + \lambda_1\lambda_2 + \lambda_2^2$$

$$\Delta_{24} = -\left[\mu_1\mu_2 + \mu_1(\lambda_1 + \lambda_2) + \mu_2\lambda_1 + \lambda_1\lambda_2 + \lambda_1^2\right]$$

After simple transformations, the inequality $p_{11}^{(1)} > p_{11}^{(2)}$ reduces to the equivalent inequality

$$\lambda_1(\lambda_1 + \mu_1 + \mu_2) > \lambda_2(\lambda_2 + \mu_1 + \mu_2)$$

For $\mu_i \gg \lambda_i$, we easily obtain the approximate condition

$$\lambda_1 > \lambda_2$$

The latter expresses the rule of choice of the repair order. The same result can be extended to a series system of n units.

Thus, for a redundant system, the priority of repair is determined by the repair speed, and, for a series system, it is determined by the failure rate.

Maximization of the MTBF Often the MTBF is of interest. We first consider a redundant system. If discipline 1 is used, the system always enters state S_{01} from state S_{00}. If discipline 2 is used, the system always enters state S_{10} from state S_{00}. In other words, in these two cases the system starts its period of successful operation from different initial states.

To find the values of interest, we should solve the system of equations corresponding to the transition graph of Figure 12.2. For discipline 1 the system of equations, in terms of the Laplace–Stieltjes transform, has the

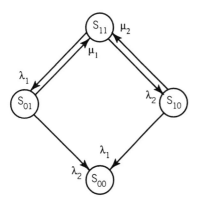

Figure 12.2. Transition graph for computing MTBF.

form

$$
\begin{pmatrix}
s + \lambda_1 + \lambda_2 & -\mu_1 & -\mu_2 & 0 \\
-\lambda_1 & s + \lambda_2 + \mu_1 & 0 & 0 \\
-\lambda_2 & 0 & s + \lambda_1 + \mu_2 & 0 \\
0 & -\lambda_2 & -\lambda_1 & s
\end{pmatrix}
\begin{pmatrix}
\varphi_{11}(s) \\
\varphi_{01}(s) \\
\varphi_{10}(s) \\
\varphi_{00}(s)
\end{pmatrix}
=
\begin{pmatrix}
0 \\
1 \\
0 \\
0
\end{pmatrix}
\qquad (12.10)
$$

where $\varphi_{ij}(s)$ is the LST of $p_{ij}(t)$.

The MTBF of the redundant system is defined as

$$
T_1 = \left[\varphi_{11}(s) + \varphi_{01}(s) + \varphi_{10}(s) \right]_{s=0}
\qquad (12.11)
$$

where $\varphi_{ij}(s)$ can be easily found from (12.10). After the substitution of all $\varphi_{ij}(s)$'s into (12.11), one obtains

$$
T_1 = \frac{\lambda_1^2 + \lambda_1\lambda_2 + \lambda_2\mu_1 + \lambda_1\mu_2 + \lambda_1\mu_1 + \mu_1\mu_2}{\lambda_1\lambda_2(\lambda_1 + \lambda_2 + \mu_1 + \mu_2)}
\qquad (12.12)
$$

For discipline 2 the system of equations has the form

$$
\begin{pmatrix}
s + \lambda_1 + \lambda_2 & -\mu_1 & -\mu_2 & 0 \\
-\lambda_1 & s + \lambda_2 + \mu_1 & 0 & 0 \\
-\lambda_2 & 0 & s + \lambda_1 + \mu_2 & 0 \\
0 & -\lambda_2 & -\lambda_1 & s
\end{pmatrix}
\begin{pmatrix}
\varphi_{11}(s) \\
\varphi_{01}(s) \\
\varphi_{10}(s) \\
\varphi_{00}(s)
\end{pmatrix}
=
\begin{pmatrix}
0 \\
0 \\
1 \\
0
\end{pmatrix}
\qquad (12.13)
$$

and one obtains

$$
T_2 = \frac{\lambda_2^2 + \lambda_1\lambda_2 + \lambda_2\mu_1 + \lambda_1\mu_2 + \lambda_2\mu_2 + \mu_1\mu_2}{\lambda_1\lambda_2(\lambda_1 + \lambda_2 + \mu_1 + \mu_2)}
\qquad (12.14)
$$

Using (12.12) and (12.14), we find that discipline 1 is preferred to discipline 2 when

$$
\lambda_1(\lambda_1 + \mu_1) > \lambda_2(\lambda_2 + \mu_2)
$$

Again, for $\mu_i \gg \lambda_i$, the approximate solution can be used

$$
\lambda_1\mu_1 > \lambda_2\mu_2
$$

Notice that for a series system both disciplines are equivalent.

12.4 DYNAMIC REDUNDANCY

Dynamic redundancy models occupy an intermediate place between optimal redundancy and inventory control models. The essence of a dynamic redundancy problem is contained in the following. Consider a system with n redundant units. Some redundant units are operating and represent an active redundancy. These units can be instantly switched to a working position without delay and, consequently, do not interrupt the normal operation of the system. These units have the same reliability parameters (e.g., exponential distribution and the same failure rate). The remaining units are on standby and cannot fail while waiting. But, as the same time, these units can be switched to an active redundant regime only at some predetermined moments of time. The total number of such switching moments is usually restricted because of different technical and/or economical reasons.

A system failure occurs when at some moment there are no active redundant units to replace the main ones which have failed. At the same time, there may be many standby units which cannot be used because they cannot be instantly switched after a system failure.

In practice, such situations can arise in different spacecraft which are traveling through the solar system. A similar situation occurs when one considers using uncontrolled remote technical objects whose monitoring and service can be performed only rarely.

It is clear that if all redundant units are switched to an active working position at an initial moment $t = 0$, the expenditure of these units is highest. Indeed, many units might fail in vain during the initial period. At the same time, the probability of the unit's failure during this interval will be small. On the other hand, if there are few active redundant units operating in the interval between two neighboring switching points, the probability of the system's failure decreases. In other words, from a general viewpoint, there should exist an optimal rule (program) of switching standby units to an active regime and allocating these units over all these periods.

Before we begin to formulate the mathematical problem, we discuss some important features of this problem in general.

Goal Function   Two main reliability indexes are usually analyzed: the probability of failure-free system operation during some specified interval of time and the mean time to system failure.

System Structure   Usually, for this type of problem, a parallel system is under analytical consideration. Even a simple series system requires a very complex analysis.

Using Active Redundant Units   One possibility is that active redundant units might be used only during one period after being switched to the system. Afterwards, they are not used further, even if they have not failed. In other words, all units are divided in advance into several independent groups, and each group is working during its own specified period of time. After this period has ended, another group is switched to the active regime. In some sense, this regime is similar to the preventive maintenance regime. Another possibility is to keep operationally redundant units in use for the next stages of operation. This is more effective but may entail some technical difficulties.

Controlled Parameters   As we mentioned above, there are two main parameters under our control: the moments of switching (i.e., the periods of work) and the number of units switched at each switching moment. Three particular problems arise: we need to choose the switching moments if the numbers of switched units are fixed in each stage; we need to choose the numbers of units switched in each stage if the switching moments are specified in advance; and, in general, we need to choose both the switching moments and the numbers of units switched at each stage.

Classes of Control   Consider two main classes of switching control. The first one is the so-called *prior rule* (*program switching*) where all decisions are made in advance at time $t = 0$. The second class is the *dynamic rule* where a decision about switching is made on the basis of current information about a system's state (number of forthcoming stages, number of standby units, number of operationally active units at the moment, etc.). We note that analytical solutions are possible only for exponentially distributed TTFs. The only possible method of analysis for an arbitrary distribution is via Monte Carlo simulation.

Prior Rule Without the Transfer of Active Units   Consider a system with n redundant units which might be turned on at N moments $t_0, t_1, \ldots, t_{N-1}$ which are determined in advance. The system is designed to operate during a period of time τ. First, let $\Delta_j = t_j - t_{j-1} = \Delta$ for all $1 \le j \le N$, that is, $\Delta = \tau/N$. The problem can be written as

$$\max_{\mathbf{x}} \left\{ \prod_{0 \le i \le N-1} r(x_i, \Delta) \,\middle|\, \sum_{0 \le i \le N-1} x_i = n \right\} \qquad (12.15)$$

where $\mathbf{x} = (x_1, \ldots, x_{N-1})$ and $r(x_i, \Delta)$ is the probability of failure-free operation at the $(i + 1)$th stage, that is, at the interval $[t_i, t_{i+1}]$.

The optimal solution of this problem is trivial: the total number of redundant units is spread over the stages uniformly: $x_i = x_{opt} = n/N$. Of course, if n/N is not an integer, then N_1 stages have x^* units and N_2 stages have x^{**} units where

$x^* = \lfloor n/N \rfloor$ where, in turn, $\lfloor n/N \rfloor$ is the integer part of n/N.

$x^{**} = \lfloor n/N \rfloor + 1$.

$N_1 = \lfloor n/N \rfloor (1 + x_i) - n$.

$N_2 = n - \lfloor n/N \rfloor x_i$.

The stages which use either x^* or x^{**} redundant units may be chosen arbitrarily.

If the problem is to determine x_i^{opt} for the specified moments t_i, the problem is formulated as

$$\max_{\mathbf{x}} \left\{ \prod_{0 \le i \le n-1} r(x_i, \Delta_i) \middle| \sum_{0 \le i \le N-1} x_i = n, \quad \sum_{0 \le i \le N-1} \Delta_i = \tau \right\}$$

where $\mathbf{x} = (x_1, \ldots, x_{N-1})$ and $\Delta_i = t_{i+1} - t_i$.

If the problem is to determine t_i^{opt} for the specified numbers x_i admissible at each stage, the problem is formulated as

$$\max_{\Delta} \left\{ \prod_{0 \le i \le n-1} r(x_i, \Delta_i) \middle| \sum_{0 \le i \le N-1} x_i = n, \quad \sum_{0 \le i \le N-1} \Delta_i = \tau \right\}$$

where $\Delta = (\Delta_1, \ldots, \Delta_{N-1})$.

The solution of both these problems can be reduced to the solution of the following system of equations:

$$\ln r(x_i, \Delta_i) = c\Delta_i$$

$$\sum_{0 \le i \le N-1} x_i = n$$

$$\sum_{0 \le i \le N-1} \Delta_i = \tau$$

where c is some dimensional constant. The solution can be easily obtained, for instance, using the steepest descent method.

Finally, the solution of the problem of simultaneously choosing $\mathbf{x} = (x_0, \ldots, x_{N-1})$ and $\mathbf{t} = (t_1, \ldots, t_{N-1})$ can be reduced to the solution of (12.15). Thus, the very first problem appears to be not quite so useless, in spite of its triviality.

If one considers the problem of maximizing the mean time to failure, a simple solution can be obtained only for the case when all Δ_i are equal. In this case the following recurrent equation can be written:

$$T_k(n_k) = \max_{0 < x_k \le k} r(x_k, \Delta)(\Delta + T_{k-1}(n_k - x_k)) \qquad (12.16)$$

The solution of the problem satisfies the condition $x_k > x_{k+1}$. This condition allows us to write the following simple algorithm for finding the optimal allocation of redundant units by stages.

If $n = 1$, then this unit must obviously be allocated to the first stage. If $n = 2$, the redundant units can be allocated in two possible ways denoted by $(2, 0)$ or $(1, 1)$. We compute the values

$$T(2,0) = \Delta r(2, \Delta)$$

$$T(1,1) = r(1, \Delta)(\Delta + \Delta r(1, \Delta)) = \Delta r(1, \Delta)(1 + r(1, \Delta))$$

and choose the maximum. Further, we develop the solution process from the vertex with the maximal value of T. Thus, from the vertex $(2, 0)$, the branch can move to $(3, 0)$ or $(2, 1)$, and from the vertex $(1, 1)$, it can move to $(2, 1)$ or $(1, 1, 1)$. This process continues until all redundant units have been used, that is, the process stops at the nth level of the solution tree.

Using Operational Units at the Next Stages   We decide in advance how many units will be used at each stage. In addition, we take into account that all units which remain operational after the previous stage may be used at subsequent stages. It is clear that if the probability of a failure-free operation is considered, an a priori uniform allocation of redundant units over all stages is unreasonable. Indeed, in this case the actual number of redundant units will, on average, increase with the number of stages.

Consider the case where all Δ_i are equal and a unit's TTF is exponentially distributed. Assume there are N stages. Then for the probability of interest we can write the following recurrent equation:

$$R_N(n, \tau) = \max_{0 \leq x_0 \leq n} \sum_{1 \leq \xi_1 \leq x_0 + \xi_0} p_{\xi_1}(x_0) R_{N-1}(n - x_0 + \xi_1, \tau - \Delta)$$

$$= \max_{0 \leq x_0 \leq n} E_{\xi_1}\{R_{N-1}(n - x_0 + \xi_1, \tau - \Delta)\}$$

where $p_{\xi_1}(x_0)$ is the probability that during a time Δ, ξ_1 units will survive from the initial number of x_0. Here and in the future ξ_k is assumed to be a binomial r.v. depending on ξ_{k-1} and x_{k-1} in the following manner:

$$p_{\xi_k}(x_{k-1} + \xi_{k-1}) = \binom{\xi_k}{x_{k-1} + \xi_{k-1}} r^{\xi_k} q^{x_{k-1} + \xi_{k-1} - \xi_k}$$

where $r = 1 - q = e^{-\lambda \Delta}$ and λ is the parameter of the exponential distribution of a unit's TTF.

A direct computation is very cumbersome, so we will provide ways of obtaining a simpler approximate solution. We note that the function $R_N(n, \tau)$ is increasing and concave in N and n and decreasing and convex in τ, and

$R_N(n, \tau) - R_N(n, \tau + \delta)$ decreases by τ and $R_N(n + \delta, \tau) - R_N(n, \tau)$ decreases by n. This allows us to write the inequality

$$\tilde{R}_N(n + E\{\xi_0\}, \tau) < \max_{0 \le x_0 \le n} \tilde{R}_{N-1}(n - x_0 + E\{\xi_1\}, \tau - \Delta) \quad (12.17)$$

where $E\{\xi_1\} = (x_0 + \xi_0)r$. If the inequality in (12.17) is changed to an equality, the use of the properties of R_N leads to the approximate solution:

$$\tilde{x}^{\mathrm{opt}} \approx \left\lfloor \frac{n}{1 + Nq} \right\rfloor + 1$$

This solution can be successfully used as an initial approximation. Notice that the actual x^{opt} is not less than $\tilde{x}^{\mathrm{opt}}$.

In addition, we strongly suspect that the following hypothesis, related to the optimal solution, is correct though we have no proof:

If

$$\sum_{1 \le j \le N} x_j > \sum_{1 \le j \le N} x_j'$$

then

$$\sum_{1 \le j \le k} x_j \ge \sum_{1 \le j \le k} x_j'$$

for all $k < N$. This structure of the solution allows us to decrease the number of enumerations because we have a restricted range over which x_k can vary.

Now consider the problem of maximizing a system's MTTF. If units can be switched at the moments $t_k = k\Delta$, then for N stages chosen a priori the recurrent equation is

$$\begin{aligned}
T_N(n) &= \max_{0 \le x_0 \le n} \sum_{1 \le \xi_1 \le x_0} p_{\xi_1}(x_0; \Delta)[\Delta + T_{N-1}(n - x_0 + \xi_1)] \\
&= \max_{0 \le x_0 \le n} \left\{ \sum_{1 \le \xi_1 \le x_0} p_{\xi_1}(x_0; \Delta) T_{N-1}(n - x_0 + \xi_0 + \xi_1) + (1 - q^{x_0})\Delta \right\} \\
&= \max_{0 \le x_0 \le n} \left[E_{\xi_1}\{T_{N-1}(n - x_0 + \xi_1)\} + (1 - q^{x_0})\Delta \right]
\end{aligned}$$

Obviously,

$$T_0(m) = \frac{1}{\lambda} \sum_{1 \le i \le m} \frac{1}{i}$$

that is, after the last switching of redundant units, the system becomes a common redundant group with active redundant units.

In this case we may also use the following inequality for an approximation:

$$T_N(n) < \max_{0 \le x_0 \le n} \left[T_{N-1}(n - x_0 + \mathrm{E}\{\xi_1\}) + (1 - q^{x_0})\Delta \right]$$

Dynamic Rule by Current State The problem of choosing the number of redundant units which have to be switched at the beginning of each stage is formulated for two cases as follows:

$$R(n; \tau) = \max_{0 \le x_0 \le n} \max_{0 \le \Delta_0} \mathrm{E}_{\xi_1}\{R(n - x_0 + \xi_1; \tau - \Delta_0)\}$$

and

$$T(n) = \max_{0 \le x_0 \le n} \max_{0 \le \Delta_0} \left[\mathrm{E}_{\xi_1}\{T(n - x_0 + \xi_1)\} + \Delta_0(1 - q^{x_0}) \right]$$

Both of these problems can be approximately solved if one replaces the mean of the function by the function of the mean arguments. Then the recurrent equations are

$$\tilde{R}(n; \tau) = \max_{0 \le x_0 \le n} \max_{0 \le \Delta_0} \tilde{R}(n - x_0 + \mathrm{E}\{\xi_1\}, \tau - \Delta_0)$$

and

$$T(n) = \max_{0 \le x_0 \le n} \max_{0 \le \Delta_0} \left[T(n - x_0 + \mathrm{E}\{\xi_1\}) + \Delta_0(1 - q^{x_0}) \right]$$

The analysis of these problems is analogous to the previous one.

12.5   TIME SHARING OF REDUNDANT UNITS

Consider a series system of n units. Each unit has an arbitrary distribution of TTF. There are standby redundant units of each type which can increase a system's reliability. In general, all standby units can be different. In practice, such a situation might be explained by the fact that the standby units may change their parameters depending on their history before their use. In general, each unit may change its reliability characteristics during operation, so the preventive unit replacements are considered to be reasonable. After some period of use, a new unit, which has replaced a used one, can spend more reliability resources than the previously used one. Thus, the problem of switching the previously used (but unfailed) unit back arises if, of course, such a possibility of reswitching exists. Each switching is assumed to be absolutely reliable and to do no harm to the system's operation. Of course, the failure of any unit in a working position leads to a system failure.

Consider two rules of unit replacement:

1. One can switch from a main unit to any redundant unit at any moment and continue to use all units until any one of them fails at the working position.
2. One can switch from a main unit to any redundant unit at any time but a unit once used is excluded; that is, each unit can be used only once and without interruption.

The first rule reminds one of a time-sharing rule in a computer: after some small interval of time, any working unit can be replaced by a standby one. Then, after some period of use, it can be returned to a standby position. Such a rule allows us to spend a unit's "reliability resource" in a rational way. The second rule can be used if a standby unit can only be used once. We call the first rule *operative switching*, and the second rule, *program switching*.

Consider three reliability indexes: the MTTF, the PFFO during a specified or random operational time, and the mean "profit" from the system until failure.

Operative Switching   Let $F_i(t)$ be the distribution of TTF of unit i and let $f_i(t)$ be the density function of this distribution. Assume that the distributions are "aging." Each $\lambda_i(t)$ is increasing in such a way that the inverse function $\lambda^{-1}(y)$ has a restricted (finite) derivative $\sigma_i = d\lambda_i(y)/dy$.

Consider some arbitrary rule of switching the redundant units into a working position and call this rule S. Let rule S include using unit i_1 during a small interval $[t_1, t_2]$ and unit i_2 during a small interval $[t_2, t_3]$, and $t_2 - t_1 = t_3 - t_2 = \Delta$. Let rule S^* differ from rule S only in the order of switching the units i_1 and i_2. Let the failure rates of these units be $\lambda_1(t_1) = \lambda_1$ and $\lambda_2(t_1) = \lambda_2$ at the moment t_1, respectively, and assume $\lambda_1 > \lambda_2$. Remember that we are considering standby redundancy; that is, only the unit in a working position changes its reliability characteristics.

Consider the residual time to failure of socket i. For rule S,

$$\xi^S = \begin{cases} \eta \text{ with probability } \lambda_1\Delta \\ \Delta + \eta \text{ with probability } (1 - \lambda_1\Delta)\lambda_2\Delta \\ 2\Delta + \xi_0 \text{ with probability } (1 - \lambda_1\Delta)(1 - \lambda_2\Delta) \end{cases} \qquad (12.18)$$

and, for rule S^*,

$$\xi^{S^*} = \begin{cases} \eta \text{ with probability } \lambda_2\Delta \\ \Delta + \eta \text{ with probability } (1 - \lambda_2\Delta)\lambda_1\Delta \\ 2\Delta + \xi_0 \text{ with probability } (1 - \lambda_2\Delta)(1 - \lambda_1\Delta) \end{cases} \qquad (12.19)$$

Here η is a conditional residual operative time under condition that failure has occurred within a considered interval Δ, and ξ_0 is a TTF after switching if no failure has occurred.

It is clear that if $\lambda_1 > \lambda_2$, then $\Pr\{\xi^S > t\} < \Pr\{\xi^{S^*} > t\}$. Notice that we use the same transposition rule for the proof which was used in the problem of the optimal technical diagnosis.

Now if $\Delta \to 0$, the rule of a unit's switching (replacement) can be interpreted in the following way. Assume that several units (even all of them), say with numbers $i_1, i_2, \ldots, i_k$, are currently located in a working position. Each unit can be used with a "partial" intensity. The intensity of use defines a unit's "aging process": the longer the unit is in use, the higher the failure rate of that unit. The rule of optimal switching of the units is such that, if the minimal failure rate equals y, then the intensity of the use of a unit i_j is determined by

$$\frac{\sigma_{i_j}(y)}{\sum_{1 \le l \le k} \sigma_{i_l}(y)}$$

This maximizes the MTTF of the system. From (11.18) and (11.19) it follows that this rule is even stronger: it maximizes the probability of a failure-free operation, even over a random time interval.

We remark that this rule is not unique for a fixed time of operation t_0. One possible solution is to find y such that all units $i_1, i_2, \ldots, i_k$ of the ith socket satisfy

$$\lambda_{i_1}(t_1) = \lambda_{i_2}(t_2) = \cdots = \lambda_{i_k}(t_k) = y$$

and

$$\sum_{1 \le l \le k} t_l = t_0$$

Thus, for a fixed time of operation, the above-described rule of choosing a priori the duration of using redundant units gives the same result as a "time-sharing" rule.

Program Switching We first consider the case where all units are identical. Let a_L denote the mean time to failure of the redundant group if there is no failure until the moment t_1 and the switching is performed at this moment, and all L remaining switches will be performed at optimal moments. The solution of the problem of choosing t_1 optimally to maximize the system's MTTF coincides with the solution of the problem of maximizing the function

$$\int_0^{t_1} f(u)u\,du + [1 - F(t_1)](t_1 + a_{L-1}) \tag{12.20}$$

The maximum of (12.20) is reached when

$$t_1 = \lambda^{-1}\left(\frac{1}{a_{L-1}}\right)$$

In an analogous way, we can find the other moments of switching. In an optimal procedure the ith switching must be performed at the moment

$$t_i = \sum_{1 \le j \le i} \lambda^{-1}\left(\frac{1}{a_{L-j}}\right)$$

where a_L is found from

$$a_L = \sum_{1 \le i \le L}\left[\prod_{1 \le j \le i-1}\left(1 - F\left(\lambda^{-1}\left(\frac{1}{a_{L-j}}\right)\right)\right)\right]\int_0^{\lambda^{-1}(1/a_{L-j})}(1 - F(u))\,du$$

$$+ \prod_{1 \le i \le L}\left(1 - F\left(\lambda^{-1}\left(\frac{1}{a_{L-j}}\right)\right)\right)\Bigg]a_0 \qquad L = 1, 2, \ldots$$

and

$$a_0 = \int_0^\infty (1 - F(u))\,du$$

CONCLUSION

Optimization problems very often appear in engineering practice. More correctly speaking, they are problems of the optimization of various parameters of mathematical models not real objects. In real life one has too many nonformalized or unknown factors which are not included in an optimization model. Thus, we should be very cautious when we consider an "optimization" in engineering design.

In spite of this warning, one must consider optimization models as a powerful tool in everyday decision making. If one does not receive reliable quantitative results, these models give reasonable qualitative answers on some important questions.

Of course, it is impossible to represent the entire spectrum of optimization problems arising in a system design. We split optimal redundancy and optimal diagnosis into two separate chapters only because they are arranged in more or less self-contained discussions. Besides, optimal redundancy is a very special branch of reliability theory with its direct practical implementation.

The list of references includes only papers concerning a sample of models for solution of optimization problems in reliability. This list is far from complete, as is the chapter, in covering varied optimization problems.

REFERENCES

Gertsbah, I. B. (1966). On optimal control of redundant units switching. *Engrg. Cybernet.*, no. 5.

Gertsbakh, I. B. (1970). Dynamics reserves: optimal control of spare items (in Russian). *Avtomat. i Vychsl. Tekhn.*, no. 1.

Gnedenko, B. V., ed. (1983). *Aspects of Mathematical Theory of Reliability* (in Russian). Moscow: Radio i Sviaz.

Konev, V. V. (1974). Optimal inclusion of standby elements. *Engrg. Cybernet.*, vol. 12, no. 4.

Pechinkin, A. V., and A. G. Tatashev (1981). On one generalized discipline of processor priority sharing. *Engrg. Cybernet.*, no. 4.

Pestov, G. G., and L. V. Usahkova (1973). Investigation of optimal strategies in dynamic redundancy. *Engrg. Cybernet.*, no. 5.

Raikin, A. L. (1978). *Reliability Theory Methods in Engineering Design.* (in Russian) Moscow: Sovietskoe Radio.

Raikin, A. L., and A. S. Mandel (1967). Construction of the optimal schedule of redundant units switching. *Automat. Remote Control*, no. 5.

Schrage, L. (1967). The queue $M/G/1$ with feedback to lower priorities queues. *Management Sci.*, vol. 13, no. 7.

Tatashev, A. G., and I. A. Ushakov (1981). A problem of reswitching standby elements after a schedule. *Engrg. Cybernet.*, vol. 12, no. 3.

Ushakov, I. A. (1965). Approximate solution of the optimal redundancy problem (in Russian). *Radiotechnika*, no. 12.

Ushakov, I. A. (1971). Approximate solution of optimal redundancy problem for multipurpose systems. *Engrg. Cybernet.* (USA), vol. 9, no. 2.

Ushakov, I. A. (1981). Methods of approximate solution of dynamic standby problems. *Engrg. Cybernet.* (USA), vol. 19, no. 2.

CHAPTER 13

HEURISTIC METHODS
IN RELIABILITY

The researcher is often faced with the problem of finding a "solution" to a problem when the problem is unsolvable. In spite of the problem's "unsolvability," a solution must be found! And even if a solution cannot be found, the designer is forced to make a practical decision. When there is no exact analytic solution and the problem is still too hard for even Monte Carlo simulation, the only possibility is to use a heuristic procedure (heuristics).

Sometimes heuristics are thought to lead to an arbitrary "solution," based only on an "I personally believe"–type of argument. We oppose such "heuristics," as we understand the term *heuristic* to be an extension of analytical methods in areas where such methods cannot be exactly proven. Sometimes we omit some specified conditions. Sometimes we make additional assumptions and are still not sure that the method of solution is correct. Sometimes we change an analyzed phenomenon description to allow for the use of available mathematical tools.

In fact, the building of a mathematical model is always a heuristic procedure itself. No mathematical model completely reflects all of the properties of a real object. We always create "an ideal image" based on a real object and then build a mathematical model for this image. Moreover, approximate calculations can be viewed as "good proven heuristics." Thus, heuristics are an inevitable part of mathematical modeling.

Below, we will introduce several heuristic approaches. Generally, they concern the constructing of models.

13.1 APPROXIMATE ANALYSIS OF HIGHLY RELIABLE REPAIRABLE SYSTEMS

Highly reliable systems are of special interest because of their practical importance. In order to obtain important results for repairable systems, we may use the limit results of point stochastic processes. In Chapter 1 we discussed two very important limit theorems: the Rényi theorem and the Khinchine–Ososkov theorem and its generalization, the Grigelionis–Pogozhev theorem.

We will now show that both of these theorems are extremely constructive and effective for a heuristic analysis of highly reliable repairable systems. We will suggest a simple heuristic method based on strong assumptions. This method, in its main meaning, reminds one of an asymptotic method and produces good results which coincide with such a method.

The idea of the proposed heuristic method is as follows. Consider a repairable system consisting of n independent units. Each unit's operation might be described with the help of an alternating stochastic process. Each such process represents a sequence of the operational and idle periods of the unit. The intersection of the idle periods of different units (i.e., their overlapping in time) may lead to the system's failure. All such possible occurrences are considered. Of course, one should take into account only events significant in a probabilistic sense.

Consider a summarized (resulting) failure flow which is formed by the failure flows of all of the system's units. The system's failure might occur if the idle periods of some specified units coincide in time. (Of course, a single unit can cause a system to fail.)

For a redundant system, we use the following procedure of thinning of this resulting process. A failure of a system appears when all units of a redundant group are in idle states at the same time. A unit's failure may "not meet" an appropriate "environmental condition," and the system as a whole does not fail in this case. (For example, the first failure of a unit of a duplication does not lead to a redundant group's failure but, at the same time, this failure creates the condition for a second unit's failure and the next failure leads to the system's failure.) Unit failures which do not lead to a system failure are excluded from future consideration.

For a series system, obviously, each unit failure leads to the system's failure. Thus, in this case we are dealing with the superposition of a point process.

For highly reliable systems, the resulting alternating process for the system can be represented as a sequence of long periods of successful operation and short periods of idle time. In turn, each alternating process can be approximated by a point process. The intensity of this "pseudo-point" process and the duration of each such "point failure" can be easily computed.

Note that, for a highly reliable system, a random TBF has a distribution which is close to exponential even if the distribution of the TTF and the distribution of the idle time for single units are other than exponential. As far as the system's idle time is concerned, for practical purposes it is enough to know only the mean value, and this value is easily computed.

Everybody understands that the main imperfection of any heuristic method usually lies in this difficulty of defining the domain where the results obtained with its help valid. But in this particular case, we can suggest a simple (and convenient!) rule: if you have obtained a reliability index with a high value, the application of the heuristic method *was correct*. It is not a bad rule because otherwise a system is improper for practical use!

The best way to explain a heuristic method in detail is to show examples of how it works.

13.1.1 Series System

Let a series system consist of n units. Each unit itself might represent a complex object. The only condition is that a unit's failure will lead to the system's failure. The ith unit has an MTTF T_i and a mean repair time τ_i. Thus, the individual availability coefficient for the ith unit is $K_i = T_i/(T_i + \tau_i)$. First of all, we should check the condition

$$\frac{\tau_i}{T_i + \tau_i} \ll \frac{1}{n} \tag{13.1}$$

If this condition holds, the next step is to calculate the system's parameters.

If n is sufficiently large (in practice, $n \gg 10$ is enough), the Khintchine–Ososkov or Grigelionis–Pogozhev theorem might be applied. In fact, the process of a unit's operation can be described as an alternating process consisting of long (on average) intervals of TBF and relatively short intervals of idle time.

To calculate the resulting system's failure rate, we use the rule

$$\Lambda_{\text{syst}} = \sum_{1 \le i \le n} \frac{1}{T_i}$$

The superposition of the recurrent point processes by the above-mentioned theorems in the limit will converge to a Poisson process as n increases. For large n, we can assume that a system's TBF approximately has an exponential distribution with parameter Λ.

The system's MRT can be calculated as a weighted average

$$\tau_{\text{syst}} = \frac{1}{\Lambda} \sum_{1 \le i \le n} \frac{\tau_i}{T_i}$$

Note that it is likely that the random repair time has a DFR-type distribution. This assumption is correct at least when all repair times have an exponential distribution.

13.1.2 Unit with Periodic Inspection

Some units of a renewal system are subjected only to a periodic inspection of their state. Thus, a failure occurring in such a unit can be found only with some delay. Such inspections can be performed periodically with either regular or random periods. Assume that, on the average, the TTF is much larger than the interval between the inspections. In this case the process of a unit's operation is as follows: after a failure occurs at a random time, the idle time appears. This time equals the waiting time for a next inspection plus the repair time. (In many practical models the repair time can be neglected.) Thus, the process reminds one of an alternating stochastic process. This is only an approximation because inspections are performed regularly.

We now investigate the distribution of time during which the system is in a state of hidden failure before detection. Consider two cases with regular and random inspection intervals θ.

1. If the inspection intervals are regular and a unit's MTTF T is much larger than θ, the moments of failure are approximately uniformly distributed within the corresponding inspection intervals. Thus, the mean waiting time (the interval of a hidden failure) on the average equals $\theta/2$. The total idle time in this case equals $\tau^* = 0.5\theta + \tau$ where τ is the renewal time (the time to repair or replace the failed unit). Let us assume this process can be considered as an alternating process (ignoring the fact that a waiting time always ends at regular time moments).

2. If inspections are performed at random times θ' and a unit's MTTF T is much larger than $\theta^* = E\{\theta'\}$, then after the failure one observes the r.v. θ' which is a residual r.v. Thus, the distribution of the waiting time might be expressed as

$$\Pr\{\theta' \le t\} = 1 - \frac{1}{\theta^*} \int_t^\infty \left[1 - F_\theta(x)\right] dx$$

As we stated above, the mean residual time depends on the kind of distribution $F_n(t)$. If the distribution of the inspection intervals is exponential, the waiting time has the same distribution as the random interval between neighboring inspections: $\theta = \theta^*$. The total idle time in this case equals $\tau^* = \theta + \tau$. We may approximate this type of unit by a common renewal unit which is described as an alternating process (with the parameters mentioned above).

13.1.3 Parallel System

At first, consider a system of two different independent renewal units. Their TTFs have exponential distributions with parameters λ_1 and λ_2 and their repair times are arbitrarily distributed with distributions $G_1(t)$ and $G_2(t)$. At the very beginning, we note that we are considering highly reliable units, that is, with availability coefficients K_1 and K_2 close to 1. Thus, the process of operating each unit might be approximately considered to be a point process. (The duration of a unit's idle time is negligibly small in comparison with the unit's TTF.)

By definition, a system fails when both units are down. So the necessary condition of a system failure after a current failure is that another unit is in a failed state. For convenience, we call a single unit failure a "system defect." The system defect might be the case of the system failure if during the system defect state another unit's failure occurs.

Thus, one observes two flows of system defects formed by both units. The first flow has parameter λ_1. The probability that a defect transforms into a system failure equals

$$q_1 = 1 - \int_0^\infty e^{-\lambda_2 t}\, dG_1(t) \tag{13.2}$$

The simple expression for (13.2) can be obtained due to the assumption of the high reliability of the system. For the exponential distribution of TTF,

$$q_1 \approx 1 - \int_0^\infty [1 - \lambda_2 t]\, dG_1(t) = \lambda_2 \tau_1 \tag{13.3}$$

Therefore, if we consider the flow of a system's failures, a part of them is determined by the flow of the system defects which, in turn, are connected with the first unit failures. The rate of this subflow is $\Lambda_1 = \lambda_1 q_1$. In an analogous way, we easily obtain the second component of a system's failure flow: $\Lambda_2 = \lambda_2 q_2$. [Here q_2 is obtained as (13.2) or (13.3) with changing subscripts.] Thus, the system failure flow is approximately a Poisson process with parameter $\Lambda_{\text{syst}} = \lambda_1 q_1 + \lambda_2 q_2$.

In fact, we may use the exponential distributions of a unit's TTF only to simplify the calculation of the probabilities q_1 and q_2. But, under the assumption of high reliability, these probabilities can be easily calculated in the general case. Consider the probability q as the probability that the residual random TTF is shorter than the random repair time:

$$\Pr\{\xi \le \theta'\} = 1 - \int_0^\infty \left[\frac{1}{T} \int_t^\infty P(x)\, dx \right] dG(t)$$

where T is the ith unit MTTF and $G(t)$ is a unit distribution of the repair time, η. Usually, we deal with IFR distributions. Then it is reasonable to consider only two limit cases for TTF: constant or exponentially distributed.

We determine the mean idle time of the system when units are completely independent. They not only fail independently but are also independently repaired. For each unit, consider the process in terms of sequences of short "impulses" of idle time. The system's idle time can be found as follows. Suppose the first unit is being repaired and the second unit fails during this time. (This corresponds to the case when "the rising edge" of the second type of impulse appears within the impulse of the first type.) The system's idle time can be found as $\tau_{syst}^1 = E[\min(\eta_2, \theta_1)]$ where θ_i is a residual idle time of the ith unit. If both distributions are exponential, then

$$\tau_{syst}^1 = \frac{1}{\dfrac{1}{\tau_1} + \dfrac{1}{\tau_2}} = \frac{\tau_1 \tau_2}{\tau_1 + \tau_2} \tag{13.4}$$

Obviously, $\tau_{syst}^2 = E\{\min(\eta_1, \theta_2)\} = \tau_{syst}^1$ because (13.4) is symmetric with respect to subscripts.

If both units have constant repair times, a similar result can be obtained with the help of the following semiheuristic arguments. Assume that $\tau_1 > \tau_2$ and consider two flows of impulses (see Figure 13.1a). For a highly reliable system, the overlapping of impulses of these two flows coincides with the following event. Consider the first flow as a flow of "enlarged" impulses of length $\tau_1 + \tau_2$ and the second flow as a point process where each point corresponds to the leading edge of an impulse (see Figure 13.1b). For a highly reliable system where the overlapping of impulses of the idle time of both flows is rare, both of these models are equivalent to finding the length of the overlapping impulses.

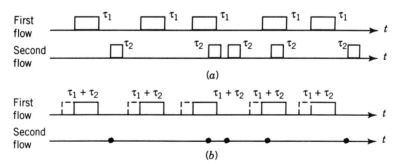

Figure 13.1. Two simultaneous flows of impulses and equivalent representation of their overlapping.

Indeed, if a point of the second flow appears in the front "time window," τ_1, this corresponds to the case where an impulse of the first flow appears inside an impulse of the second flow. The probability that a point of the second flow appears in this time window is $p_1 = \tau_2/(\tau_1 + \tau_2)$. The mean time of impulse overlapping equals $\tau_2/2$ (see Figure 13.2a).

The appearance of a point of the second flow within the interval $\tau_1 - \tau_2$ corresponds to the case where τ_1 completely covers τ_2. The last case of the point appearance in the end part of impulse τ_1 corresponds to an uncomplete covering τ_2 by τ_1. The probability of this event equals $p_2 = (\tau_1 - \tau_2)/(\tau_1 + \tau_2)$. The mean time of overlapping for this case is τ_2 (see Figure 13.2b).

Finally, if a point of the second flow appears in the end part of impulse τ_1, this corresponds to the partial overlapping of these two impulses. The probability of this event equals $p_3 = \tau_2/(\tau_1 + \tau_2)$. The mean time of overlapping for this case is $\tau_2/2$ (see Figure 13.2c).

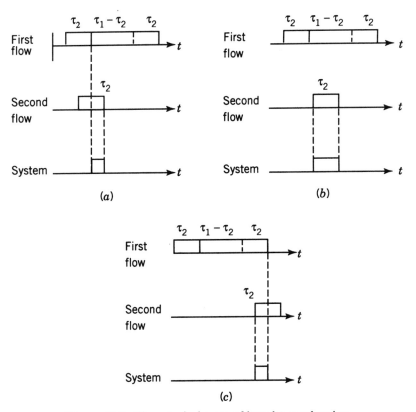

Figure 13.2. Three typical cases of impulse overlapping.

The system's idle time can be found as the weighted average

$$T_{\text{syst}} = \frac{\tau_2}{2}(p_1 + p_3) + \tau_2 p_2$$

$$= 2\frac{\tau_2}{\tau_1 + \tau_2}\frac{\tau_2}{2} + \frac{\tau_1 - \tau_2}{\tau_1 + \tau_2}\tau_2 = \frac{\tau_1\tau_2}{\tau_1 + \tau_2} \tag{13.5}$$

Because the degenerate and exponential distributions are boundary distributions in the class of all IFR distributions, it follows that the final expression

$$T_{\text{syst}} = \frac{\tau_1\tau_2}{\tau_1 + \tau_2}$$

is valid for all such distributions of renewal times.

If units are dependent through the repair process, for example, because only one repair facility is available for two units, the system's mean idle time can be found in a similar way with minor corrections. If a system's failure occurs because of a system defect caused by the first unit, and the serving discipline is FIFO, then

$$T_{\text{syst}}^1 = E\{\theta_1\}$$

where θ_1 is a residual value of η_1. For an exponential distribution of θ_1, this value equals τ_1. If the repair time is almost constant, this value equals $\tau_1/2$. In an analogous way

$$T_{\text{syst}}^2 = E\{\theta_2\}$$

for an exponential distribution and a constant repair time equal to τ_2 and $\tau_2/2$, respectively.

The system's idle time is defined as before

$$T_{\text{syst}} = T_{\text{syst}}^1 \frac{\tau_1}{\tau_1 + \tau_2} + T_{\text{syst}}^2 \frac{\tau_2}{\tau_1 + \tau_2} \tag{13.6}$$

where $\tau_i/(\tau_1 + \tau_2)$ is the probability that the system has failed during the ith unit repair. If the units are different, the unit with the smaller expected repair time may have an earlier priority of repair if both units have failed. In this case

$$T_{\text{syst}}^1 = \min(E\{\theta_1\}, E\{\eta_2\})$$

if the system failure develops after the first unit's failure, and

$$T_{\text{syst}}^2 = \min(E\{\theta_2\}, E\{\eta_1\})$$

if the system failure develops after the second unit's failure. Of course, for exponential distributions, a unit with the smallest MRT has absolute priority.

All of these simple considerations are possible, without invoking queuing theory methods, only because we are considering highly reliable systems and, consequently, we are assuming that the possibility of overlapping events is effectively negligible.

The availability coefficient of such a redundant system consisting of two units equals

$$K = \frac{1}{1 + \Lambda_{\text{syst}}\tau_{\text{syst}}} = \frac{1}{1 + \lambda_1\lambda_2\tau_1\tau_2} \tag{13.7}$$

If there are several system units whose failure may be considered as a system defect, the problem can be solved sequentially.

13.1.4 Redundant System with Spare Unit

Consider a system of two identical independent renewal units where the redundant unit is in a standby regime. Assume that this unit can be immediately switched to an operating state when the main unit fails. We assume that both units—main and standby—are identical because this is the most frequently encountered practical case (e.g., spare parts). Assume that the distributions of the TTF, $F(t)$, and the repair time, $G(t)$, are arbitrary, and that, as before, the MTTF T is much larger than the MRT τ. The flow of system defects has parameter λ. A system defect develops into a system failure if the unit which has replaced the failed one fails before the failed unit has been repaired. Thus,

$$q = \int_0^\infty F(t)\, dG(t)$$

Under the assumption of high reliability, we can write

$$q \approx F(\tau)$$

The system failure rate is

$$\Lambda_{\text{syst}} = \lambda F(\tau) \tag{13.8}$$

To facilitate the calculation of the system's mean idle time, we make the assumption that $F(t)$ has a first derivative at $t = 0$. Then, in a sufficiently small interval of time $[t, t + \Delta]$, this function can be approximated by a linear function; that is, the conditional density function in this interval is very close to uniform. From this assumption, it follows that the mean idle time equals $\tau/2$. The availability coefficient of such a system is

$$K = \frac{1}{1 + \Lambda_{\text{syst}} \tau_{\text{syst}}} = \frac{1}{1 + \frac{1}{2}\lambda\tau F(\tau)}$$

13.1.5 Switching Device

Usually, a switching device works as an interface between redundant units and a "technical environment": the input and output signals go through this device. But sometimes switches are necessary only during the process of switching. In this case, if a switching device is not needed during its repair, a system does not "feel" the failure. Thus, a system failure occurs only when the main unit fails during the switch's restoration.

Let the switch's failure rate be λ_s and let its mean repair time be τ_s. As above, assume that the moment of the main unit's failure has a conditional distribution which is uniform in the interval τ_s. Thus, the system idle time equals

$$\tau_{\text{syst}} = \frac{\tau_s}{2}$$

The probability that a switching device failure leads to a system failure is $q \approx \tau_s/T$, where T is the main unit's MTTF. Then the system failure rate, caused by the switch failure, is

$$\Lambda_{\text{syst}} = \frac{\lambda_s \tau_s}{T}$$

Example 13.1 Consider a duplicate system, that is, a system of two identical units. The failure rate of the main unit equals λ_1 and its mean repair time is τ_1. The redundant unit is in an underloaded regime. Its failure rate equals λ_2 and its mean repair time is τ_2. The redundant unit is monitored continuously, but there are false signals indicating failure. The rate of such signals is λ_f. In this case the redundant unit is considered to have failed until a test shows that the signal was false. The idle time of the redundant unit connected with this check is τ_f.

The main unit is only partially monitored. A part of the unit, characterized by a failure rate $\alpha\lambda$, is monitored continuously, and a failure is found instantly. The remaining part of the main unit, characterized by a failure rate $(1 - \alpha)\lambda$, is monitored only periodically with a constant period t^*. Thus, from a failure until the beginning of a current periodic test, the system is in a state of hidden failure.

The switching device is not necessary. Thus, the system fails because of the switch only if the main unit fails during the switch's repair. The switch's MTTF is T_s and the MRT is τ_s. Only one repair facility is available. The serving discipline is FIFO.

Solution. The solution of this problem by means of "classical" reliability or queuing theory methods is an extremely difficult task. The use of the following heuristic method allows us to solve the problem with no difficulty.

First, let us assess all failure situations. As we mentioned above, some events differ from others not only by the set of failed units but also by the order of their appearance. In Table 3.1 we use the following notation of events: $M_{1-\alpha}$ is the failure of the nonmonitored part of the main unit; M_α is the failure of the monitored part of the main unit; R is the failure of the redundant unit; F is the false signal about the failure of the redundant unit; and S is the failure of the switch. As before, the average residual time τ^r depends on the distribution of corresponfing r.v. θ..

With the help of the expressions in Table 3.1, we can easily obtain different reliability indexes of the system:

$$\Lambda_{syst} = \Lambda^1_{syst} + \Lambda^2_{syst} + \Lambda^3_{syst}\Lambda^4_{syst} + \Lambda^5_{syst}$$

$$P(t) \approx e^{-\Lambda_{syst}t}$$

$$\tau_{syst} \approx \frac{1}{\Lambda_{syst}} \sum_{1 \le i \le 5} \Lambda^i_{syst}\tau^i_{syst}$$

$$K \approx \frac{1}{1 + \sum_{1 \le i \le 5} \Lambda^i_{syst}\tau^i_{syst}}$$

In conclusion, we repeat: if the resulting reliability indexes obtained with the help of this method are good, you can trust them; if not—forget about a reliability evaluation and begin to improve the system or discard it!

TABLE 13.1

Situation with Ordered Events	Rate	Idle time
$M_\alpha\,R$	$\Lambda^1_{syst} = \alpha\lambda_1\lambda_2\tau_1$	$\tau^1_{syst} = \tau^r_1$
$R\,M_\alpha$	$\Lambda^2_{syst} = \alpha\lambda_1\lambda_2\tau_2$	$\tau^2_{syst} = \tau^r_2$
$F\,M_\alpha$	$\Lambda^3_{syst} = \alpha\lambda_1\lambda_f\tau_f$	$\tau^3_{syst} = \tau^r_f$
$M_{1-\alpha}$	$\Lambda^4_{syst} = (1 - \alpha)\lambda_1$	$\tau^4_{syst} = t_k/2$
$S\,M_\alpha$	$\Lambda^5_{syst} = \alpha\lambda_1\lambda_s\tau_s$	$\tau^5_{syst} = \tau^r_s$

13.2   TIME REDUNDANCY

Many technical systems possess a property termed *time redundancy*. A failure of such a system does not necessarily lead to a failure of the operation which the system performs. For example, a system's failure leads to the failure of its operation if the idle time duration is longer than some specified value. All conveyer systems with intermediate storage are of this type.

Sometimes a system has a redundant time to perform its operation; in this case at least once during a total permissible period a failure-free operation time must exceed a specified value. An example of such a system is a defense missile system which can launch several missiles sequentially until a target has been destroyed.

Some systems require that their failure-free operation time (collected during a permissible period of operational performance) exceed a specified value. For example, a computer with a restarting regime needs some time to fulfill the performance of a task.

The reader can find more complex cases where one considers series–parallel or parallel–series systems and also more sophisticated "time collecting" procedures in Cherkesov (1974) and Kredentser (1978).

We will illustrate a heuristic approach on some simple examples.

13.2.1   Gas Pipeline with Underground Storage

We consider a simple gas pipeline system with underground storage which allows a user to use gas from storage during a pipeline breakdown.

Let η be the pipeline's random idle time with distribution $G(t) = \Pr\{\eta \le t\}$ and let ξ be its TTF with distribution $R(t) = \Pr\{\xi \le t\}$. The storage volume equals V. The speed of the storage expenditure (after a pipeline failure) equals α, and its speed of refilling equals ρ. This process of expenditure and refilling of the storage is depicted in Figure 13.3. For simplicity, we assume that the storage begins refilling immediately after the pipeline's repair.

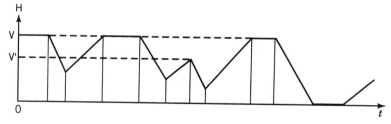

Figure 13.3. Example of a process of expenditure and refilling the storage (immediately after the pipeline's repair).

The system's failure occurs when the user does not obtain gas (the storage becomes empty). It is clear that because of the storage, a user does not "feel" the short idle times of the pipeline.

Assume that pipeline failures occur "not too often" and the probability of the storage's exhaustion during a pipeline's repair is "small enough." (The meaning of the expressions in quotation marks will be explained below.) Also assume that, on the average, the duration of TBF is much larger than the duration of the pipeline's repair.

Under these assumptions, one can consider the process of the pipeline's disruption occurrences as a renewal stochastic process. The appearance of the system failures can be considered as a "thinning" (sifting) procedure because a pipeline's disruption rarely leads to the system's failure. In other words, the process of system failures might be approximated by a Poisson process. Note that for the acceptance of this hypothesis, the probability of developing a pipeline failure in a system failure should be less than 0.1.

Let us denote $T = E\{\xi\}$, $\tau = E\{\eta\}$, and $q = 1 - G(V/\alpha)$. The system MTTF equals

$$T_{syst} = \frac{T + \tau}{1 - G\left(\dfrac{V}{\alpha}\right)}$$

For the PFFO, one can write

$$P_{syst} \approx e^{-t/T_{syst}} \tag{13.9}$$

This expression gives an upper bound because we actually assume that the storage is refilled instantaneously. The smaller the probability q, the better is the bound. One can easily obtain a lower bound. Assume that any pipeline failure, which appears during the refilling, leads to the system's failure. This bound is lower because, as a matter of fact, not each failure during refilling leads to the system's failure. The probability of a system's failure under this assumption equals

$$q^* = 1 - G\left(\frac{V}{\alpha}\right) + \int_0^x R\left(\frac{\alpha x}{\rho}\right) dG(x)$$

Obviously, the probability q^* will be larger if we write

$$q^* = \left[1 - G\left(\frac{V}{\alpha}\right)\right] + G\left(\frac{V}{\alpha}\right) R\left(\frac{V}{\rho}\right) \tag{13.10}$$

that is, if we consider that each pipeline failure always has a maximal duration equal to V/ρ. Expression (13.10) can be rewritten as

$$q^* = 1 - G\left(\frac{V}{\rho}\right)\left[1 - R\left(\frac{V}{\alpha}\right)\right] \qquad (13.11)$$

Now we can write a lower bound for the system's MTTF

$$T^*_{\text{syst}} = \frac{T + \tau}{q^*}$$

and for the system's PFFO

$$P^*_{\text{syst}} \approx e^{-t/T^*_{\text{syst}}} \qquad (13.12)$$

Note that in practical cases the MTTF and MTBF coincide. Of course, these bounds can be improved in the above-described manner.

13.2.2   Oil Pipeline with Intermediate Reservoirs

As an extension of this problem, consider an oil pipeline which has several intermediate reservoirs (storage) between an oil field and a customer. Let the oil pipeline have n pipe links with their individual reservoirs (see Figure 13.4). Denote the maximal volume of the kth storage by V_k, its nominal filling (in a so-called stationary regime) by v_k, the nominal pipeline capacity by α, the MTTF for the kth pipeline's link by T_k, and the repair time distribution for the kth link by $G_k(t)$.

When some link fails, all reservoirs between a failed link and the oil field are working in a regime of collecting oil. (Because of technological reasons, the oil extraction should not be stopped completely and, at the same time, excessively intensive extraction can destroy the oil field.) At the same time, all reservoirs between a failed link and the customer are working in a regime which guarantees a nominal customer supply. The system fails when one of these two operational functions cannot be provided.

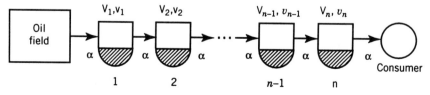

Figure 13.4. Oil pipeline with an individual reservoir at each link.

If the kth link has failed, then during time

$$\tau_k' = \frac{1}{\alpha'} \sum_{1 \leq i \leq k} (V_i - v_i) \qquad (13.13)$$

the extracted oil will not fill the corresponding reservoirs. (We use α' instead of α because, in the case of pipeline failure, the oil extraction can be slightly decreased.) At the same time, during time

$$\tau_k'' = \frac{1}{\alpha} \sum_{k+1 \leq i \leq n} v_k$$

oil from the reservoirs, cut from the oil field, can supply the customers. Thus, a system failure caused by the kth link's failure will be observed if its duration is longer than

$$\tau_k = \min(\tau_k', \tau_k'')$$

The system failure rate caused by the kth link's failure can be written as

$$\lambda_k^{\text{syst}} = \frac{1}{T_k^{\text{syst}}} = \frac{G_k(\tau_k)}{T_k}$$

where T_k is the kth link's MTTF.

The system's failure rate is defined as

$$\Lambda_{\text{syst}} = \sum_{1 \leq i \leq n} \lambda_i^{\text{syst}}$$

The PFFO can be found as usual with the help of an exponential distribution. This case is made simple for illustrative purposes. If there is a single customer, the best way is to arrange one reservoir whose nominal regime is "half-empty–half-full," that is, $V_k = 2v_k$. Another solution is to have an empty reservoir of volume $V/2$ close to the oil field and a entirely full reservoir of volume $V/2$ close to the customer.

A real need for intermediate reservoirs arises if customers are distributed along the pipeline (see Figure 13.5). In this case expressions for the reliability indexes are not formally more difficult. Denote by α_k an intensity of consumption by the kth customer from the kth reservoir. We can use (13.13) and substitute there

$$\tau_k' = \frac{\displaystyle\sum_{1 \leq i \leq k} (V_i - v_i)}{\alpha - \displaystyle\sum_{1 \leq i \leq k} \alpha_i^*}$$

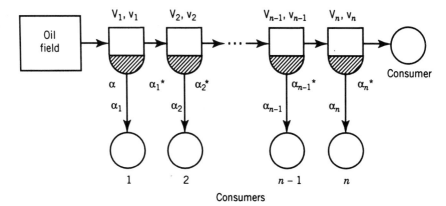

Figure 13.5. Pipeline along which customers are distributed.

and

$$\tau_k'' = \frac{\displaystyle\sum_{k+1 \le i \le n} v_i}{\alpha_n^* + \displaystyle\sum_{k+1 \le i \le n} \alpha_i}$$

All the remaining expressions (for the system's MTTF and PFFO) can be written in the above-presented form. Several related problems are considered in more detail in Rudenko and Ushakov (1989).

13.3 BOUNDS FOR THE MTTF OF A TWO-POLE NETWORK

Consider a highly reliable two-pole network. Using the lower Esary–Proschan bound and the upper Litvak–Ushakov bound, we can write combined bounds of the form

$$\prod_{1 \le k \le M} \left(1 - \prod_{i \in B_k} q_i\right) \le E\{\Phi(\mathbf{X})\} \le \min_{1 \le i \le W} \prod_{k \in b_i} \left(1 - \prod_{j \in B_k^i} q_j\right)$$

where the following notation from Chapter 9 are used: M is the number of all simple cuts of the network; B_k is a set of indexes of units of the kth simple cut; W is the number of different sets of non-intersected cuts; b_i is the ith complete set of non-intersected cuts; B_k^i is a set of indexes of units of the kth simple cut of the ith complete set of non-intersected cuts. It is clear that in both cases we use the expressions for series connections (or parallel connections) of network cuts. For the Esary–Proschan bound, cuts are dependent but we will ignore this fact for approximate expressions. This leads to negligible errors if the network is highly reliable.

The probability q_i may be considered as the availability of the ith unit (link) of the network

$$q_i = \frac{\tau_i}{T_i + \tau_i}$$

where, as usual, T_i is the MTTF of the ith unit and τ_i is its repair time.

If all units belonging to the same cut are independent in terms of their failure and repair, then we can write for an exponential distribution of repair times that the effective idle time for the kth cut can be found as

$$\theta_k = \left(\sum_{i \in B_k} \tau^{-1} \right)^{-1}$$

(We emphasize that the links' independence also means their independence in repair, that is, the number of repair facilities is large and they are available to each failed link.) Note that this result is a good approximation for an arbitrary distribution of repair time.

For the kth cut, the availability coefficient is

$$\kappa_k = \prod_{i \in B_k} q_i$$

At the same time, the availability coefficient of the kth cut can be written as

$$\kappa_k = \frac{1}{1 + \theta_k \Lambda_k}$$

where Λ_k is the failure rate of the kth cut. For highly reliable network units, we can write

$$\Lambda_k \approx 1 - \frac{\kappa_k}{\theta_k}$$

As mentioned above, the flow of failures of a highly reliable redundant system can be approximated by a Poisson process. Considering a highly reliable system as a whole, we assume that the idle times of different cuts of the network almost never overlap. Thus, for a series connection of cuts, one can write approximately

$$\frac{1}{T_{\text{up}}} = \max_{1 \le i \le W} \sum_{k \in b_i} \Lambda_k^{(i)} \le \Lambda \le \sum_{1 \le k \le M} \Lambda_k = \frac{1}{T_{\text{low}}}$$

By our previous assumptions, the failure flow of the network as a whole should be close to Poissonian. This allows us to write the expression for the PFFO:

$$e^{-t_0/T_{low}} \leq P(t_0) \leq e^{-t_0/T_{up}}$$

13.4 DYNAMIC REDUNDANCY

Here we present an approximate solution for the dynamic redundancy problem when there is a possibility of transferring units to the next stage. (For the problem statement and terminology we refer to Chapter 10.) This solution concerns the case where the number of switching stages and the number of redundant units are both large. In this case the following simple solution may be suggested.

We have shown in Chapter 10 that the number of redundant units which are switched at a stage of the operational process can only decrease at the following stages. Moreover, the number of redundant units at a stage decreases smoothly. It seems that a uniform prior allocation of the redundant units between stages is not too bad. So now the problem is to find how to choose the number of units which should work at each stage.

Consider again the case where all units are identical and independent and each of them has an exponentially distributed TTF with parameter λ. If the duration of each stage equals t and the number of operating units in n, then the average number of failed units at each stage equals λnt. Using a Poisson distribution to describe the distribution of the (random) number of failed units at each stage v, we must note that if the system operates successfully, the number of failed units cannot exceed n. In other words, we should consider a distribution truncated from the right. Hence, the actual distribution must be written as

$$\Pr(k = v/v \leq n) = \frac{1}{\Pr\{v \leq n\}} \frac{(\lambda nt)^k}{k!} e^{-\lambda nt} \qquad (13.14)$$

Such a dynamic redundant system should be highly reliable, that is, $\lambda nt \ll 1$. If so, the (random) number of failed units will practically always be less then n and we might neglect the truncation in (13.14). Of course, this correction essentially changes nothing in the numerical results but makes the solution possible via a standard of the Poisson distribution.

It is clear that a system's reliability increases with n. But, at the same time, the more redundant units are used at a stage, the more of them fail during this stage, and, consequently, the faster spare units of the system will be exhausted.

Let the total number of spare units for the system during an operating period T be N. If a single stage's duration is t, then the total number of stages is $m + T/t$.

We assume that a system's successful operation during the specified period T and the expenditure of spare units are independent events. Under the assumption that the system is highly reliable, this is almost correct and we can write the approximate expression of the system's successful operation as

$$P_{\text{syst}}(T) \approx \left[P_n(\lambda nt) \right]^m P_n(\lambda nT) \tag{13.15}$$

where $P_n(A)$ is the Poisson distribution

$$P_n(A) = \sum_{1 \le k \le n} \frac{A^k}{k!} e^{-A}$$

The simplicity of (13.15) allows us to use different methods of one-dimensional optimization (e.g., binary search, Golden section, and Fibonacci search).

13.5 OBJECT WITH REPAIRABLE MODULES AND AN UNRENEWABLE COMPONENT STOCK

18.5.1 Description of the Maintenance Support System

The designer usually uses standardized modules to enhance the performance of a fast repair. To increase an object's repairability, there should be a minimal number of different types of modules. If the designer uses a very restricted number of standardized modules, they are used ineffectively. Hence, the designer is always faced with a compromise: to decrease the total number of spare modules (by a high level of standardization) or to make an object more compact (by the use of less standardized modules).

Failed modules are sent to a repair shop. The repair shop needs spare components of various types to repair the failed modules. These components are kept in the repair shop's stock. The components themselves are assumed to be nonrepairable. The process of module repair at the repair shop continues until at least one type of the spare components in stock has been exhausted. This means that a failed module cannot be repaired because there are no available components of some type to replace a failed component of the same type in the module to be repaired.

The spare components stock is refilled in a specified time period t. After repair, a module is assumed to be completely new. Different types of modules can be repaired independently. Thus, this maintenance support system (MSS) consists of standby modules and the repair shop with a stock of spare components which are needed to repair the failed modules.

In practice, any failed module replacement by a standby module takes a finite amount of time, Usually, this time is substantially smaller than the time for repair. At any rate, we should emphasize that we are speaking about a decrease in the idle time rather than about an instant replacement. Sometimes, in engineering practice, such objects are considered to be repaired systems with a standby redundancy. It is up to the reader to decide what kind of mathematical model is best in each specific case.

The failure of the object may occur if there is no standby module of some given type for replacement (e.g., if all of them were sent to the repair shop and none had returned). Another reason for the object failure might be deficiency of components needed for performance of the repair at the repair shop.

These two above-described situations may be the cause of a failure but they do not lead to an object's failure with necessity. Indeed, the object can contain redundant modules in its structure and a deficiency of spare components still does not mean object failure. In other words, we consider the reliability of an MSS rather than the reliability of an object.

For simplicity of presentation, assume that all TTFs are distributed exponentially. The time of module replacement and repair may have a general form of distribution but its mean must be very small in comparison with the MTTF. The number of standby modules of each type and the number of various spare components at the workshop are such that the probability of

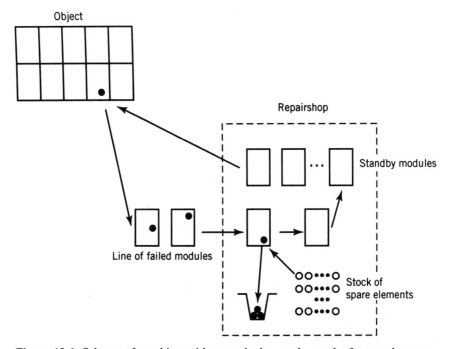

Figure 13.6. Scheme of an object with a repair shop and a stock of spare elements.

failure of the object is very small. The structure of the MSS with the served objects is depicted in Figure 13.6.

13.5.2 Notation

For simplicity, we establish the following notation.

m is the number of different types of modules, and j is the current module type, $1 \leq j \leq m$.

M_j is the number of operating modules of the jth type in the object as a whole.

M is a vector denoting the total set of different modules in the object:

$$\mathbf{M} = (M_1, \ldots, M_m)$$

y_j is the number of standby modules of the jth type.

Y is a vector denoting the total set of different standby modules:

$$\mathbf{Y} = (y_1, \ldots, y_m)$$

Λ_j is the failure rate of one module of the jth type.

a_j is the jth module replacement rate (a_j is the inverse of the mean time of replacement, A_j, of the module).

b_j is the jth module repair rate at the repair shop (b_j equals the inverse of the mean time of an actual repair, B_j, of the module).

REMARK. The actual repair time consists of the time of delivery of the failed module to the repair shop, the waiting time in line, and the returning time to the object's warehouse from the repair shop. Sometimes one also needs to include the time to search for the failed module.

n is the number of different types of components used in the various modules.

N_i is the total number of components of the ith type which are installed in the object.

N is a vector denoting the total set of different components in all modules of the object:

$$\mathbf{N} = (N_1, \ldots, N_n)$$

x_i is the number of spare components of the ith type in the repair shop stock for the total period t.

X is a vector denoting the total set of different spare components in the repair shop stock:

$$\mathbf{X} = (x_1, \ldots, x_n)$$

λ_i is the failure rate of the ith unit.

$\Pi_y\{\alpha\}$ is a Poisson d.f. with parameter α:

$$\Pi_y(\alpha) = \sum_{1 \le i \le y} \pi_i(\alpha)$$

where

$$\pi_i(\alpha) = \frac{\alpha^i}{i!} e^{-\alpha}$$

13.5.3 Probability of Successful Service

Remember that we are focusing on a successful supply of the object rather than on a successful operation of the object. For simplicity, assume that failed modules of different types are repaired independently. For a very reliable object, such an assumption leads to negligible errors.

We wish to find the probability of a successful supply R. This probability can be approximately presented as the product of two probabilities

$$R \approx R_1(\mathbf{M}, \mathbf{Y}; t) R_2(\mathbf{N}, \mathbf{X}, t) \tag{13.16}$$

Here the multipliers are defined as:

$R_1(\mathbf{M}, \mathbf{Y}; t)$ is the probability of a successful operation of the object during time t, if the object consists of **M** operating modules and there are **Y** standby modules.

$R_2(\mathbf{N}, \mathbf{X}; t)$ is the probability that during time t there is no deficit of spare components of any type in the stock of the repair shop, if the stock consists of **N** operating components and there are **X** spare ones.

Expression (13.16) is not, strictly speaking, correct because all of its multipliers are dependent. Indeed, the more module failures that occur, the more spare components are used. Nevertheless, for highly reliable objects, this dependence can be neglected.

Consider the first term of (13.16). The assumption of an independent repair of each type of module allows us to write

$$R_1(\mathbf{M}, \mathbf{Y}; t) = \prod_{1 \le j \le m} R_{1j}(M_j, y_j, t) \tag{13.17}$$

where $R_{1j}(M_j, y_j; t)$ is the corresponding probability for the jth type of module. Each of these probabilities for $j = 1, \ldots, m$ can be found by standard methodology (see Chapter 4). But we use an approximation based on the Rényi theorem of the thinning (sifting) of a renewal point process.

Each current failure of some particular jth-type module can cause an object to fail if, during its repair time B_j, all y_j standby modules of the same type have also failed. Such an event occurs with probability

$$Q_{1j}(M_j, y_j; B_j) = 1 - \Pi_{y_j}(B_j M_j \Lambda_j) \approx \pi_{y_j+1}(B_j M_j \Lambda_j) \qquad (13.18)$$

where $Q_{1j} = 1 - R_{1j}$. Here $M_j \Lambda_j$ is the total failure rate of the jth-type module, $B_j M_j \Lambda_j$ is the parameter of the corresponding Poisson distribution, and $\pi_{y_j}(B_j M_j \Lambda_j)$ is the probability that, during the module repair time of the jth type, y_j failures of this type have occurred (this leads to an object's failure caused by this particular type of module.

Notice that the repair time is random. The precise expression is

$$Q_{1j}(M_j, y_j; t) = \int_0^\infty \pi_{y_j+1}(\tau M_j \Lambda_j) \, dF_{B_j}(\tau) \qquad (13.19)$$

where $F_{B_j}(t)$ is the distribution of repair time. Above we used a substitution: instead of the mean of a function of the r.v. we used a function of the mean value argument. For a highly reliable system, however, this substitution does not lead to substantial errors. Moreover, for rough calculations one sometimes may regard B_j as the delivery time for a spare module from the repair shop.

Using the Rényi theorem, the total failure rate of object failures caused by the jth type of module can be written as

$$\tilde{\Lambda}_j = \pi_{y_j}(B_j M_j \Lambda_j) M_j \Lambda_j \qquad (13.20)$$

Indeed, we consider the event flow (the flow of the system's failures because of the jth type of module) with parameter $M_j \Lambda_j$. Each such event might cause an object's failure with probability π. Thus, the approximate expression for the probability $R_{1j}(M_j, y_j; t)$ is

$$R_{1j}(M_j, y_j; t) = e^{-\tilde{\Lambda}_j t} \qquad (13.21)$$

For the object as a whole, (13.17) can be rewritten as

$$R_1(\mathbf{M}, \mathbf{Y}; t) = \prod_{1 \le j \le m} e^{-\tilde{\Lambda}_j t} = \exp\left(-t \sum_{1 \le j \le m} \tilde{\Lambda}_j\right) = e^{-t\tilde{\Lambda}} \qquad (13.22)$$

where

$$\tilde{\Lambda} = \sum_{1 \le j \le m} \tilde{\Lambda}_j$$

The second term of (13.16) relates to the event that, during period t, there is no deficit of spare components for the repair of failed modules at the workshop. If spare components of some kind have been exhausted, each failure of a component of this type leads to the exclusion of one of the spare modules from circulation because it is impossible to repair it. The exact solution to this problem is practically hopeless. We propose an approximate method giving the lower bound for the probability.

We introduce the value z_i equal to the smallest number of modules of some type in which components of the ith type are installed. We are not interested in a particular type of module for the approximate solution, we only note that there are modules with such a property. Then we assume that when there are no spare components of the ith type for repair, any forthcoming failure involves a component of the ith type. It is clear that a new failure may not actually be connected with this type of spare unit. Moreover, we assume that such a failure occurs in the "weakest" type of module which is characterized by the value z_i. Thus, each forthcoming failure leads to the exclusion of one of these modules of the above-mentioned type. Under these assumptions, if z_i failures have occurred, there is no standby module of this type.

Of course, this yields a lower bound on the probability of interest. Indeed, we consider an artificial case where all failures of the ith component occur in the "most vulnerable" module. (In general, one should be very cautious in the use of bounds for an optimization problem!) Hence, if we assume that all operating components are independent:

$$R_2(\mathbf{N}, \mathbf{X}; t) = \prod_{1 \le i \le n} \left[1 - \Pi_{x_i + z_i}(t N_i \lambda_i) \right] \approx \prod_{1 \le i \le n} \pi_{x_i + z_i + 1}(t N_i \lambda_i) \right] \quad (13.23)$$

13.5.4 Availability Coefficient

We first emphasize that we are to ignore any internal redundancy of the object. Thus, in some sense, we consider a system with a series structure. A reliability index for the kth object can be written as

$$K = K_1 K_2 K_3 \qquad (13.24)$$

where the multipliers K_j mean the following:

K_1 is determined by the module's replacement time. If one considers that such a replacement takes negligible time, then K_1 equals 1.

K_2 is determined by the idle time which occurs because there are no available standby modules of any type (this may occur if all modules of the required type are at the repair shop).

K_3 is determined by the idle time at the end of period t because of the exhaustion of spare components of any type at the repair shop. (As a consequence of this, it is impossible to repair some modules.)

The total failure rate of an object of the jth type of module can be expressed as

$$\tilde{\Lambda}_j = \Lambda_j M_j$$

After each such failure, the system is idle during the replacement time. This leads to the following availability coefficient:

$$K_j = \frac{1}{1 + B_j \tilde{\Lambda}_j}$$

Finally, we may write

$$K_1 \approx \prod_{1 \leq j \leq m} K_j \qquad (13.25)$$

For practical engineering calculations, we often neglect the time for a module's replacement. If such an assumption is reasonable, we can consider an ideal standby redundancy with repair (see Chapter 4). As a matter of fact, it is not always reasonable to do this.

The term K_2 in (13.24) depends on the repair regime at the individual repair shop and on the type of repair time distribution. The delivery time of a failed module to the repair shop and back to the object also has to be taken into account. Assuming that a repair shop is located directly at an object's site, we may neglect these two components of the idle time.

Table 13.2 contains expressions of the object's idle time for two important cases: (1) the repair time is constant (a degenerate distribution) and (2) the repair time is distributed exponentially.Let us briefly explain the values in

TABLE 13.2 The mean residual time of repair (the object waiting time)

Number of repair facilities	Type of the repair time	
	Exponential	Constant
1	A_j	
		$A_j/(y_j + 1)$
no restrictions	$A_j/(y_j + 1)$	

this table. Assume that the repair time A is constant and the failure flow is Poisson. Suppose y failures occur during period A. They divide this period of time into $y + 1$ random parts. All of these new r.v.'s are i.i.d., and so the mean of each equals $A/(y + 1)$.

Consider the case where an object's failure is caused by the jth-type module. This may occur if the main module of this type has failed and, during its repair, y_j more failures occur. As we found above, the system's idle time is supposed to be approximately equal to $A_j/(y_j + 1)$.

This is an approximation because we neglect the possibility of the appearance of more than y failures in the period. Of course, we might write the exact expression of the type

$$\tau = A_j \sum_{1 \le k < \infty} \frac{1}{y_j + k} \frac{(\Lambda_j A_j)^{y_j + k}}{(y_j + k)!} e^{-\Lambda_j A_j}$$

But it is not reasonable to do this for an approximate analysis. We have made a number of disputable assumptions (Poisson failure flow, constant repair time, independent of some dependent events, etc.), and so why not make another which is not less reasonable?

If the repair time is distributed exponentially and there is only one repair facility, then, by the memoryless property of the exponential distribution, the residual time has the same distribution with the same mean equal to A_j. If there are enough repair facilities to simultaneously and independently repair all failed modules of any type, then we observe the minimal value of $y_j + 1$ exponentially distributed random variables. The mean of this value equals $A_j/(y_j + 1)$. Denote the repair time by $\tilde{A}_j$. The appropriate value might be chosen from Table 13.2.

Note that such periods of idle time appear with a rate Λ_j determined above. We assume that all modules of different types are independently repaired. This gives us an opportunity to write an expression for the second term of (13.24)

$$K_2 = \prod_{1 \le j \le m} \frac{1}{1 + \tilde{\Lambda}_j \tilde{A}_j} \tag{13.26}$$

The third term of (13.24) is determined by the shortage of spare components during a specified interval t. We remind the reader again of the property of a conditional distribution of arrival times of a fixed number of events in the Poisson process: all intervals between arrivals have the same distribution.

For the ith type of unit, we can write

$$K_{3i} = \Pi_{x_i}(\tilde{\lambda}_i t) + \sum_{z > x_i} \pi_z(\tilde{\lambda}_i t) \frac{x_i + 1}{z + 1} \tag{13.27}$$

where $\tilde{\lambda}_i$ is the resulting failure rate of the ith type of component:

$$\tilde{\lambda}_i = \lambda_i N_i$$

Notice that in this case the first component failure occurs inside the time interval t, so that inside this interval, $N_i + 1$ failures are observed before the object begins to "feel" the spare components deficit. Thus, if there are z failures, $z > x$, then the first $x + 1$ of them are considered to be intervals during which there is no deficit of spare components.

Again, if

$$\pi_{x_i+1}(\tilde{\lambda}_i t) \gg \sum_{z > x_i} \pi_z(\tilde{\lambda}_i t)$$

expression (13.27) can be rewritten as

$$K_{3i} \approx \Pi_{x_i}(\tilde{\lambda}_i t) + \frac{x_i + 1}{x_i + 2}\pi_{x_i+1}(\tilde{\lambda}_i t) \approx 1 - \pi_{x_i+1}(\tilde{\lambda}_i t)\frac{1}{x_i + 2} \quad (13.28)$$

For highly reliable systems, we might assume that the probability of a deficit of any type of spare unit is sufficiently small. Thus, we can neglect the possibility of the deficit of several types of spare components at the same moment. This assumption is very important because all idle times of the latter type appear at the very end of the interval t. Consequently, these intervals are strictly dependent:

$$K_3 > \prod_{1 \le i \le n} K_{3i}$$

If spare components are delivered in a nonperiodic manner, we can construct an approximate solution of the above-described type.

13.6 TERRITORIALLY DISPERSED SYSTEM OF OBJECTS WITH INDIVIDUAL MODULE STOCK, GROUP REPAIR SHOPS, AND HIERARCHICAL WAREHOUSE OF SPARE UNITS

13.6.1 Description of the Maintenance Support System

Consider a set of objects which are dispersed territorially. Each object has its own stock of standby modules. Failed modules are sent to a repair shop. The repair shop serves several objects. A spare unit's delivery to the repair shop allows for the continuous repair of failed modules. The repair shop needs spare units of different types to replace failed units during the repair. These units are nonrepairable. The repair of modules can continue until the depletion of the stock of spare units. Let such a maintenance system using an individual stock of repairable modules and the common (group) repair shop with a stock of spare units be called a *maintenance support system* (MSS).

We assume that the failure of an individual stock occurs if there are no available spare modules of some type when a failure of a module of the same type has occurred. This might not lead to the object failure. Thus, we again (as we did in Section 13.5) consider the ability of the MSS to supply the object with spare parts.

The exhaustion of spare units of at least one type at the repair shop is also considered to be an MSS failure, although this means only that a potential disability of repair is starting. If there is a failure of the module because of a unit of another type, the repair can be performed successfully. Below we will show how to take this fact into account.

The exhaustion of spare modules of some type in the stock of the repair shop leads only to an increase in the repair time because of the waiting time for the appropriate module to be repaired and returned.

Assume that all TTFs are distributed exponentially, the mean repair times are relatively small, and the number of spare modules and spare units are such that the probability of failure of any object is very small.

From the above-given description of the MSS, one can try to construct a Markov model of the investigated system of several technical objects and the MSS for their maintenance. But any attempts to construct and practically use even such simple models are doomed to defeat. Real practical problems,

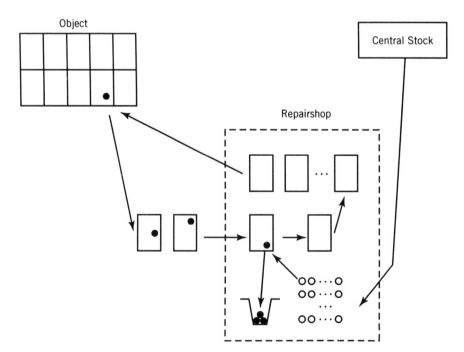

Figure 13.7. Scheme of an object with a repair shop supplied from the central stock of spare elements.

involving a large number of different modules and units, seem to be completely unsolvable in any exact analytical way.

We shall construct a heuristic model for the complex case described above.

13.6.2 Object with an Individual Repair Shop and a Hierarchical Stock of Spare Units

The structure of the hierarchical supply system and the object subjected to service is presented in Figure 13.7 The supply system consists of an individual repair shop with its own stock of spare units, and a central stock which periodically supplies the individual warehouses. During the specified period of system operation, the central stock refills the individual stock several times.

The failure of the system can be caused by the following circumstances:

- There are no spare modules of some type, and a failure of this particular type of module has occurred.
- During an interval between the neighboring refillings of the individual stock, there are no available spare units of some type and, at the same time, such a module has failed, which needs this type of unit.
- During the total period of time during which we consider the MSS and objects, a central stock is not able to supply a given individual stock with an appropriate number of spare units.

First, let us introduce the notation used in this section.

m is the number of different types of modules.

M_{kj} is the number of operating modules of the jth type in the kth object.

$\mathbf{M}_k$ is the vector denoting the total set of different modules in the kth object:

$$\mathbf{M}_k = (M_{k1}, \dots, M_{km})$$

Y_{kj} is the number of spare modules of the jth type in the kth object.

$\mathbf{y}_k$ is the vector denoting the set of different spare modules of the kth object in the warehouse:

$$\mathbf{y}_k = (y_{k1}, \dots, y_{km})$$

Λ_j is the failure rate of one module of the kth type.

a_{kj} is the intensity of replacement of the jth module in the kth object (a_{kj} equals the inverse of the mean time of replacement of the module, A_{kj}).

b_{kj} is the intensity of repair of the jth module in the kth object (b_{kj} equals the inverse of the mean time of the actual repair of the module, B_{kj}).

REMARK. The actual repair time consists of the delivery time of a failed module to a repair shop, waiting time in a line, and time to return the module to the object's individual stock.

n is the number of different types of units.

N_{ki} is the number of operating units of the ith type in the kth object.

$\mathbf{N}_k$ is the vector denoting the complete set of units in all modules in the kth object:

$$\mathbf{N}_k = (N_{k1}, \ldots, N_{kn})$$

x_{ki} is the number of spare units of the ith type in the kth in the warehouse of the kth object for one refilling period.

$\mathbf{x}_k$ is the vector denoting the set of spare units of all types in the kth object for the period:

$$\mathbf{x}_k = (x_{k1}, \ldots, x_{kn})$$

X_i is the number of spare units of the ith type in the central warehouse for the total period of the inventory system's operation (at least before a current refilling of the central warehouse).

$\mathbf{X}$ is a vector denoting the set of spare units of all types in the central stock:

$$\mathbf{X} = (X_1, \ldots, X_n)$$

λ_i is the failure rate of one unit of the ith type.

t is the period of refilling the kth object's stock with spare units from the central stock.

T is the total period of the inventory system's operation (at least before a current refilling of the central warehouse).

k is the number of refilling periods:

$$k = T/t$$

Now we start by writing the goal functions of interest.

1. Probability of successful supplying the kth object with spare modules and spare units during time T.

Again, we do not focus on the successful operation of the object, but rather on its successful supply. To simplify the explanations and to obtain reasonable formulas, assume that modules of each type can be repaired independently. Without this assumption, we need to solve a large system of equations corresponding to a multichannel queuing system with nonhomogeneous demands. Such an expression would essentially be useless for the purposes of optimization. Note once more that such an assumption leads to negligible errors for a highly reliable system, and makes all calculations much easier.

The probability R can be approximately represented as a product

$$R = R_1(M_k, y_k; T) R_2(N_k, x_k; kt) R_3(N_k, X, T) \qquad (13.29)$$

where

$R_1(M_k, y_k; T)$ is the probability of a successful operation of the kth object during a period of time T if this consists of M_k operating modules of different types and there are y_k spare modules in the object's stock.

$R_2(N_k, x_k; kt)$ is the probability that during all k periods between the kth object's stock refilling there will be no deficiency of spare units of any type if the kth object contains N_k different types of units and there are x_k spare units in the object's stock.

$R_3(N_k, X, T)$ is the probability that during the period of time T all $k - 1$ refillings of the kth object's stock will be successful (i.e., there will be no deficiency of spare units of any type in the kth object's stock during time T if the kth object contains N_k different types of units and there are X spare units in the central stock).

Of course, (13.29) is not, strictly speaking, a correct formula. In fact, all multipliers there are dependent: the more modules that fail (and, consequently, the less the system availability), the more spare units are spent. But we can neglect this dependence for highly reliable systems.

Now let us write approximate expressions for the multipliers of (13.29). We obtain the probability $R_1(M_k, y_k; T)$ after several auxiliary steps. Each failure of the module of the jth type generates a potentially dangerous situation: the next failure may cause a system failure. If the mean time of actual repair of the jth module of the kth object is A_{kj}, the probability that a failure of the jth module will lead to a system failure can be written as a cumulative Poisson probability

$$P_j = 1 - \Pi_{y_{kj}}\{A_{kj} M_{kj} \Lambda_j\} \qquad (13.30)$$

where

$$\Pi_{y_{kj}}\{A_{kj} M_{kj} \Lambda_j\} - = \sum_{0 \le s \le y_{kj}} \frac{(A_{kj} M_{kj} \Lambda_j)^s}{s!} e^{-A_{kj} M_{kj} \Lambda_j}$$

is the probability that during the repair of the first failed module of the jth type there will be less than y_{kj} failures of modules of this type.

REMARK. One should mention that the time of repair is a random value and the above-mentioned probability, strictly speaking, has to be found as

$$P_j = \int_0^\infty \Pi_{y_{kj}}\{tM_{kj}\Lambda_j\}\, dF_{A_{kj}}(t)$$

where $F_{A_{kj}}(t)$ is the distribution of the random repair time. But, for a highly reliable system, we can neglect this fact and use an approximation.

The parameter of the flow of failures of the jth module is $M_{kj}\Lambda_j$. Using the Rényi theorem, we can find the failure rate of the kth object caused by the jth-module failure:

$$\tilde{\lambda}_{kj} = \left[1 - \Pi_{y_{kj}}\{A_{kj}M_{kj}\Lambda_j\}\right] M_{kj}\Lambda_j \tag{13.31}$$

Consider a highly reliable system. For the object as a whole, the failure rate caused by exhausting at least one type of module equals

$$\tilde{\lambda}_k = \sum_{1 \le j \le m} \tilde{\lambda}_{kj}$$

On the basis of the Rényi theorem, we can write the approximation

$$R_1(M_k, y_k, T) \approx e^{-\tilde{\lambda}_k T} \tag{13.32}$$

Now let us obtain $R_2(N_k, x_k; T)$. This is the probability that during k periods of length t there will be no deficiency of spare units of the kth object's stock. This probability includes the corresponding probabilities for all different types of units. For spare units of the ith type during a single period t, the probability equals

$$p_i(t) = \Pi_{x_{ki}}\{tN_k\lambda_i\} \tag{13.33}$$

Obviously,

$$R_2(N_k, x_{ki}, kt) = \left[\sum_{1 \le i \le n} p_i(t)\right]^k \tag{13.34}$$

Finally, let us consider R_3. First of all, we must note that two events are dependent: an expenditure of spare units on a period t and an expenditure of spare units on a total period T. Indeed, if excessively many spare units, for instance, are spent at the first period t, then an expected total number of spare units spent on a total period T will exceed the mean number calculated before the first stage. But, fortunately, for highly reliable systems the approximation (13.27) is a reasonable practical result.

The probability of a successful supply of the kth object's stock with an ith type of spare unit equals

$$R_{3i} = \Pi_{X_i}(\lambda_i(T - t) N_{ki}) \qquad (13.35)$$

and, obviously,

$$R_3 = \prod_{1 \leq i \leq n} R_{3i} \qquad (13.36)$$

2. Availability coefficient of the kth object

Again, note that we consider a series system, that is, a system without inner redundancy. (More correctly, we assume that the maintenance system's duty is to keep an object in its initial state.) The availability coefficient for the kth object can be written as

$$K_k = K_{1k} K_{2k} K_{3k} K_4 \qquad (13.37)$$

where

K_{1k} is the component of the system's availability coefficient determined by the time of replacement of failed modules for operating ones.

K_{2k} is the component of the system's availability coefficient determined by the idle time caused by the absence of available spare modules of some type.

K_{3k} is the component of the system's availability coefficient determined by the idle time at the end of the time period t caused by the absence of spare units in the repair shop's stock.

K_4 is the component of the system's availability coefficient determined by the idle time at the end of the total period T caused by the absence of spare units in the central stock.

For the kth object all modules of the jth type have a total failure rate equal to

$$\tilde{\Lambda}_{kj} = \Lambda_j M_{kj} \qquad (13.38)$$

This type of module has an availability coefficient

$$K_{kj} = \frac{1}{1 + B_{kj}\tilde{\Lambda}_{kj}} \qquad (13.39)$$

For a highly reliable system, we can neglect the possible overlapping of idle times caused by the failures of different modules. This allows us to write

$$K_{1k} = \prod_{1 \leq j \leq m} K_{kj} \qquad (13.40)$$

REMARK. For most engineering calculations one neglects the replacement time, considering instead a standby redundancy with repair. In fact, this is not always reasonable to do.

The component K_{2k} depends on the repair regime at the repair shop and on the type of distribution of the repair time.

REMARK. One should take into account the time of delivery to and return from the repair shop. Assuming that a repair shop is located at the investigated object, one can neglect this time in comparison with the repair time.

Table 13.3 contains expressions of the object's idle time for two important cases: (1) the repair time is constant and (2) the repair time is distributed exponentially. Let us briefly explain the values in this table. Assume that y_{kj} failures occur during the repair time of the main unit equal to constant A_{kj}. The last failure causes a failure of the object. We know that these failures form a Poisson process. Thus, they divide A_{kj} into y_{kj+1} random parts. All of these new random values are identically distributed (uniformly), so the mean value of each of them equals $A_{kj}/(y_{kj} + 1)$. The first failed module (a module at the internal structure of an object) is repaired by time A_{kj}. Under the assumption of a constant repair time, it does not matter if there are one or many repair facilities: the earliest repaired module is that which failed first.

If the repair time is distributed exponentially and there is only one repair facility, then, by the memoryless property, the residual repair time of any module is also exponentially distributed with the same mean equal to A_{kj}. If there are enough repair facilities, all failed modules might be repaired instantly. When y_{kj} failures have occurred, one observes an object failure. The system's idle time begins from this moment. The time until the completion of the first repair is equal to the minimum of the values of $y_{kj} + 1$ exponentially distributed random variable is determined the system idle time. The mean of this value equals $A_{kj}/(y_{kj} + 1)$. Below we denote the repair time by $\tilde{A}_{kj}$.

Now note that an object's failure rate equals λ_{kj} as determined above. We will assume that all modules of different types are independent in terms of

TABLE 13.3   The Mean Object's Idle Time

Number of Repair Facilities	Type of Repair Time Distribution	
	Exponential	Constant
1	A_{kj}	$A_{kj}/(y_{kj} + 1)$
Y_{kj}	$A_{kj}/(y_{kj} + 1)$	

their repair. This allows us to write an expression for the second component
of (13.37)

$$K_{2k} = \sum_{1 \le j \le m} \frac{1}{1 + \tilde{\lambda}_{kj}\tilde{A}_{kj}} \tag{13.41}$$

The third component of the product (13.40) is determined by the absence
of spare units during a time interval t. We again use the property of a
Poisson process: arrival moments within a "time window" divide this interval
as independent uniformly distributed r.v.'s.

For the ith type of unit, we can write

$$K_{3ki} = \Pi_{x_{ki}}(\lambda_i N_{ki} t) + \sum_{z > x_{ki}} \pi_z(\lambda_i N_{ki} t) \frac{x_{ki} + 1}{z + 1} \tag{13.42}$$

Again, if

$$\pi_{x_{ki}+1}(\lambda_i N_{ki} t) \gg \sum_{z > x_{ki}+2} \pi_z(\lambda_i N_{ki} t)$$

expression (13.42) can be rewritten as

$$K_{3ki} = \Pi_{x_{ki}}(\lambda_i N_{ki} t) + \frac{x_{ki} + 1}{x_{ki} + 2} \sum_{z > x_{ki}} \pi_z(\lambda_i N_{ki} t)$$

$$= 1 - \left[1 - \Pi_{x_{ki}}(\lambda_i N_{ki} t) \frac{1}{x_{ki} + 2} \right]$$

For highly reliable systems we may assume that the probability of a
deficiency of any type of spare unit during the time interval t is very small.
This means that the probability of an occurrence of a deficit of several types
of spare units is negligibly small. This assumption is very important because
all idle times appear at the very end of the time interval t; that is, all such
failures are not independent.

To make it clear, let us consider a simple case where there are only two
types of spare units, say x_s and x_r. For these two types of spare units, we can
write an approximation

$$K_3 = P_r P_s + Q_r P_s \frac{x_r + 1}{x_r + 2} + P_r Q_s \frac{x_s + 1}{x_s + 2}$$

$$+ Q_r Q_s \min\left\{ \frac{x_r + 1}{x_r + 2}, \frac{x_s + 1}{x_s + 2} \right\} \tag{13.44}$$

In this case for highly reliable systems one can write an approximate expression

$$K_3 \approx P_r P_s + Q_r P_s \frac{x_r + 1}{x_r + 2} + P_r Q_s \frac{x_s + 1}{x_s + 2}$$

$$\approx 1 - \left\{ Q_r \left[1 - \frac{x_r + 1}{x_r + 2} \right] + Q_s \left[1 - \frac{x_s + 1}{x_s + 2} \right] \right\}$$

where P_i is a compact notation for $\Pi_{x_{ki}}(\lambda_i N_{ki} t)$, $Q_i = 1 - P_i$, and $i = r$ or s. Using the same arguments as above, we have

$$K_{3k} = 1 - \sum_{1 \le i \le n} \left[1 - \Pi_{x_{ki}}(\lambda_i N_{ki} t) \right] \left[1 - \frac{x_{ki} + 1}{x_{ki} + 2} \right] \qquad (13.45)$$

Similar arguments lead us to an expression for K_4, with appropriate changes of parameters:

$$K_4 = 1 - \sum_{1 \le i \le n} \left[1 - \Pi_{X_i}(\lambda_i (T - t) N_{ki}) \right] \left[1 - \frac{X_i + 1}{X_i + 2} \right] \qquad (13.46)$$

Thus, all terms of (13.37) are determined.

13.6.3 Set of *K* Objects with Repair Shops and a Central Stock with Spare Units

In this case an entire system subjected to service consists of a set of *K* objects. Each of these objects has its own individual repair shop with a stock of spare units. The hierarchical supply system includes a central stock which periodically refills all of the individual stocks. The structure of the investigated supply system and the served objects is presented in Figure 13.8.

The failure of the *k*th object might be a consequence of the following events:

- There are no spare modules of some specified type in the *k*th stock, and the current failure is caused by a module of this type.
- During the interval between refilling, *t*, there are no available spare units of a given type in the *k*th stock, and the module currently failed needs a unit of this particular type for its repair.
- Over the entire interval *T* of the system's operation, the central stock cannot supply the *k*th individual warehouse with an appropriate number of spare units.

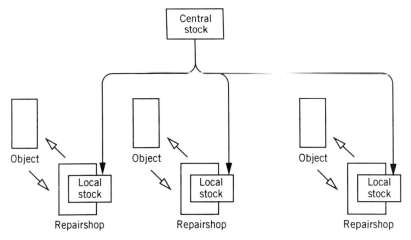

Figure 13.8. Several objects with repair shops supplied from the central stock of spare elements.

First, let us introduce some additional notation.

K is the number of different objects.

$N_{(i)}$ is the total number of operating units of the ith type installed in each of K objects.

$\mathbf{N}_*$ is the vector

$$\mathbf{N}_* = \left(N_{(1)}, \ldots, N_{(n)} \right)$$

where n is the number of different types of units; that is, this vector shows the complete set of units installed in all modules of all objects

$$\mathbf{N}_* = \left\{ \sum_{1 \le k \le K} N_{k1}, \ldots, \sum_{1 \le k \le K} N_{kn} \right\}$$

$\mathbf{N}$ is the vector

$$\mathbf{N} = (N_1, \ldots, N_K)$$

where N_k, $k = 1, \ldots, K$, is in turn, a vector with components defined above, that is, N_k is the set of units of the ith type in the kth object.

$\mathbf{x}$ is the vector

$$\mathbf{x} = (x_1, \ldots, x_K)$$

where x_k, $k = 1, \ldots, K$, is, in turn, a vector with components defined above; that is, x_k is the set of spare units of the ith type in the stock of the kth object.

$\mathbf{M}$ is the vector

$$\mathbf{M} = (M_1, \ldots, M_K)$$

where M_k, $k = 1, \ldots, K$, is, in turn, a vector, each component of which is the set of different modules installed in the kth object.

y is the vector

$$\mathbf{y} = (y_1, \ldots, y_K)$$

where y_k, $k = 1, \ldots, K$, is, in turn, a vector, each component of which is the set of spare modules of different types in the stock of the kth object.

z is the total number of refilling: $z = (T/t) - 1$.

Now we will define the goal functions.

1. R^*: The probability that all K objects will be successfully supplied with spare modules and spare units during the entire system's operation T.
2. R_k: The probability that the kth object will be successfully supplied with spare modules and spare units during the entire system's operation T.
3. K^*: The availability coefficient of the MSS (the probability that at some moment in time there is no deficiency in the supply of all K objects).
4. K_k: The availability coefficient of the kth object (the probability that at some moment in time there is no deficiency in the supply of the kth object).
5. $\tilde{K}$: The average availability coefficient of all objects (the average of all K_k).

Again, we consider the probability of successfully supplying the set of objects rather than the PFFO of these objects. All assumptions concerning the independent repair of failed modules that we considered before are preserved. We also continue to assume that all units operate with a high reliability.

We now consider the above-mentioned reliability indexes. We will refer to the previous section for intermediate results where needed.

1. Probability of successfully supplying all objects (R^*)

This can be approximately represented as a product of three probabilities

$$R^* = R_1(M, y; T) R_2(N, x; zt) R_3(N_*, X; T) \qquad (13.47)$$

where

$R_1(M, y; T)$ is the probability of successfully operating all K objects during time T.

$R_2(N, x; zt)$ is the probability that there is no deficiency of spare units of any type for any object during all refillings.

$R_3(N_*, X; T)$ is the probability that during the entire period T there will be no deficiency of spare units of any type in the central stock for refillings of all individual stocks.

Again, we have to mention that (13.47) is not strictly correct because its multipliers are dependent.

We now write an approximate expression for the multipliers in (13.47). For the term $R_1(M, y; T)$, we use (13.32)

$$R_1(M, y; T) = \prod_{1 \le k \le K} R_1(M_k, y_k; T) = \exp\left(-T \sum_{1 \le k \le K} \tilde{\lambda}_k \right) \quad (13.48)$$

where all necessary notation and intermediate formulas were written above. For $R_2(N, x; zt)$ we use (13.34)

$$R_2(N, x; zt) = \prod_{1 \le i \le K} R_2(N_i, x_i; zt) \quad (13.49)$$

Finally, we consider R_3. Again note that the approximation (13.47) gives good practical results for highly reliable systems.

For the ith type of spare unit, the probability of a successful supply of the kth individual stock during z refillings is similar to expression (13.46), but with partially changed parameters

$$R_{3i} = \Pi_{X_i}\big(\lambda_i (T - t) N_{(i)} \big) \quad (13.50)$$

and, obviously,

$$R_3 = \prod_{1 \le i \le n} R_{3i} \quad (13.51)$$

2. Probability of successfully supplying the kth object (R_k)

This probability can be approximately represented as a product of three probabilities

$$R^* = R_1(M_k, y_k; T) R_2(N_k, x_k; zt) R_3(N_*, X; T) \quad (13.52)$$

where

$R_1(M_k, y_k; T)$ is the probability of a successful operation of the kth object during time T.

$R_2(N_k, y_k; zt)$ is the probability that in all z periods t between neighboring refillings there will be no deficiency of any spare units in the stock of the kth object.

$R_3(N_*, X; T)$ is the probability that during the entire period T there will be no deficiency of any spare units in the central stock for refillings of the stock of the kth object.

We now write approximate expressions for the multipliers in (13.52). The term $R_1(M_k, y_k; T)$ coincides with (13.43). The term $R_2(N_k, x_k; zt)$ coincides with (13.45). Finally, consider $R_3(N_*, \mathbf{X}; T)$. Again, we note that the expenditure of spare units in each period t and the expenditure during the total period T are dependent events. At any rate, (13.52) is a good approximation for highly reliable systems.

For the kth object we can write

$$R_3(n_*, \mathbf{X}; T) = R_3(n_*, \mathbf{X}; T) + \left[1 - R_3(n_*, \mathbf{X}; T)\right]\left[1 - \frac{1}{K}\right]$$

$$= 1 - \frac{1}{K}\left[1 - R_3(n_*, \mathbf{X}; T)\right] \tag{13.53}$$

3. Availability coefficient of the supply system and all objects (K^*)

The reliability index $\hat{K}^*$ for all K objects can be written as

$$K^* = K_1 K_2 K_3 K_4 \tag{13.54}$$

where

K_1 is the component of the system's availability coefficient determined by the time of replacement of the failed modules by operating modules in all K objects.

K_2 is the component of the system's availability coefficient determined by the idle time associated with the absence of available spare modules of some type of at least at one object (this occurs when an object has failed and there is no available module of the required type; the object is idle until the repair's completion).

K_3 is the component of the system's availability coefficient determined by the idle time at the end of the time period t caused by a deficiency of spare units in at least one warehouse.

K_4 is the component of the system's availability coefficient determined by the idle time at the end of the total period T caused by a deficiency of spare units in the central warehouse.

Let us write the expressions for each of the terms of (13.54). Term K_1 can be expressed through K_{1k} obtained in (13.40) as

$$K_1 = \prod_{1 \leq k \leq K} K_{1k} \tag{13.55}$$

Term K_2 depends on the repair regime at each individual repair shop and on the repair time distribution. Formula (13.41) is used for calculating K_{2k}.

Thus, K_2 can be expressed as

$$K_2 = \prod_{1 \le k \le K} K_{2k} \qquad (13.56)$$

The third term of (13.54) is determined by a shortage of spare units during the time interval t in at least one stock. For one object we determined K_{3k} in (13.45). For all objects, under the assumption of high reliability, we can approximately write

$$K_3 = \prod_{1 \le k \le K} K_{3k} \qquad (13.57)$$

The expression for K_4 practically coincides with (13.46), with an appropriate change of parameters,

$$K_4 = 1 - \sum_{1 \le i \le n} \left[1 - \Pi_{X_i}\left(\lambda_i(T - t) \sum_{1 \le k \le K} N_{ki} \right) \right] \left[-\frac{X_i + 1}{X_i + 2} \right] \qquad (13.58)$$

Thus, all terms of (13.54) are determined.

4. Availability coefficient of the kth object (K_k)

Again, we do not take into account the possibility of internal redundancy within an object. For the kth object this reliability index can be written as

$$K_k = K_{1k} K_{2k} K_{3k} K_4 \qquad (13.59)$$

where

K_{1k} is the component of the system's availability coefficient determined at the time of the failed modules' replacement by operating modules.

K_{2k} is the component of the system's availability coefficient caused by the absence of available spare modules.

K_{3k} is the component of the system's availability coefficient determined by the idle time at the end of the time period t caused by a deficiency of spare units in the warehouse at the repair shop.

K_4 is the component of the system's availability coefficient determined by the idle time at the end of the entire period T caused by a deficiency of spare units in the central warehouse.

Let us consider the first term of (13.59). The expression for K_{1k} is completely equivalent to (13.30). The component K_{2k} depends both on the repair regime at the kth workshop and also on the type of repair time distribution. We take the same arguments into account because, in this case, the expression for K_{2k} coincides with (13.41). The third component of the product (13.59) is determined by a shortage of spare units during the time interval t. The expression for K_3 coincides with (13.45).

The last component K_4 reflects the influence of a deficiency in the central stock on the kth object. We assume that no individual stock has a priority in supplying. Under these conditions the kth chosen object meets a deficiency in spare units with probability equal to $1/K$. (We again assume that for a highly reliable MSS we can neglect the possibility that more than one object has a deficiency.) Note that if such a deficiency occurs, we observe it during the last period t within the entire period T. Thus, the expression for K_4 can be written as

$$K_4 = K_4 + (1 - K_4)\left(\frac{1}{K}\right) = 1 \tag{13.60}$$

where K_4 is determined in (13.46).

Thus, all terms for (13.59) are obtained.

5. Average availability coefficient $(\tilde{K})$

With (13.59) it is very simple to write the desired expression for $\tilde{K}$:

$$\tilde{K} = \prod_{1 \leq k \leq K} K_k \tag{13.61}$$

Thus, all of the different reliability indexes of the system are obtained. We can see that the assumptions relating to a high level of system reliability were essential in all respects.

13.7 CENTRALIZED INVENTORY SYSTEM

Consider an inventory system which serves k objects. Each object is considered as a single unit with standby redundant units. In practice, we can consider any group of similar units as such an object. Each working unit fails at a random time ξ that has an arbitrary distribution $F(t)$. The number of standby units at each object is restricted. A stock of standby units might be refilled from a central storage facility. As soon as the number of standby units at some object decreases to m, an invoice for n new units is sent to a central storage facility. We call this rule an (m, n) policy.

Let us consider a period of time T and assume that the central storage facility has N spare units at the beginning of this period. If a central stock cannot satisfy a claim for spare units from some object, we will consider this as an inventory system failure. (Of course, an object can work successfully after sending the claim.)

The delivery of an order of requested spare units from central stock to an object takes a time τ. If during this time all m standby units are spent, an object failure has occurred. We call such failures "failures of the second type." The probability of a successful operation of the inventory system will be denoted by R.

Consider a process supplying a single object. If N is large, centralized inventory control plays an important role in reliability improvement. Increasing m leads to a decreasing probability of failure of the second type. On the other hand, it increases the probability of a failure of the first type. This can be explained as follows: some object sends its claim to the central stock, but is rejected. This case is considered as an inventory system failure even though, in all probability, all of the remaining objects have enough standby units (including those which are waiting for delivery).

Approximate Solution   The "order level" m must be chosen in such a way that the probability of failure during τ is small:

$$\Pr\left\{ \sum_{1 \le i \le m} \xi_i < \tau \right\} = F^{*m}(\tau) = \varepsilon(m)$$

where F^{*m} denotes an m-ordered convolution and $\varepsilon(m)$ denotes a small value depending on m.

But, at the same time, the larger m, the more unused standby units remain in the system up to the moment of a failure of the first type. Thus, up to the moment of a first type of failure, the number of spent units will not be N, but $N - Z$ where Z is the random number of unused units in the system. Thus,

$$R = \sum_{\forall j} \Pr\{Z = j\} \sum_{0 \le l \le [(1/m)(N-j)]} (1 - \varepsilon(m))^l \Pr\left\{ \sum_{1 \le i \le k} \gamma_i^*(T) = l \right\} \quad (13.62)$$

where [**a**] is the integer part of **a**, and $\gamma^*(T)$ is the number of events of a renewal process with d.f. F_0^{*m} for the intervals between neighboring events:

$$F_0^{*m}(t) = \Pr\left\{ \frac{\sum_{1 \le i \le m} \xi_i < t}{\sum_{1 \le i \le m} \xi_i > \tau} \right\} = \begin{cases} 0 & 0 \le t \le \tau \\ \dfrac{F^{*m}(t) - F^{*m}(\tau)}{1 - F^{*m}(\tau)} & t > \tau \end{cases}$$

$$(13.63)$$

Let us denote the mean and the variance of the distribution (13.63) by ma_* and σ_*^2. From renewal theory one knows that

$$E\{\gamma^*(T)\} \approx \frac{T}{ma_*} \tag{13.64}$$

and

$$\text{Var}\{\gamma^*(T)\} \approx \frac{T\sigma_*^2}{m^2 a_*^3} \tag{13.65}$$

If $\varepsilon(m) \ll 1$, then the truncated moments (13.64) and (13.65) differ from their nontruncated counterparts by a value of order $O(\varepsilon)$. Thus,

$$E\{\gamma^*(T)\} \approx \frac{T}{ma} \tag{13.66}$$

and

$$\text{Var}\{\gamma^*(T)\} \approx \frac{T\sigma^2}{m^2 a^3} \tag{13.67}$$

where a and σ^2 are the mean and variance of a nontruncated distribution.

We might assume that the remaining spare units at each object are distributed approximately uniformly between 1 and m. Thus, the r.v. Z has mean $E\{Z\} = (m + 1)/2$ and $\text{Var}\{Z\} = (m^2 - 1)/12$. In the system as a whole, the total number of unused units N equals the sum of $k - 1$ uniform r.v.'s. Then $E\{Z\} = (k - 1)(m + 1)/2$ and $\text{Var}\{Z\} = (k - 1)(m^2 - 1)/12$. In practice, each operating unit is considered to be a new one. Strictly speaking, this is true only when a unit's TTF has an exponential distribution. If one considers an almost constant TTF (e.g., a normal r.v. with an extremely small coefficient of variance), then $E\{Z\} = (k - 1)m/2$. For an "aging" r.v. ξ, the following bounds for the mean are valid:

$$(k - 1)m/2 \le E\{Z\} \le (k - 1)(m + 1)/2$$

Consider the superposition of k renewal processes and calculate the number of spare units which were not used at $k - 1$ objects. The distribution of this sum can be approximated by a normal distribution. Substituting this approximation into (13.62) and using integration instead of summation, we obtain

$$R(m) \approx \exp\left[-\frac{kt}{ma}F^{*M}(\tau)\right]\Phi\left(\frac{N - \dfrac{kt}{a} - \mathrm{E}\{Z\}}{\sqrt{\dfrac{kT\sigma^2}{a^3} + \mathrm{Var}\{Z\}}}\right) \quad (13.68)$$

where $\Phi(x)$ is a standard normal distribution. The expression (13.68) has an obvious explanation. The first multiplier reflects the influence of failures of the second type and increases with increasing m. The value of $(kT)/(ma)$ is the mean number of claims sent before a failure of the second type. $F^{*m}(\tau)$ is the probability of the latter failure. The second multiplier reflects the influence of failures of the first type and decreases with increasing m.

Using essentially the same argument, one can easily adapt expression (13.68) to the case where $n > m$ and τ_i and F_i are different. In this case different objects are characterized by different values of n_i and m_i. Under the assumption that the r.v. τ has a d.f. $G(t)$,

$$\varepsilon(m) = \mathrm{Pr}\left\{\sum_{1 \le i \le m} \xi_i < \tau\right\} = \int_0^\infty F^{*m}(x)\, dG(x)$$

13.8 HEURISTIC SOLUTION OF OPTIMAL REDUNDANCY PROBLEM WITH MULTIPLE RESTRICTIONS

In many practical situations the optimal redundancy problem can be solved approximately. This situation arises when the object, subjected to an improvement in reliability, is not well formalized for a deep mathematical analysis or the available statistical data are too uncertain. For instance, at the first stage of a system's design, we need to evaluate the designed system in very rough terms. We consider heuristic procedures based on the SDM. These procedures are described in a more verbal way than before, because our goal is to illustrate a general approach, rather than to describe a specific algorithm in detail.

13.8.1 Sequence of One-Dimensional Problems

An approximate solution can be obtained based on the solutions of M one-dimensional problems with a simultaneous checking of all of M restrictions. Consider a series system consisting of n units. Each unit is characterized by M types of restrictions: cost, size, weight, and so on. For convenience, we will call each of these restrictions a "cost" and will use subscripts to distinguish different types of "costs." Redundant units are used to improve a system's reliability.

Thus, one must solve the following general problem:

$$\max_{\mathbf{X}} \left\{ R(\mathbf{X}) | C_j(\mathbf{X}) \le C_j^0 \right\}$$

(Here we use the notation of Chapter 10.)

Consider the starting moment of the computation procedure where no redundant units are used. The initially available quantities for each type of "cost" are denoted by C_j^0, $1 \le j \le M$.

At the first stage, we use the SDM, taking into account only one type of cost at a time, say the jth type. Calculate the relative increments $g_{(j)i}^N(x_i^N)$ as described in Chapter 10. [In this case, the subscript (j) denotes the type of restricting factor, the subscript i denotes the type of unit, and the superscript N denotes the number of the current step.] In correspondence with the SDM, add an appropriate unit to the system at each step. Consider the Nth step of the process constructed for the jth type of cost. Suppose $\mathbf{X}_{(j)}^N = (x_{(j)1}^N, \dots, x_{(j)n}^N)$ redundant units have been installed in the system. The process continues until step N, when, at least for one type of restriction,

$$\sum_{1 \le i \le n} c_{ij} x_{(j)i}^N \le C_j^0 \le \sum_{1 \le i \le n} c_{ij} x_{(j)i}^{N+1} \qquad (13.69)$$

where $1 \le j \le M$. We stop when (13.69) is satisfied. Denote the conditional (marginal) optimal solution for this case by

$$\mathbf{X}_{(j)}^{\mathrm{opt}} = \left(x_{(j)1}^{\mathrm{opt}}, \dots, x_{(j)n}^{\mathrm{opt}} \right)$$

In a such way we find all M solutions, one for each restricting factor. The best current vector $\mathbf{X}_{(j)}^{\mathrm{opt}}$ is treated as a candidate for a quasi-optimal solution. The best approximate solution in this case can be found as

$$\mathbf{X}^{\mathrm{opt}} = \left\{ \mathbf{X}_k : \max_{1 \le j \le M} R_{\mathrm{syst}} \left(\mathbf{X}_j^{\mathrm{opt}} \right) \right\}$$

Suppose that we use the procedure for some specific type of cost, say j, and the procedure has stopped because of a violation of the restriction of the same type. In this case we might say that the current solution is accurate (with the accuracy determined by the SDM, of course). Indeed, this means that all of the remaining restrictions are not critical at all: during the solution we spent other resources nonoptimally but, at any rate, no restrictions were violated.

13.8.2 Using the Most Restrictive Current Resource

In the previous case we found the solution in a "corner" vertex of the domain of restrictions. The question arises: how can one find some intermediate solution? We can exploit ideas based on the use of the most constrained restriction at each step of the solution process to construct a simple heuristic procedure for finding an approximate solution of the optimal redundancy problem.

But what does the "most constrained restriction" mean? Suppose that at any current step of the process, each unit might be used with an "approximately equal" probability. This means that, on average, each step of the SDM will require an expenditure of

$$\hat{C}_j = \frac{1}{n} \sum_{1 \le i \le n} c_{ij}$$

units of resource of the jth type (a "jth cost"). If so, we find that there will be about $k_j = C_j^0/c_j$ steps before the jth restriction is violated. If k_j is the smallest value for all j, $1 \le j \le M$, it is intuitively clear that the restriction C_j^0 might be considered as the most constraining.

If so, we should choose an intuitively appealing procedure: as a first step try to solve the one-dimensional problem for the jth type of cost for which $\hat{k}_j$ is smallest. After this first step, the best unit has been installed into the system in correspondence with the SDM. In other words, it corresponds to

$$j = \left\{ j : k_j = \min_{1 \le j \le M} \frac{C_j^0}{\sum_{1 \le i \le n} c_{ij}} \right\}$$

Then one finds a unit of type i such that

$$i = \left\{ i : g_i(x_i) = \max_{1 \le i \le n} \frac{\ln R_i(x_i + 1) - \ln R_i(x_i)}{c_{ij}} \right\}$$

where j has been chosen above. Here $R_i(x_i)$ is the reliability index of the ith redundant group with x_i redundant units, and at the very first step of the procedure all x_i's are 0.

After this step new restrictions C_j^1 are formed

$$C_1^1 = C_1^0 - c_{i1}, \ldots, C_M^1 = C_M^0 - c_{iM} \tag{13.70}$$

Since we have found the "best" unit for the "most constrained restriction," other restrictions are ignored. This could lead to an overspending of some other types of resources. Thus, at the second step of the process, it may be reasonable to change the type of resources for optimization processing.

The procedure described for the first step is repeated for the second step with a new set of restrictions (13.70). This allows us to control the most critical current restrictions. The stopping rule again coincides with (13.69).

13.8.3   Method of "Reflecting Screen"

Another heuristic procedure uses an idea which might be considered to be a composite of the two previously presented ideas. Let us use an arbitrary type of cost, say j, for the one-dimensional SDM. At the same time, we control the violation of all of the remaining restrictions. The intermediate stopping rule is a violation of any of the restrictions. If a first violation is observed for the restriction of the same type j which was used for the optimization procedure, the problem is assumed to be accurately solved. But if the first violation is observed for a cost of type j^*, this means that during the procedure this type of resource has been spent in an ineffective way. (Of course, it may also happen because of a heavy restriction on the cost of this type.) What can one do? Common sense tells us that we should find a way to release some of the most critical types of resources. For this purpose, we can exclude the units with the highest cost of type j^* from the obtained solution $\mathbf{X}_j^{\mathrm{opt}}$. For this we can use an inverse-directed SDM: the process should move back, releasing those units which are "most expensive" as measured by the cost j^*.

From this new point, say $\mathbf{X}^*$, the process starts again in a forward direction, but now we use a cost j^* for controlling the SDM. If a new restriction is violated, the procedure might be continued in the same manner until an appropriate solution is obtained.

13.9   MULTIFUNCTION SYSTEM

Consider a complex system designed for N different operations or functions which might be performed simultaneously. Each function is performed by some subsystem. These subsystems which form the system as a whole may have some common parts: some units might belong to several subsystems. These functions might vary by importance and, consequently, the reliability requirements for the corresponding subsystems might also be different.

This problem might be considered as an inverse to the optimal redundancy problem with several cost restrictions. A strict formulation of the problem is as follows:

$$\min_{\mathbf{X}} \left\{ C(\mathbf{X}) \mid \prod_{i \in G_j} R_i(x_i) \geq R_j^0, \, 1 \leq j \leq N \right\} \qquad (13.71)$$

where G_j is the set of the subscripts of units which form the jth subsystem and N is the number of different subsystems of the system:

$$X = \bigcup_{1 \leq j \leq N} X_j$$

and $X_j = \{x_i, i \in G_j\}$. By assumption, $G_j \cap G_k \neq \emptyset$ at least for some j and k. (Otherwise, the problem is trivial because it decomposes into N independent one-dimensional optimization problems.)

This problem can be solved by exact methods, but for practical purposes it is enough to use a heuristic method. It is described as follows:

1. For each subsystem solve an independent optimal redundancy problem using the SDM

$$\min_{X_j} \{C(X_j) | R_j(X_j) \geq R_j^0\} \qquad (13.72)$$

Denote an optimal solution of the one-dimensional problem (13.72) as

$$X_{(j)}^* = \{x_{(j)i}^*, i \in G_j\}$$

REMARK. We have chosen the SDM as the particular method of optimization because, for any current kth step of the optimization process, there is a strict majorization:

$$X^{(K-1)} \leq X^{(K)} \leq X^{(K+1)} \qquad (13.73)$$

Note that for an exact algorithm (e.g., dynamic programming), condition (13.73) may not hold. Thus, the proposed algorithm gives only an approximate solution.

2. Divide the complete set of system units into nonintersecting subsets $g_j, g_{jk}, \ldots, g_{jk \ldots w}$. Here we use the following notation for the subscripts: g_j is the set of units belonging only to G_j, g_{jk} is the set of units belonging to both G_j and G_k, and so on. We also use the general notation g_α where α is a subset of subscripts.

3. Choose an arbitrary unit, say i. Denote the subset which includes this particular unit as $g_{\alpha(i)}$. If this set represents an intersection of several subsystems G_i, we order all subsystems by their "importance," using the values of $x_{(j)i}^*$, $j \in \alpha(i)$. The larger $x_{(j)i}^*$, the more important is the subsystem. If $x_{(j')i}^* = x_{(j'')i}^*$ for some j' and j'', we try another unit i' from $g_{\alpha(i)}$ and check the preference condition again. It may happen that $x_{(j')i'}^*$ and $x_{(j'')i'}^*$ are unequal. Note that, as a matter of fact, the vectors $X^{(K)}$ change monotonically during the SDM procedure, so different vectors may have some equal components. After such a check, all subsystems G_k which participate in $g_{\alpha(i)}$ are ordered linearly.

4. This procedure continues until all subsystems are ordered. The ordering of the system's subsystems may not necessarily be linear. A tree of the partially ordered preferences might be drawn. This tree may have several separated roots; that is, in graph-theoretic terms, the graph represents a "forest." If all subsystems are independent (no intersection between subsystems), the forest is represented by its set of roots.

5. Consider the root of the preference tree (if there are several roots, then pick anyone). Let this correspond to the subsystem G_{j_1}. For each unit of this subsystem which has a maximal priority, we take the values of $x^*_{(j_1)i}$ as the final solutions. Afterwards, the subsystem G_{j_1} is excluded from further analysis. Now the preference tree (or its fragment if there were several separated trees) is separated into two or more subtrees.

6. The values of the above-fixed solutions, $x^*_{(j_1)i}$, relate to other subsystems which intersect G_{j_1}. Denote these subsystems by $G_{(j_1)s}$. Thus, $G_{(j_1)s} \cap G_{j_1} \neq \varnothing$. Denote this intersection by $g_{(j_1)s}$. Because the subsystem G_{j_1} is the most critical among all of its "neighbors," the use of $x^*_{(j_1)i}$'s for them can only improve them; that is, the specified requirements on reliability will not be violated. One understands that this circumstance allows one to reduce the needed number of other redundant units in the subsystem $G_{(j_1)s}$'s.

   Consider some specific subsystem $G_{(j_1)s}$. One needs to use redundant units only for those subsystem units which belong to

$$g'_{(j_1)s} = G_{(j_1)s} \setminus G_{(j_1)} = G_{(j_1)s} \setminus g_{(j_1)s}$$

To determine the needed x_i's for the units from $g'_{(j_1)s}$, we must:

- Calculate the probability of a successful operation for the part of the subsystem $G_{(j_1)s}$ which intersects the subsystem $G_{(j_1)}$:

$$R_{j_1s} = \prod_{i \in g_{j_1s}} R_i(x_i)$$

- Revise the restriction on the cost for the remaining part of the subsystem $G_{(j_1)s}$, that is, for $g'_{(j_1)s}$,

$$\tilde{R}^0_j = \frac{R^0_j}{R_{(j_1)s}}$$

- Use the SDM to find x^*_i for all $i \in g'_{(j_1)s}$ as should be done for the independent system.

7. Continue the procedure until all solutions are found.

REMARK. Note that the solution obtained with the help of this procedure is approximate. But practical experience in solving such problems has shown that this approximation is very close to the optimal solution and, as a matter of fact, they almost always coincide.

13.10 MAXIMIZATION OF MEAN TIME TO FAILURE

The maximization of the MTTF is important for many systems. For example, the outcome of some particular system may depend on the total useful operating time and any short interruption is harmless. (Usually, this corresponds to a case of spare redundancy.)

At the same time, the standard exact methods of discrete optimization in this case cannot be used at all or they produce results which are quasi-optimal. We assume that in such situations, it is reasonable to use a good heuristic.

Consider a series system of n independent units. To improve a system's MTTF, we can use standby redundancy. Consider the case where the number of spare units of each type x_i is reasonably large. Such a situation appears when the system is supposed to operate over a long period of time without any outside monitoring or when remote supply centers are difficult to access.

Suppose we know the MTTF T_i, the standard deviation σ_i, and the cost c_i for each unit of the ith type. Under the assumption that $x_i \gg 1$, the random time until exhaustion of the stock of spare units of each type approximately has a normal distribution.

Thus, the random TTF of x_i units of the ith type (used consequently) is

$$\xi_{\text{syst}}(x_i) = \sum_{1 \le j \le x_i} \xi_i^j$$

where ξ_i^j is the jth TTF of the random value of ξ_i and x_i is the number of units in the redundant group. The mean equals

$$E\{\xi_{\text{syst}}(x_i)\} = x_i T_i \tag{13.74}$$

and the variance is

$$\sigma_{\text{syst}}^2 = \text{Var}\{\xi_{\text{syst}}(x_i)\} = \sigma_i^2 x_i \tag{13.75}$$

Let $\overline{\Phi}(T, \sigma)$ denote the complementary function for the normal distribution function $\Phi(T, \sigma)$, that is, $\overline{\Phi}(T, \sigma) = 1 - \Phi(T, \sigma)$.

If we observe a series system of n redundant groups, then

$$P_{\text{syst}}(t) = \prod_{1 \le i \le n} \overline{\Phi}\left(x_i T_i, \sigma_i \sqrt{x_i}\right) \tag{13.76}$$

It is clear that from $T_{\text{syst}} = T_0$ it follows that $x_i T_i \geq T_0$ must be held for each i, $1 \leq i \leq n$; that is, a lower bound can be written as

$$T_{\text{syst}} < \min_{1 \leq i \leq n} x_i T_i \qquad (13.77)$$

Now consider a particular case when all units are identical with an MTTF $T_i = T$ and a standard deviation $\sigma_i = \sigma$. Each redundant group has x units. Then, for the moment $t = xT$, we can easily calculate

$$P_{\text{syst}}(xT) = \left[\overline{\Phi}\left(0 | xT, \sigma\sqrt{x}\right) \right]^n \qquad (13.78)$$

For a relatively large value of x, we may assume that the mean and the median of this distribution are close. (As a matter of fact, there is a left-hand skewness, and the median is larger than the mean.)

Now we come to an important point of this heuristic approach: why not to replace the mean by the median? Historically, people used means, we suppose, not because of their "extraordinary" physical sense but because of their mathematical simplicity. Indeed, everybody knows that the mean of the sum of r.v.'s is equal to the sum of the means, that the LST allows one to obtain easily the distribution moments (including the mean), and so on. All of these are good reasons . . . . (The authors of this very book use the mean as one of the main reliability parameters for the same reason!)

If a practical engineer were asked: "How is the mean located within a sample?" the most probable answer would be: "The mean divides the sample into approximately two equal parts, one on the left, and another on the right." This answer appears because most engineers tend to draw the bell-shaped density function which corresponds to the normal distribution. In this particular (though very common and useful) case, the mean coincides with the median. Note that the process we are considering represents the sum of a large number of r.v.'s (a sequential replacement of standby spare units) which leads to a normal distribution.

For the problem under consideration, let us choose the median as a reliability measure from the very beginning. This substitution of one reliability index by another is appropriate if the mean and the median are relatively close to each other.

For the case (12.5) we can easily find the median M:

$$M = \left\{ t : \overline{\Phi}\left(0 | xT, \sigma\sqrt{x}\right) = (0.5)^{-n} \right\} \qquad (13.79)$$

From (13.79) the following convenient heuristic rule can be derived: the median of the system TTF is not less than t if x_i^o for each ith redundant group is chosen with the help of the rule:

$$x_i^o = \min_{1 \le i \le n} \left\{ x_i : x_i - b_n \sigma_i \sqrt{x_i} \le t \right\} \qquad (13.80)$$

Here the coefficient b_n depends only on the number of redundant groups in the system and can be found from the solution of the equation

$$\Phi(b_n) = 1 - 0.5^{-n} \qquad (13.81)$$

We must mention that this rule does not give an accurate value of the median for the system, but rather allows us to obtain a good "quasi-optimal" solution.

A practical rule for the solution of the problem:

$$\min_{\mathbf{X}} \left(C(\mathbf{X}) | T(\mathbf{X}) \ge T_0 \right) \qquad (13.82)$$

can be expressed as follows:

1. For all i, $1 \le i \le n$, find $x_i^{(0)}$ according to the rule

$$x_i^{(0)} = \frac{T_0}{T_i} \qquad (13.83)$$

2. With the help of a standard table of the normal distribution, determine b_n using (13.80).
3. For all i, $1 \le i \le n$, find the first approximation of a possible optimal value of x_i:

$$x_i^{(1)} = x_i^{(0)} + \frac{1}{T_i} b_n \sigma_i \sqrt{x_i^{(0)}} \qquad (13.84)$$

The vector of the solutions for the system is denoted by

$$\mathbf{X}^{(1)} = \left(x_1^{(1)}, \ldots, x_n^{(1)} \right) \qquad (13.85)$$

REMARK. The solution (13.85) usually suffices for practical purposes. For a more accurate solution, it is possible to use a variant of the "greedy" algorithm described below.

4. Calculate $T(\mathbf{X}^{(1)})$:

$$T(\mathbf{X}^{(1)}) = \int_0^\infty \prod_{1 \le i \le n} \overline{\Phi}_i \left(t | T_i x_i \sqrt{x_i^{(1)}}, \sigma_i \sqrt{x_i^{(1)}} \right) dt \qquad (13.86)$$

5a. If $T(\mathbf{X}^{(1)}) < T_0$, calculate $T(\mathbf{X}_i^{(1)})$ with the use of a formula similar to (13.86). Hence $\mathbf{X}_i^{(1)} = \mathbf{X}^{(1)} + e_i$ where e_i is the unit vector: $e_i = (0, 0, \ldots, 1_i, \ldots, 0)$, that is,

$$\mathbf{X}_i^{(1)} = \left(x_1^{(1)}, \ldots, x_i^{(1)} + 1, \ldots, x_n^{(1)} \right)$$

6a. Calculate

$$\delta_i = \frac{1}{c_i}\left[T(\mathbf{X}_i^{(1)}) - T(\mathbf{X}^{(1)})\right] \qquad (13.87)$$

7a. For the redundant group with index i_1 where

$$i_1 = \{i: \max \delta_i\} \qquad (13.88)$$

replace x_{i_1} with $x_{i_1} + 1$.

Repeat this procedure until $T(\mathbf{X}^{(k)}) \geq T_0$. The vector $T(\mathbf{X}^{(k)})$ so determined is an approximate solution.

5b). If $T(\mathbf{X}^{(1)}) \geq T_0$, calculate $T(\hat{\mathbf{X}}_i^{(1)})$ with the use of the formula similar to (13.86). Here $\hat{\mathbf{X}}_i^{(1)} = \mathbf{X}^{(1)} - e_i$, that is,

$$\hat{\mathbf{X}}_i^{(1)} = \left(x_1^{(1)}, \ldots, x_i^{(1)} - 1, \ldots, x_n^{(1)}\right)$$

6b. Calculate

$$\hat{\delta}_i = \frac{1}{c_i}\left[T(\mathbf{X}^{(1)}) - T(\hat{\mathbf{X}}_i^{(1)})\right]$$

7b. For the redundant group with index i_1 where

$$i_1 = \{i: \min \hat{\delta}_i\}$$

replace x_{i_1} with $x_{i_1} - 1$.

Repeat the procedure until $T(\mathbf{X}^{(k)}) \leq T_0$. The vector $T(\mathbf{X}^{(k-1)})$ is an approximate solution.

If redundant groups have more than one main unit, the reader can develop a similar procedure.

Now consider an inverse optimal redundancy problem

$$\max_{\mathbf{X}} \left(T(\mathbf{X})|C(\mathbf{X}) \leq C_0\right) \qquad (13.89)$$

As mentioned above, the optimal solution should satisfy the approximate condition

$$x_1 T_1 \approx x_2 T_2 \approx \cdots \approx x_n T_n \qquad (13.90)$$

At the same time, we expect that

$$\sum_{1 \le i \le n} c_i x_i \approx C_0 \tag{13.91}$$

Thus, the first approximation of the optimal solution can be obtained by combining (13.90) and (13.91) to obtain

$$x_i^* \approx I\left(\frac{C_0}{\displaystyle\sum_{1 \le i \le n} c_i \frac{T_0}{T_i}}\right) \tag{13.92}$$

where $I(x)$ is the integer part of x.

The solution (13.92) is good enough for most practical purposes, but could be improved. Below we present an SDM algorithm for finding an optimal solution. This algorithm is similar to that was described above.

1. Find $x_i^{(0)}$ for all i, $1 \le i \le n$, by the rule (13.92). The vector of these rough solutions is denoted by

$$\mathbf{X}^{(0)} = \left(x_1^{(0)}, \dots, x_n^{(0)}\right) \tag{13.93}$$

2. Calculate $T(\mathbf{X}_i^{(0)})$ and $T(\hat{\mathbf{X}}_i^o)$ with the use of a formula similar to (13.86). Here $\mathbf{X}_i^{(0)} = \mathbf{X}^{(0)} + e_i$, $\hat{\mathbf{X}}_i^{(0)} = \mathbf{X}^{(0)} - e_i$, and e_i is the uniform vector.

3. Calculate

$$\delta_i = \frac{1}{c_i}\left[T(\mathbf{X}_i^{(0)}) - T(\mathbf{X}^{(0)})\right] \tag{13.94}$$

and

$$\hat{\delta}_i = \frac{1}{c_i}\left[T(\mathbf{X}^{(0)}) - T(\hat{\mathbf{X}}_i^{(0)})\right] \tag{13.95}$$

4. Note that, in general, we use only the integer parts in (13.92), so not all resources C_0 will be spent. For the redundant group with i_1,

$$i_1 = \{i: \max \delta_i\} \tag{13.96}$$

$x_{i_1} := x_{i_1} + 1$. Set the new tentative solution $\mathbf{X}^{(1)} = \mathbf{X}^{(0)} + e_{i_1}$.

5. Check the restriction $C(\mathbf{X}^{(1)}) \le C_0$. If the restriction has not been violated, continue the procedure. If the restriction has been violated, go to step 6.

6. For the redundant group with $\hat{i}_1$,

$$\hat{i}_1 = \left\{ i: \min \hat{\delta}_i \right\} \tag{13.97}$$

$x_{\hat{i}_1} := x_{\hat{i}_1} - 1$. Recall $\mathbf{X}^{(1)} = \mathbf{X}^{(0)} - e_{\hat{i}_1}$.

7. Check the restriction and continue the procedure, if necessary.

If each redundant group has more than one main unit, a similar procedure can be developed with the use of the approximate results given in Chapter 3.

CONCLUSION

Heuristic methods are not exceptional in reliability theory. Moreover, we would like to emphasize that almost all applied research concerning modeling represents heuristic methods. Indeed, any model is a reflection of the researcher's understanding of a real object based on his or her background, experience, and intuition.

This chapter is not followed by references or exercises. This can be explained by the fact that this chapter is a collection of case studies rather than an attempt to present a comprehensive overview on the subject. All of the material in this chapter reflects the authors' experience in research and consulting in industry.

We wrote in the Introduction that the book is open for additions. Now we reiterate this: this chapter on heuristic methods needs more practical approaches in various reliability applications. We wait to hear from the readers.

Closing this chapter, we hope that the number 13 did not force the reader to avoid it. We understand that this chapter is of almost no interest for a pure mathematician. But, at the same time, we would like to ask everybody: what is a model design in general? It is an assumption embedded in an assumption and covered with an assumption pretending to be a reflection of a real object. Seriously, any mathematical model is a reflection of our guessing about a real object, that is, heuristic!

In conclusion, we recall a well-known fable about Christopher Columbus. According to the story, once Columbus entered the crew's space and found the sailors rolling an egg. He asked what were they doing. They answered that they were trying to arrange a temporary equilibrium of the egg by rolling it. "What's the problem?" Columbus asked, and he hit the egg on the table surface. The broken egg stably stood on the table. A sailor replied, "I could have done the same ... ." Columbus answered, "You could have, but I did!"

This fable might be a good analogy for many engineers in their attempts to solve so-called "unsolvable problems." Do not always try to follow a strong formulation of the problem. The problem itself can be corrected. Moreover, most "unsolvable problems" have a deficiency in their formulations.

Some references concerning the subject can be found below.

REFERENCES

Cherkesov, G. N. (1974). Reliability of Technical System with Time Redundancy. Moscow: Sovietskoe Radio.

Genis, Ya. G., and I. A. Ushakov (1984). Optimization of the Reliability of Multipurpose Systems. Soviet Journal of Cybernetics and Computer Sciences (USA), vol. 22, no. 3.

Genis, Ya. G., and I. A. Ushakov (1988). Estimation of Reliability of Redundant Restorable System with Known Test and Switching Characteristics. Automation Remote Control (USA), vol. 49, no. 6.

Kredentser, B. P. (1978). Prediction of Reliability of Systems with Time Redundancy (in Russian). Kiev: Naukova Dumka.

Rudenko, Yu. N., and I. A. Ushakov (1985). Models and Methods for Investigation of Energy-System Reliability. Power Engineering: Journal of the USSR Academy of Sciences (USA), vol. 23. no. 5.

Rudenko, Yu. N., and I. A. Ushakov (1989). Reliability of Power Systems (Russian). 2nd ed. Novosibirsk: Nauka.

Ushakov, I. A. (1980). An Approximate Method of Calculating Complex Systems with Renewal. Engineering Cybernetics (USA), vol. 18, no. 6.

Ushakov, I. A. (1981). Methods of Approximate Solution of Dynamic Standby Problems. Engineering Cybernetics (USA), vol. 19, no. 2.

Principal Books in Russian

Barzlovich, Ye. Yu., and Kashtanov, V. A. (1975). *Maintenance Management under Incomplete Information* (in Russian). Soviet Radio, Moscow.

Belyaev, Yu. K. (1975). *Probabilistic Methods of Sampling Control* (in Russian). Nauka, Moscow.

Bolotin, V. V. (1971). *Application of Probabilistic Methods and Reliability Theory in Construction Design* (in Russian). Stroiizdat, Moscow.

Buslenko, N. P. (1978). *Monte Carlo Simulation of Complex System* (in Russian). Nauka, Moscow.

Buslenko, N. P., Kalashnikov, V. V., and Kovalenko, I. N. (1973). *Lectures on Theory of Complex Systems* (in Russian). Sovietskoe Radio, Moscow.

Cherkesov, G. N. (1974). *Reliability of Technical Systems with Time Redundancy* (in Russian). Sovietskoe Radio. Moscow.

Dzirkal, E. V. (1981). *Setting and Estimation of Reliability Requirements for Complex Systems*. Sovietskoe Radio, Moscow.

Gadasin, V. A., and Ushakov, I. A. (1975). *Reliability of Complex Control Information Systems* (in Russian). Sovietskoe Radio, Moscow.

Gnedenko, B. V., ed. (1983). *Aspects of Mathematical Theory of Reliability* (in Russian). Authors: Barzilovich, Ye. Yu., Belyaev, Yu. K., Kashtanov, V. A., Kovalenko, I. N., Solovyev, A. D., and I. A. Ushakov, Radio i Svyaz, Moscow.

Konyonkov, Yu. K., and Ushakov, I. A. (1975). *Aspects of Reliability of Electronic Equipment under Mechanical Stress*. Sovietskoe Radio, Moscow.

Korolyuk, V. S., and Turbin, A. F. (1982). *Markov Renewal Processes in Reliability*. Naukova Dumka, Kiev.

Kovalenko, I. N. (1980). *Analysis of Rare Events for Reliability and Performance Effectiveness of Complex Systems Investigation*, Sovietskoe Radio, Moscow.

Kozlov, B. A. (1969). *Redundancy with Renewal*. Sovietskoe Radio, Moscow.

Kozlov, B. A., and Ushakov, I. A. (1966). *Brief Handbook on Electronic Equipment Reliability Estimations*, Sovietskoe Radio, Moscow.

Kozlov, B. A., nad Ushakov, I. A. (1975). *Handbook on Reliability of Automation and Electronic Equipment*. Sovietskoe Radio, Moscow.

Pavlov, I. V. (1982). *Statistical Methods of Reliability and Performance Estimation of Complex Systems on Basis of Experimental Data* (ed. I. A. Ushakov). Sovietskoe Radio, Moscow.

Pashkovsky, G. S. (1981). *Aspects of Failure Searching and Finding for Electronic Equipment* (ed. by I. A. Ushakov). Sovietskoe Radio, Moscow.

Pollyak, Yu. G. (1971). *Monte Carlo Simulation with Computers*. Sovietskoe Radio, Moscow.

Raykin, A. L. (1978). *Elements of Reliability Theory* (ed. by I. A. Ushakov). Sovietskoe Radio, Moscow.

Reinschke, K., and Ushakov, I. A. (1988). *Application of Graph Theory in Reliability Analysis*. Radio i Svyaz, Moscow.

Rubalsky, G. B. (1977). *Inventory Control under Stochastic Requests* (ed. by I. A. Ushakov). Sovietskoe Radio, Moscow.

Rudenko, Yu. N. and Cheltsov, M. B. (1974). *Reliability of and Redundancy in Electric Power Systems*. Nauka, Novosibirsk.

Rudenko, Yu. N., and Ushakov, I. A. (1989). *Reliability of Power Systems* (ed. by B. V. Gnedenko). Nauka, Novosibirsk.

Ushakov, I. A. (1969). *Methods of Solving Optimal Redundancy Problems under Restrictions*. Sovietskoe Radio, Moscow.

Ushakov, I. A., ed. (1985). Reliability of Technical Systems: Handbook. Radio i Svyaz, Moscow.

Volkovich, V. L., A. F. Voloshin, V. A. Zaslavsky, and I. A. Ushakov (1992). *Models and Methods of Optimization of Complex Systems Reliability*. Kiev: Naukova Dumka.

GENERAL REFERENCES

Abdel-Hameed, M. S., et al. (1984). *Reliability Theory and Models*. San Diego: Academic.

Anderson, R. T. (1990). *Reliability Centered Maintenance*. New York: Elsevier.

Asher, H., and H. Feingold (1984). *Repairable Systems Reliability*. New York: Marcel Dekker.

Barlow, R. B., and F. Proschan (1965). *Mathematical Theory of Reliability*. New York: Wiley.

Barlow, R. B., and F. Proschan (1981). *Statistical Theory of Reliability and Life Testing: Probability Models*, 2nd ed. Silver Spring, MD: To Begin With.

Bazovsky, I. (1961). *Reliability Theory and Practice*. Englewood Cliffs, NJ: Prentice-Hall.

Becker, P. W., and F. Jensen (1977). *Design of Systems and Circuits for Maximum Reliability and Maximum Production Yield*. New York: McGraw-Hill.

Dhillon, B. S., and H. Reiche (1985). *Reliability and Maintainability Management*. New York: Van Nostrand Reinhold.

Dhillon, B. S., and C. Singh (1981). *Engineering Reliability: New Techniques and Applications*. New York: Wiley.

Gertsbakh, I. B. (1977). *Models of Preventive Maintenance*. Amsterdam: North-Holland.

Gertsbakh, I. B., and Kh. B. Kordonsky (1969). *Models of Failure*. Berlin: Springer.

Gnedenko, B. V., Yu. K. Belyaev, and A. D. Solovyev (1969). *Mathematical Methods of Reliability Theory*. San Diego: Academic.

Henley, E. J., and H. Kumamoto (1981). *Reliability Engineering and Risk Assessment*. Englewood Cliffs, NJ: Prentice-Hall.

Jardine, A. R. S. (1973). *Maintenance, Replacement and Reliability*. London: Pitman.

Kapur, K. C., and L. R. Lamberson (1977). *Reliability in Engineering Design*. New York: Wiley.

Kececiogu, D. (1991). *Reliability Engineering Handbook*. Englewood Cliffs, NJ: Prentice-Hall.

Kozlov, B. A., and I. A. Ushakov (1970). *Reliability Handbook*, New York, Holt, Rinehart, and Winston.

Lloyd, D. K., and M. Lipov (1962). *Reliability Management, Methods and Mathematics*. Englewood Cliffs, NJ: Prentice-Hall.

O'Connor, P. (1990). *Practical Reliability Engineering*, 3rd. ed. New York: Wiley.

Osaki, S. (1984). *Stochastic System Reliability Modeling*. Singapore: World Scientific.

Polovko, A. M. (1968). *Fundamentals in Reliability Theory*. San Diego, CA: Academic.

Ravichandran, N. (1990). *Stochastic Methods in Reliability Theory*. New York: Wiley.

Roberts, N. H. (1964). *Mathematical Models in Reliability Engineering*. New York: McGraw-Hill.

Ross, S. (1980). *Introduction to Probability Models*, 2nd ed. New York: Academic.

Sandler, G. H. (1964). *System Reliability Engineering*. New York: McGraw-Hill.

Shoman, M. L. (1968). *Probabilistic Reliability: An Engineering Approach*. New York: McGraw-Hill.

Sundararajan, C. (1991). *Guide to Reliability Engineering*. New York: Van Nostrand Reinhold.

Tillman, F. A., H. Ching-Lai, and W. Kuo (1980). *Optimization of System Reliability*. New York, Marcel Dekker.

Tobias, P. A., and D. C. Trindade (1986). *Applied Reliability*. New York: Van Nostrand Reinhold.

Ushakov, I. A., ed. (1994). *Handbook of Reliability Engineering*. New York, Wiley.

Villemeur, A. (1992). *Reliability, Availability, Maintainability and Safety Assessment*, vol. 1, *Methods and Techniques*. New York: Wiley.

Villemeur, A. (1992). *Reliability, Availability, Maintainability and Safety Assessment*, vol. 2, *Assessment, Hardware, Software and Human Factors*. New York: Wiley.

Vinogradov, O. G. (1991). *Introduction to Mechanical Reliability: A Designer's Approach*. New York: Hemisphere.

Zacks, S. (1992). *Introduction to Reliability Analysis: Probability Models and Statistical Models*. New York: Springer.

INDEX

515